Lasers and Optical Engineering

P. Das

Lasers and Optical Engineering

With 391 Illustrations

Springer-Verlag
New York Berlin Heidelberg London
Paris Tokyo Hong Kong Barcelona

P. Das
Electrical, Computer, and
    Systems Engineering Department
Rensselaer Polytechnic Institute
Troy, NY 12180-3590
U.S.A.

Cover illustration: Photograph of the optical processor for real-time spectrum analysis. This figure appears on p. 315 of the text.

Library of Congress Cataloging-in-Publication Data
Das. P.
    Lasers and optical engineering / P. Das.
        p.   cm.
    Includes bibliographical references.
    ISBN 0-387-97108-4
    1. Lasers. 2. Optics. I. Title.
    TA1677.D37   1990
    621.36′6—dc20   89-26312
                                                                    CIP

Printed on acid-free paper.

Typeset by Asco Trade Typesetting Ltd., Hong Kong.
Printed and bound by R.R. Donnelley & Sons, Harrisonburg, Virginia.
Printed in the United States of America.

9 8 7 6 5 4 3 2 1

ISBN 0-387-97108-4 Springer-Verlag New York Berlin Heidelberg
ISBN 3-540-97108-4 Springer-Verlag Berlin Heidelberg New York

To

Virginia, Andrea, Joshua and the memory of Susama
and Upendra

Preface

A textbook on lasers and optical engineering should include all aspects of lasers and optics; however, this is a large undertaking. The objective of this book is to give an introduction to the subject on a level such that under-graduate students (mostly juniors/seniors), from disciplines like electrical engineering, physics, and optical engineering, can use the book. To achieve this goal, a lot of basic background material, central to the subject, has been covered in optics and laser physics. Students with an elementary knowledge of freshman physics and with no formal courses in electromagnetic theory should be able to follow the book, although for some sections, knowledge of electromagnetic theory, the Fourier transform, and linear systems would be highly beneficial.

There are excellent books on optics, laser physics, and optical engineering. Actually, most of my knowledge was acquired through these. However, when I started teaching an undergraduate course in 1974, under the same heading as the title of this book, I had to use four books to cover the material I thought an electrical engineer needed for his introduction to the world of lasers and optical engineering. In my sabbatical year, 1980–1981, I started writing class notes for my students, so that they could get through the course by possibly buying only one book. Eventually, these notes grew with the help of my undergraduate and graduate students, and the final result is this book.

It is a pleasure to thank Janet Tomkins for typing the class notes, over and over again. Without her patience and efforts, this book would not have been possible. Also, I would like to thank many of my students who helped improve the manuscript by criticizing, finding mistakes and correcting them, and editing and writing projects reports which I have freely used.

Contents

Some Fundamental Constants

q = electronic charge = 1.6022×10^{-19} C.
m = electron rest mass = 9.106×10^{-31} kg.
h = Planck constant = 6.626×10^{-34} J s.
K = Boltzmann constant = 1.381×10^{-23} J/K.
μ_0 = permeability of free space = $4\pi \times 10^{-7}$ H/m.
ε_0 = permittivity of free space = 8.8542×10^{-12} F/m.
C = velocity of light in vacuum = $1/\sqrt{\mu_0 \varepsilon_0}$ = 2.9979×10^8 m/s.
Z_0 = characteristic impedance of free space = $\sqrt{\mu_0/\varepsilon_0}$ = 120π Ω = 377 Ω.
eV = electronvolt = 1.602×10^{-19} J.
$KT = 25 \times 10^{-3}$ eV for $T = 300$ K.

Decimal Prefixes

tera = 10^{12} cente = 10^{-2}
giga = 10^9 milli = 10^{-3}
mega = 10^6 micro = 10^{-6}
kilo = 10^3 nano = 10^{-9}
hecto = 10^2 pico = 10^{-12}
deca = 10^1 femto = 10^{-15}
deci = 10^{-1} atto = 10^{-18}

1 eV electron transition corresponds to $\lambda = 1.2394$ μm.
1 μm = 10,000 Å, 1 Å = 10^{-10} m.

Chronology of Optical Discoveries*

1637 Laws of refraction.
1666 Discovery of diffraction.
Light dispersion by prism—Newton.
1669 Double refraction.
1675 Determination of speed of light—Roemer.
1690 Huygens' wave theory.
1704 Newton's *Optics*.
1720 Three-color copper plate printing.
1727 Light images with silver nitrate.
1758 Achromatic telescope.
1790 Ultraviolet rays discovered.
1800 Infrared rays discovered.
1801 Discovery of interference of light waves.
1808 Discovery of polarization of light.
1814 Discovery of Fraunhofer black lines in the sun's light spectrum.
1823 Faraday—laws of electromagnetism.
Discovery of silicon.
1827 Ohm's law.
1828 Electromagnet.
1832 Principles of induction—Faraday.
1837 Electric motor—Davenport.
1838 Photography—Daguerre.
1842 Doppler effect discovered.
1864 Electromagnetic theory—Maxwell.
1866 Dynamite–Nobel.
1869 Angstrom.
1875 Telephone.
1885 Transformer.

---

* From R. Buckminster-Fuller, *Critical Path*, St. Martin's Press, New York, 1981.

1887   Electromagnetic waves—Hertz.
        Michelson–Morley experiment.
1893   Motion picture machine.
1895   Wireless telegraphy.
        X-ray—Roentgen.
1896   Zeeman effect.
1897   Electron discovered.
1900   Quantum theory—Planck.
1905   Special theory of relativity—Einstein.
1915   General relativity theory—Einstein.
1922   Radar.
1924   Wave mechanics—De Broglie.
1932   X-ray diffraction.
1948   Holography—Gabor.
        Xerography—Carlson.
        Transistor—Bardeen, Brattain, and Shockley.
1954   Solar cell.
1954   Maser—Towne.
1960   Laser demonstrated—Maiman.

Introduction

Until recently, optics had not been considered as part of the electrical engineering curriculum, even though the fundamental laws for electromagnetic waves, which include those in optics, are governed by Maxwell's equations. The main reason for this was the absence of coherent optical sources such as klystrons or magnetrons for microwaves, or oscillators for lower frequencies. However, the invention of the laser has changed this situation and, as expected, there has been enormous activity in using the optical frequency region for conventional electrical engineering applications, such as optical communication, laser radar, and optical signal processing. These applications are in addition to those traditionally belonging to optics, such as photography, spectroscopes, microscopes, telescopes, etc., which generally use incoherent light. The "optical revolution" in electrical engineering is not only fueled by the availability of the laser, but also by other technical developments such as integrated optics, fiber-optics, acousto-optics, electro- and magneto-optics, Fourier optics, and a phenomenal need for parallel computation; hence optical computing, systolic arrays, photodetector arrays and charge coupled device (CCD), charge injection device (CID), focal plane arrays, GaAs technology, very high speed integrated circuits, and the overall desire of society to perform real-time signal processing with greater speed and higher bandwidth.

To understand the role played by optics in electrical engineering, often referred to as photonics, opto-electronics, electro-optics, optronics, etc., we consider two topics, optics and devices, shown in compact form as trees with branches in Figs. 1 and 2. The optics tree root includes work by Maxwell, Fresnel, and Fraunhofer as the fundamentals of physical optics or wave optics. Abbé introduced the fundamental concepts of Fourier optics, augmented by Zernicke in his applications to phase contrast microscopy. Fourier optics was developed further by Maréchal, Tsujiuchi, O'Neill, and Lohman, who successfully applied more of the conventional electrical engineering techniques in traditional one-dimensional time domains to two-dimensional space domains.

Gabor also used the concept of spatial-frequency multiplexing and demultiplexing in holography which, after the invention of the laser, was further

Fig. 1. Optics tree.

refined by Leith, Cutrona, Palermo, Porcello, and Van der Lught and applied to synthetic aperture radar, matched filtering, and pattern recognition. Integrated optics, optical fiber propagation, and crystal optics are some of the other branches of the optics tree. Note that the tree is still growing very rapidly and, from the point of view of application, some of the branches (e.g., optical fibers) might overshadow some of the other branches with respect to engineering applications in volume, mostly because of the eventual replacement of most telephone lines by optical fibers. Nonlinear optics is also an important branch which has important applications in phase conjugation.

For different applications such as optical communication, optical data and signal processing, image processing, etc., we need to implement these using different devices as depicted in the devices tree shown in Fig. 2. The

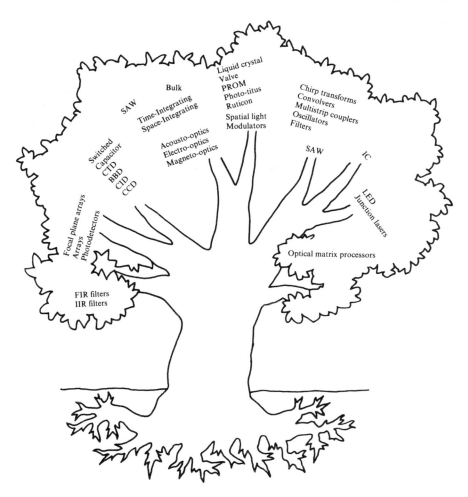

Fig. 2. Devices tree.

important devices are acousto-optic, electro-optic, or magneto-optic devices. These devices can be used as real-time correlators, convolutors, or matched filters, etc., or as optical matrix processors. For the conversion of light signals, we have spatial light modulators (SLMs) which include a host of devices, such as liquid crystal valves, e-beam potassium dihydrogen phosphate (KDP), Pockel's readout optical modulator (PROM) photo-titus, strain-based SLMs using ceramic ferroelectrics, SLMs using deformable surface tubes, and ruticon and membrane light modulators. Of these, the acousto-optic devices have so far, been, the most useful in actual applications because of their large bandwidth operations.

For any application, of course, we need a light source and light detectors. The source is usually a light emitting diode (LED) or a junction laser, or other lasers such as He–Ne or argon. Of course, the LED and junction lasers have the advantage of compactness and higher efficiency. The detectors are photo-detector, photodetector arrays, focal plane arrays, and sometimes photo-multipliers and ordinary photographic film if electronic output is not needed. For the electronic output case, the detectors also need amplifiers and other conventional electronics for further processing and display. For the array output, we also need CCDs or digital circuits to manipulate the large amount of data coming from the arrays.

Because of the importance of acousto-optic devices and the potential for integrating an acousto-optic device, SAW device, laser source, and photo-detector array, CCD and digital circuits on a GaAs substrate, we also need to consider all of these technologies separately (and together) for the system level problems.

Why Photonics?

Electronics deals with electrons and the manipulation of their flow to perform a useful function. In photonics we use clever combinations of photons, and their flow and conversion to electrons, in conjunction with the usual elec-tronics. Examples of photonics used in everyday life are growing rapidly: compact disc players, video recording machines, laser printers, laser-scanned check-out counters in supermarkets, robot vision, laser processing of materials in factories, laser diagnostics, and surgery in hospitals are some examples. Fiber-optic telephone and cable TV cables, lasers for missile defense (Star War), laser fusion, and laser ranging and guiding are other examples. Conven-tional optical equipment like cameras, microscopes, telescopes, etc., are also encountering photonic modifications, as the outputs of many of these devices are detected by area photodetectors or scanned to obtain electronic images which can be processed further.

Thus it is natural that at the present time an electrical engineer should have some knowledge of optics. The main objective of this book is to provide this basic knowledge of photonics to undergraduate students in a one- or two-semester course. Unfortunately, to limit the size of the book, all the applica-tions mentioned in this Introduction will not be covered adequately in this book. However, it is hoped that once the reader understands the fundamentals he will read reference books and articles for further details or for a particular application.

Bird's Eye View and Guide

The book is divided into four parts, excluding the Introduction. These are:

  I. Geometrical Optics.
 II. Wave Optics.
III. Lasers.
IV. Applications.

"Geometrical Optics" deals with situations where the wave nature of light is disregarded. This is generally true when aperture size is very large compared to the wavelength of light. This part starts with Snell's laws and its matrix formulation. This is then used to develop lens formulas, concepts of image formation, and optical instruments like microscopes and telescopes, although practical details of these instruments are discussed in Part IV. At the beginning, paraxial approximation is used for mathematical simplification. This is later extended to the exact matrices and a discussion on aberration. This part also includes a discussion on apertures and stops, radiometry, and photometry.

In "Wave Optics", the wave nature of light is introduced through Maxwell's equations. Using elementary arguments of spherical waves and superposition due to linearity, the diffraction formula is derived. This diffraction formula forms the backbone for the discussion on Fresnel and Fraunhofer diffraction and Fourier transforming properties of spatial signals, for which this part might also be called "Fourier Optics." Special emphasis is given to the concept of *spatial* signals and their use in image processing and holograms. Interference is considered as an extension to the diffraction formula and the Fabry–Perot interferometer is extensively discussed. The final section in this part, under the heading "Physical Optics," includes discussions on the following topics without any rigorous discussion: optical tunneling, reflection and transmission coefficients, polarization, phase and group velocity, light propagation in anisotropic solids, double refraction, polarizers, and electro-, acousto- and magneto-optic interactions in matter.

The "Laser" part starts with a discussion on feedback oscillators leading to the Fabry–Perot laser. Gaussian beam optics are considered in detail as well as their use in laser cavity modes and their properties. The physics of light amplification includes Einstein's coefficients for stimulated and spontaneous emission, and the derivation of the threshold population inversion density. The four-level laser is considered in detail, including power output and optimum reflectivity of the output mirror. Other topics covered include mode locking, Q-switching, and a detailed discussion on different lasers such as gas lasers, solid state lasers, dye lasers, semiconductor lasers, and free electron lasers.

The last part, "Applications", discusses first the practical details of optical instruments such as cameras and binoculars, including their lens design. Fiber-optics and integrated optics are considered next, where guided waves are introduced and their properties and usefulness in different applications are mentioned. The next section discusses optical signal processing devices such as modulators, deflectors, and correlators, both for time and spatial signals. Some applications of these are also included in this optical signal processing section including optical matrix processors. The section on laser applications includes both industrial, military, and medical applications. The final section, entitled "Recent Advances" discusses optical interconnection, optical logic, and "Star War".

Guide

I have successfully used the following sections in a one-semester course, mostly for electrical engineering upperclass undergraduates:

Part I   Sections 1.1–1.8.
Part II   Sections 2.1–2.12.
Part III   Sections 3.1–3.10.

The subdivisions naturally result in three quizzes. This lecture course has generally been supplemented with laboratory demonstrations of diffraction, interference, holograms, and lasers.

The rest of the book can be used easily as a second-semester undergraduate course for students who are interested further.

References

Each part has a reference list at the end. These references were used by the author in preparing this text. The reader will find further clarification or detailed derivation and discussion of the topic in these references. At the end of the book there are additional references which the author feels will be useful to the reader. For new topics, such as optical matrix processors or optical computing, the number of references are quite large. The author hopes that these references will be helpful to the reader interested in a particular topic.

[1]   J.W. Goodman, The optical data processing family tree, *Optics News*, **10**, 25–28, 1984.
[2]   P. Das, *Optical Signal Processing*, Springer-Verlag, 1990.

PART I

Geometrical Optics

1.1. Fundamentals of Geometrical Optics

The subject of geometrical optics will be developed on the basis of Snell's laws. It will be assumed that Snell's laws are experimental fact. However, as any student of electromagnetic theory knows, Snell's laws can be derived from Maxwell's equations. This derivation can be found in the book, *Optical Signal Processing* by Das, which deals with other topics and applications relevant to this subject but not covered in this book.

Snell's Laws

(1) Law of Rectilinear Propagation. In homogeneous media, light rays propagate in straight lines.

Before we go further, it is worthwhile to define a "light ray", which is shown in Fig. 1.1.1. The line AB is the line of constant phase for a light wave, or in three dimensions it will be a plane of constant phase. This plane is also generally known as the wavefront. The line CD is the light ray which is normal to this plane of constant phase with an arrow indicating in which direction the wave is propagating. Actually, a ray belongs to a wavefront which is infinite in size. A finite-sized wavefront will have many rays—this will be discussed later in "Wave Optics". In this section, we consider that the wavefront sizes are much larger than the wavelength.

1.1.1. Discussion of Waves

In general, a one-dimensional wave, $E(x, t)$, is mathematically represented by the form

$$E(x, t) = Ae^{j(\omega t - kx)} \qquad (1.1.1)$$

in complex notation. Actually, we consider either

$$E(x, t) = \mathrm{Re}[Ae^{j(\omega t - kx)}] = A\cos(\omega t - kx), \qquad (1.1.2)$$

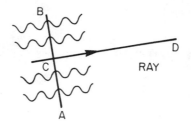

Fig. 1.1.1. Relationship between a light ray and a constant phase wavefront.

or

$$E(x, t) = \text{Im}[Ae^{j(\omega t - kx)}] = A \sin(\omega t - kx), \qquad (1.1.3)$$

the real or imaginary parts of the exponential. It is observed that the maximum value of the wave amplitude is A. The expressions (1.1.1)–(1.1.3) are all periodic in both x and t, as shown in Fig. 1.1.2. For a fixed time, the wave motion, as a function of the distance x, is shown in Fig. 1.1.2(a). (Figure 1.1.2(a) is intentionally drawn at an angle to indicate that the direction of wave motion is denoted by the x direction.) It is observed that the period is

$$\lambda = \frac{2\pi}{k}, \qquad (1.1.4)$$

where λ is the wavelength and k is the propagation constant or wave number. At a fixed point in space, the light wave goes through a periodic motion, as shown in Fig. 1.1.2(b). It is found that the time period, T, is given by

$$T = \frac{2\pi}{\omega} = \frac{1}{f}, \qquad (1.1.5)$$

where ω is called the radian frequency and f is just the frequency. Thus

$$\omega = 2\pi f. \qquad (1.1.6)$$

To describe the wave motion, we consider how the wavefront is moving or, equivalently, how the planes of constant phase are moving. The plane of

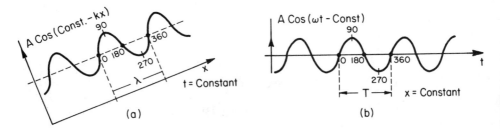

Fig. 1.1.2. Plot of the one-dimensional wave equation: (a) for $t = $ constant and (b) for $x = $ constant.

constant phase is defined by

$$\omega t - kx = \text{const.} \tag{1.1.7}$$

Thus the wavefronts are perpendicular to the x-axis for this case or lie in the yz-plane. The velocity of a wave (strictly speaking, *phase* velocity not *group* velocity) can be found as follows:

$$\omega \Delta t - k \Delta x = 0,$$

or

$$\left. \frac{\Delta x}{\Delta t} \right|_{\text{const. phase}} = \frac{\omega}{k} = v_{\text{p}} = f\lambda.$$

(The group velocity is a very important quantity, and will be shown later to be given by

$$v_{\text{g}} = \frac{d\omega}{dk}. \tag{1.1.8}$$

If ω is a linear function of k, then $v_{\text{p}} = v_{\text{g}} = $ independent of f or λ. However, if v_{p} is a function of ω or k, the phase velocity may not be equal to the group velocity.)

Using the relationship (1.1.7), we can easily rewrite (1.1.1) as

$$E(x, t) = A e^{j(\omega t - kx)}$$
$$= A e^{j\omega(t - x/v)}$$
$$= A e^{-jk(x - vt)}$$
$$= A e^{j2\pi f(t - x/v)}$$
$$= A e^{-j(2\pi/\lambda)(x - vt)}$$
$$= A e^{j2\pi(ft - x/\lambda)}. \tag{1.1.9}$$

All of these different forms are equivalent and useful.

Let us now take a specific example. We know that a light wave travels with the velocity, $v = 3 \times 10^8$ m/s. For green light, the wavelength is $\lambda = 6000$ Å $= 0.6$ μm. Thus, for this light wave

$$f = \frac{v}{\lambda} = 5 \times 10^{14} \text{ Hz.}$$

A light wave is an electromagnetic wave, a term which includes ordinary alternating current (a.c.) at 60 Hz, radio waves, TV-waves, microwaves, infrared waves, light waves, X-rays, and γ-rays. It is amazing that all the electromagnetic waves obey the equations described in (1.1.2)–(1.1.4). Table 1.1.1 gives a synopsis of these waves with their frequency ranges, their wavelengths, and some applications.

Further mathematical details of wave motion can be found in Section 2.1.

Table 1.1.1

Name	f	λ	Source	Applications
Direct current	0^+		battery	power supplies, d.c. motor
Ultra low frequency	1 Hz	3×10^8 m	electronic	submarine communication
Powerline	50 Hz	6×10^6 m	hydroturbine	powerline—motors
Audio	0–30 kHz		electronic	stereo
	10 kHz	$\to$ 30 km	oscillator	
Ultrasound	30–400 kHz		electric oscillator	ultrasound NDT, burglar alarm
Radiofrequency	500 kHz–1.5 MHz		electronic	AM radio
	1 MHz	$\to$ 300 m	oscillator	
Radiofrequency/ ultra high frequency	2–1000 MHz		electronic oscillator	C.B., FM radio, TV, other comm. systems
Microwave	1 GHz $\to$ 100 GHz		TWT, magnetrons	comm., satellite
	10 GHz	1 cm		
Near infrared	10^{11}–4.3×10^{14}	3000 μm	glow bar	infrared imaging
Infrared		$\to$0.7 μm	IRASER	robotic vision
Visible	14.3×10^{14} to	0.7 μm	lamp, laser	optical signal processing
Optical	5.7×10^{14}	$\to$0.4 μm	fluorescence	optical computing
Ultraviolet	5.7×10^{14} to	0.4 μm	laser	photolithography
	10^{16}	$\to$0.3 μm	fluorescence	material processing, laser fusion
X-ray	10^{16}–10^{19}	300 Å–0.3 Å	X-ray tubes	NDT, X-ray imaging, tomography
γ-ray	10^{19}–above	0.3 Å–shorter	radioactive source	imaging and tomography

1.1.2. Snell's Laws (*continued*)

(2) Law of Reflection. A light ray incident at an interface between two different homogeneous isotropic media is partially reflected and partially transmitted. The reflected ray lies in the plane of incidence and is determined by the incident ray and the normal to the surface. As shown in Fig. 1.1.3, the angle of incidence, θ_i (the angle the incident ray makes with the normal) is equal to the angle of reflection, θ_r (the angle between the reflected ray and the normal to the surface).

(3) Law of Refraction. The transmitted ray, or the refracted ray shown in Fig. 1.1.3, also lies in the plane of incidence and makes an angle, θ_T, with respect to the normal n. This angle is also called the angle of refraction. The law of refraction states that

$$n_i \sin \theta_i = n_T \sin \theta_T, \qquad (1.1.10)$$

where n_i and n_T are the refractive indices of the incident and transmitted media, respectively. Later, it will be discussed that this refractive index of a media is related to the velocity of the wave in the following fashion (here we take free

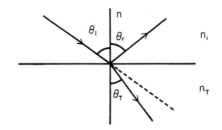

Fig. 1.1.3. Schematics for reflection and refraction. (*Note:* n is normal to the interface, not refractive index in this figure.)

space, i.e., $n_i = 1$):

$$n_T = \left(\frac{v_i}{v_T}\right)n_i = \frac{c}{v_T},$$ (1.1.11)

and

$$v_T = f\lambda_T,$$

where c is the velocity of light in vacuum and is equal to 3×10^8 m/s. (v_i and v_T are the velocities of light in the incident and transmitted media, respectively.) Thus, when we say that the refractive index of flint glass is 1.6, it means that the light velocity in flint glass is 1.875×10^8 m/s. Table 1.1.2 gives the values of the refractive index for some useful materials.

In general, the refractive index is a function of light wavelength, or the optical medium is dispersive. For example, if white light (remember white light

Table 1.1.2*

Material	Refractive index (for sodium D 589 nm; gases at normal temperature and pressure)
Vacuum	1.0
Air	1.000292
Water	1.3336[†]
Fused quartz	1.46
Glass, spectacle crown	1.523
Flint glass (light)	1.575
Flint glass (heaviest)	1.890
Carbon disulphide	1.64
Methylene iodide	1.74
Diamond	2.42
Polystyrene	1.59
Ethyl alcohol	1.36

* For other materials, see *Handbook of Chemistry of Physics*, CRC Press.

† Note that the dielectric constant of water at low frequencies (a few kilohertz) is 81 and not $(1.33)^2$.

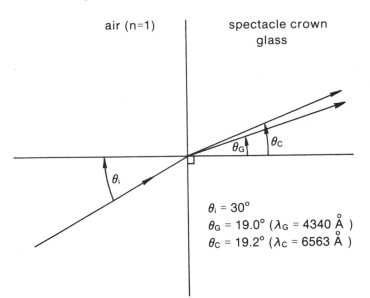

Fig. 1.1.4. Dispersion in crown glass for two wavelengths.

is a mixture of all the different colors of light, each having different wavelengths and frequencies) is incident on spectacle crown glass at an angle of 30°, then the different colors will disperse or will be refracted at different angles, as shown in Fig. 1.1.4.

We have also mentioned a word before, "isotropic". This means that the light velocity in the material is not dependent on the direction of propagation. This is true only in an amorphous material. The crystalline media is, in general, anisotropic and Snell's laws become rather complex. This will be discussed in Section 2.12.5.

Paraxial Approximation

If θ_i and θ_T are rather small, then we can approximate

$$\sin \theta_i \approx \theta_i, \tag{1.1.12}$$

and

$$\sin \theta_T \approx \theta_T.$$

(Remember that θ_i and θ_T must be in radians and not in degrees, a common source of error for students.) The law of refraction under paraxial approximation becomes

$$n_i \theta_i = n_T \theta_T \tag{1.1.13}$$

and this simplified version will be used often in this book. It is obvious why we call it paraxial approximation—because θ_i and θ_T are small and the light rays more or less move parallel to the normal, which is defined as the optical axis.

Some Interesting Points About Geometrical Optics

(i) Snell's laws do not tell us how much light is reflected and how much is refracted. It can be determined by starting from Maxwell's equations and solving for the proper boundary-value problem.

(ii) If $n_T < n_i$, then we can define θ_c (the critical angle) for total internal reflection. This θ_c is given by

$$n_i \sin \theta_c = n_T \sin 90° = n_T,$$

or

$$\sin \theta_c = \frac{n_T}{n_i} \quad \text{or} \quad \theta_c = \sin^{-1}\left(\frac{n_T}{n_i}\right).$$

For $\theta_i > \theta_c$, all the rays are reflected at the boundary.

1.2. Matrix Formulation of Geometrical Optics

Let us denote from now on the z-axis as the optical axis. Under paraxial approximation, most of the rays we will be interested in are at very small angles to this optical axis. This is shown in Fig. 1.2.1, where a light ray propagating through a homogeneous medium is shown. At point A on the optical axis, the light ray can be completely specified by x_1 and θ_1. x_1 is the vertical distance of the ray from the optical axis at A, and θ_1 is the angle the ray makes with the optical axis. Thus all the rays, lying in the plane of this light ray and the optical axis, can be represented by a value of x and a value of θ for each point along the optical axis. For point B, the position of the same light ray can be denoted by (x_2, θ_2). Thus the propagation of the light ray from point A to point B on the optical axis (a distance of D on the optical axis) can be considered as the transformation of (x_1, θ_1) to (x_2, θ_2). This is shown in Fig. 1.2.1(b), where (x_1, θ_1) is the input component vector X_1 and (x_2, θ_2) is the output vector X_2. The space between the optical axis points A and B is the optical system. However, using Snell's first law, the rectilinear propagation of light rays, we find that

$$x_2 = 1 \cdot x_1 + \theta_1 D,$$
$$\theta_2 = 0 \cdot x_1 + 1 \cdot \theta_1. \tag{1.2.1}$$

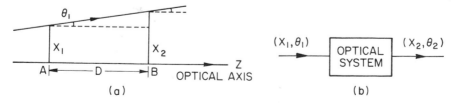
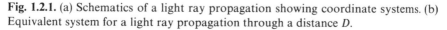

Fig. 1.2.1. (a) Schematics of a light ray propagation showing coordinate systems. (b) Equivalent system for a light ray propagation through a distance D.

Remember that in paraxial approximation, $\sin \theta \approx \tan \theta \approx \theta$. Equation (1.2.1) can be rewritten in matrix form

$$\begin{pmatrix} x_2 \\ \theta_2 \end{pmatrix} = \begin{pmatrix} 1 & D \\ 0 & 1 \end{pmatrix} \begin{pmatrix} x_1 \\ \theta_1 \end{pmatrix}, \tag{1.2.2}$$

or, symbolically,

$$X_2 = T(D)X_1, \tag{1.2.3}$$

where $T(D)$ is the (2×2) matrix given by

$$T(D) = \begin{pmatrix} 1 & D \\ 0 & 1 \end{pmatrix} \tag{1.2.4}$$

and is known as the translation matrix. It is important to note that the determinant of the translation matrix is 1.

1.2.1. Some Properties of Matrices

A matrix M with a (2×2) array element is defined as

$$M = \begin{pmatrix} M_{11} & M_{12} \\ M_{21} & M_{22} \end{pmatrix}. \tag{1.2.5}$$

A column matrix has a (2×1) array and is also known as a vector. For example,

$$X = \begin{pmatrix} x_1 \\ x_2 \end{pmatrix}. \tag{1.2.6}$$

A matrix equation between two vectors Y and X, denoted by

$$Y = MX, \tag{1.2.7}$$

or

$$\begin{pmatrix} y_1 \\ y_2 \end{pmatrix} = \begin{pmatrix} M_{11} & M_{12} \\ M_{21} & M_{22} \end{pmatrix} \begin{pmatrix} x_1 \\ x_2 \end{pmatrix}, \tag{1.2.8}$$

really means the following two equations, describing the linear transformations between (y_1, y_2) and (x_1, x_2):

$$\begin{aligned} y_1 &= M_{11}x_1 + M_{12}x_2, \\ y_2 &= M_{21}x_1 + M_{22}x_2. \end{aligned} \tag{1.2.9}$$

The multiplication implied in (1.2.8) can be written, using the summation convention (sum over the repeated indices),

$$y_i = \sum_{j=1}^{2} M_{ij}x_j \qquad i = 1, 2. \tag{1.2.10}$$

Let us consider the situation where there is another matrix equation between

Y and Z

$$Z = NY,$$

or

$$\begin{pmatrix} z_1 \\ z_2 \end{pmatrix} = \begin{pmatrix} N_{11} & N_{12} \\ N_{21} & N_{22} \end{pmatrix} \begin{pmatrix} y_1 \\ y_2 \end{pmatrix}. \tag{1.2.11}$$

It is obvious that Z and Y must be related by the matrix equation

$$Z = NY = NMX = QX, \tag{1.2.12}$$

where

$$Q \equiv NM, \tag{1.2.13}$$

or

$$\begin{pmatrix} Q_{11} & Q_{12} \\ Q_{21} & Q_{22} \end{pmatrix} = \begin{pmatrix} N_{11} & N_{12} \\ N_{21} & N_{22} \end{pmatrix} \begin{pmatrix} M_{11} & M_{12} \\ M_{21} & M_{22} \end{pmatrix}$$

$$= \begin{pmatrix} (N_{11}M_{11} + N_{12}M_{21}) & (N_{11}M_{12} + N_{12}M_{22}) \\ (N_{21}M_{11} + N_{22}M_{21}) & (N_{21}M_{12} + N_{22}M_{22}) \end{pmatrix}. \tag{1.2.14}$$

The equation for the multiplication of the two (2×2) matrices (1.2.10) can be easily derived from (1.2.9), (1.2.11), and (1.2.12).

Note that

$$NM \neq MN, \tag{1.2.15}$$

or, in other words, matrix multiplication is not commutative.

The transformations between Z, Y, and X need not be limited to three matrices. We can have a final matrix M composed of six matrices multiplied together

$$M = M_1 M_2 M_3 M_4 M_5 M_6. \tag{1.2.16}$$

Using (1.2.5), we define the determinant of a matrix M as

$$\det M = M_{11}M_{22} - M_{21}M_{12}. \tag{1.2.17}$$

It can be easily shown that (from (1.2.13))

$$\det Q = (\det N)(\det M)$$
$$= (\det M)(\det N). \tag{1.2.18}$$

Also, since the determinant is a scalar quantity, the order of this product does not matter. This readily extends, for (1.2.16), to

$$\det M = (\det M_1)(\det M_2)(\det M_3)(\det M_4)(\det M_5)(\det M_6). \tag{1.2.19}$$

1.2.2. The Translational Matrix

Let us consider the propagation of a ray to a distance $D_1 + D_2$ from A to C on the optical axis. As shown in Fig. 1.2.2, there are two ways we can make

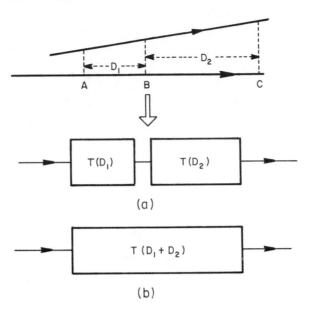

Fig. 1.2.2. Propagation of a ray through distance $D_1 + D_2$: (a) one equivalent system, $T(D_1)T(D_2)$ and (b) alternative system, $T(D_1 + D_2)$.

the equivalent optical system for this case. In the first case, we can consider two optical systems in tandem to represent the propagation by the distance $D_1 + D_2$. The equivalent optical system matrix is given by

$$T(D_2)T(D_1) = \begin{pmatrix} 1 & D_2 \\ 0 & 1 \end{pmatrix}\begin{pmatrix} 1 & D_1 \\ 0 & 1 \end{pmatrix}. \tag{1.2.20}$$

Second, we could use the translation by a distance $(D_1 + D_2)$. The system matrix in this case is given by

$$T(D_1 + D_2) = \begin{pmatrix} 1 & D_1 + D_2 \\ 0 & 1 \end{pmatrix}. \tag{1.2.21}$$

It is easy to show that

$$T(D_1 + D_2) = T(D_2)T(D_1), \tag{1.2.22}$$

as expected, because it should not matter whether we consider the propagation through empty or homogeneous space by a distance $D_1 + D_2$ on the optical axis, or through a propagation of distance D_1 and then through a distance D_2. A note on sign convention: if the ray is below the optical axis, the "x" is negative. Also, θ is positive if measured counterclockwise from the optical axis; otherwise, it is negative.

1.2.3. The Matrix for Refraction

Let us consider the refraction of a ray through an interface between two media having refractive indices n_1 and n_2, respectively. As shown in Fig. 1.2.3, the radius of curvature of the interface is R, with the origin "O" on the optical axis. We are interested in finding the final "X_2" as the ray passes through the interface. This can be done through the refraction matrix, $R(R)$, defined as

$$X_2 = R(R)X_1, \tag{1.2.23}$$

where we need to determine the elements of the $R(R)$ matrix. As the refraction does not change the position of the ray—only its angle—we immediately note that

$$x_2 = x_1. \tag{1.2.24}$$

The angle of incidence, α_1, can be written as

$$\alpha_1 = \theta_1 + \theta_0,$$

where θ_0 is the angle subtended by the optical axis and the radius drawn from the origin to the point where the ray intersects the interface, P. Thus

$$\theta_0 \approx \frac{x_1}{R}. \tag{1.2.25}$$

The angle of refraction, α_2, is given by

$$\begin{aligned}
\alpha_2 &= \theta_0 - (-\theta_2) \\
&= \theta_0 + \theta_2.
\end{aligned} \tag{1.2.26}$$

(Note that the angle of the refracted ray, as drawn, is negative.) Using Snell's law of refraction

$$n_1 \alpha_1 = n_2 \alpha_2,$$

or

$$n_1(\theta_1 + \theta_0) = n_2(\theta_0 + \theta_2),$$

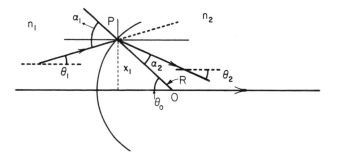

Fig. 1.2.3. Schematics of equivalent matrix for refraction.

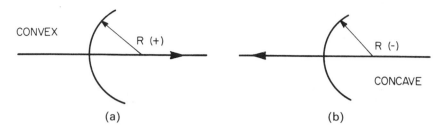

Fig. 1.2.4. Sign convention for radius of curvature of refracting surfaces: (a) convex surface and (b) concave surface.

or

$$\theta_2 = -\frac{P}{n_2}x_1 + \frac{n_1}{n_2}\theta_1, \qquad (1.2.27)$$

where $P = (n_2 - n_1)/R$, and is defined as the power of the refractive surface to bend a ray or the bending power of the surface. Thus,

$$x_2 = 1 \cdot x_1 + 0 \cdot \theta_1,$$

$$\theta_2 = -\frac{P}{n_2}x_1 + \frac{n_1}{n_2}\theta_1,$$

or

$$R(R) = \begin{pmatrix} 1 & 0 \\ -P/n_2 & n_1/n_2 \end{pmatrix}. \qquad (1.2.28)$$

Therefore, the determinant of the refraction matrix is n_1/n_2. Again, we must be careful about the sign convention. Figure 1.2.4(a) shows a convex surface where R is positive. Figure 1.2.4(b) is for negative R and for a concave surface. Note, in Fig. 1.2.4, that the direction of the optical axis points the direction in which the light rays are propagating. A simple way to remember the sign convention is: If the arrow on the radius (from the origin to the interface) and the arrow on the optical axis are in opposite directions, only the "R" is positive.

It is interesting to note that if the interface is flat (that is, if its radius of curvature is infinite), then the bending power is

$$P(R \to \infty) = \frac{n_2 - n_1}{R} = 0.$$

In that case, the refraction matrix is given by

$$R(R = \infty) = \begin{pmatrix} 1 & 0 \\ 0 & n_1/n_2 \end{pmatrix}.$$

Sign Conventions for Geometrical Optics

x (height of object measured from the optical axis)
x is negative if below the axis, and positive if above.

θ (angle of ray measured with reference to the optical axis)
θ is positive if measured counterclockwise from the optical axis, and negative if measured clockwise.

R (radius of refractive surface)
R is positive if antiparallel with the ray, and negative if parallel with the ray.

1.2.4. Matrix for a Simple Lens

A lens is an optical device, which we shall see later has many useful properties. Two of these properties are the focusing of rays and the imaging of objects. Using combinations of lens we can build equipment like binoculars, cameras, microscopes, and telescopes. In general, a simple lens is a piece of glass with two refractive surfaces having radii of curvature R_1 and R_2, as shown in Fig. 1.2.5. The thickness of the lens at the optical axis is "d". For a thin lens this "d" is considered to be approximately zero. The equivalent system model for the lens is shown in Fig. 1.2.5(b), where we need to find out the values of the lens' matrix "M" elements. This M matrix represents the property of the ray, from the point where it is incident on the front surface of the lens until it just exits from the back surface of the lens. This propagation of the ray, X_1, through the lens consists of the following three distinct operations:

(i) refraction through the surface having a radius of curvature R_1;
(ii) the translation of the ray through a distance "d" on the optical axis;
(iii) the final refraction through the surface having a radius of curvature $|R_2|$ where R_2 is negative.

If we denote the initial ray by X_1, and the intermediate rays by X_2 (refraction through R_1) and X_3 (translation by "d"), and the final ray by X_4 (refraction

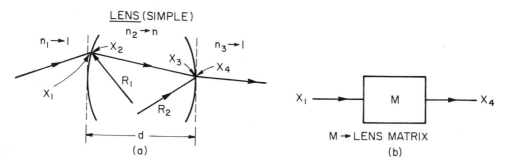

Fig. 1.2.5. Lens formation using two refracting surfaces: (a) actual lens with coordinates and (b) equivalent matrix representation.

through R_2), we can write

$$X_2 = R(R_1)X_1,$$

$$X_3 = T(d)X_2, \tag{1.2.29}$$

$$X_4 = R(R_2)X_3,$$

or

$$X_4 = R(R_2)X_3$$

$$= R(R_2)T(d)X_2$$

$$= R(R_2)T(d)R(R_1)X_1. \tag{1.2.30}$$

Noting that the equivalent lens matrix, M, is defined by

$$X_4 = MX_1, \tag{1.2.31}$$

we obtain

$$M = R(R_2)T(d)R(R_1), \tag{1.2.32}$$

or

$$M = \begin{pmatrix} 1 & 0 \\ -P_2/n_3 & n_2/n_3 \end{pmatrix} \begin{pmatrix} 1 & d \\ 0 & 1 \end{pmatrix} \begin{pmatrix} 1 & 0 \\ -P_1/n_2 & n_1/n_2 \end{pmatrix}. \tag{1.2.33}$$

Here n_1 is the refractive index of the incident media, n_2 is the refractive index of the lens material, n_3 is the refractive index of the final media, and

$$P_1 = \frac{n_2 - n_1}{R_1},$$

and   (1.2.34)

$$P_2 = \frac{n_3 - n_2}{R_2}.$$

By carrying out the matrix multiplication we obtain

$$M = \begin{pmatrix} 1 - P_1 d/n_2 & n_1 d/n_2 \\ -(P_2/n_3)(1 - P_1 d/n_2) - (P_1/n_3) & (n_1/n_3)(1 - P_2 d/n_2) \end{pmatrix}. \tag{1.2.35}$$

The above expression gives the elements of the thick lens' equivalent matrix. For a thin lens, $d \to 0$, and this simplifies to

$$(M)_{\text{thin}} = \begin{pmatrix} 1 & 0 \\ -(P_1 + P_2)/n_3 & n_1/n_3 \end{pmatrix} = \begin{pmatrix} 1 & 0 \\ -1/f & n_1/n_3 \end{pmatrix}, \tag{1.2.36}$$

where

$$\frac{1}{f} = \frac{P_1 + P_2}{n_3}. \tag{1.2.37}$$

f is generally known as the focal length, the only quantity which characterizes the thin lens. Its significance will be understood in the next section. However, we note that for a thin lens, if the refractive indices of the incident and the final media are same, i.e., $n_3 = n_1 = 1$, then

$$(M)_{\text{thin}} = \begin{pmatrix} 1 & 0 \\ -1/f & 1 \end{pmatrix} = M(f), \tag{1.2.38}$$

and

$$\frac{1}{f} = \frac{P_1 + P_2}{n_3},$$

or

$$\frac{1}{f} = \frac{1}{n_3}\left(\frac{n_2 - n_1}{R_1} + \frac{n_3 - n_2}{R_2}\right),$$

or

$$\frac{1}{f} = \left(\frac{n_2 - 1}{R_1}\right) + \left(\frac{1 - n_2}{R_2}\right),$$

or

$$\frac{1}{f} = (n_2 - n_1)\left(\frac{1}{R_1} - \frac{1}{R_2}\right)$$

$$= (n_2 - n_1)\left(\frac{1}{|R_1|} + \frac{1}{|R_2|}\right). \tag{1.2.39}$$

The last expression is generally known as the lens designer's formula and is obtained by noting that "R_2" is negative. Also, for $n_3 = n_1 = 1$, the equivalent matrix for a compound lens simplifies to

$$M = \begin{pmatrix} 1 - P_1 d/n_2 & d/n_2 \\ -P_2(1 - P_1 d/n_2) - P_1 & 1 - P_2 d/n_2 \end{pmatrix}. \tag{1.2.40}$$

An interesting case for (1.2.40) is when $R_1 = R_2 = \infty$, but $d \neq 0$. That is, for a flat piece for material with a refractive index of n_2, the M matrix is given by

$$M = \begin{pmatrix} 1 & d/n_2 \\ 0 & 1 \end{pmatrix}.$$

1.3. Image Formation

The concept of image formation is a very important concept in optics. As shown in Fig. 1.3.1, let us consider a point source of light, S, and an optical

Fig. 1.3.1. General optical system showing the image formation of the source(s), through the optical system having system matrix, M, and image, I.

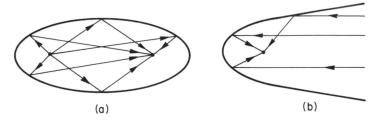

(a)   (b)

Fig. 1.3.2. Perfect image formation in: (a) ellipsoidal and (b) parabolic mirror.

system, M. Let us refer to the image of the source S through the optical system M by I. The condition of perfect image formation is that *all* the light rays emanating from the source, either must converge to the point I after passing through the optical system or appear to diverge from the image point, I. If the rays actually meet at the image point they are called *real* images, otherwise they form *virtual* images. As we know that the point source radiates in all directions equally, we immediately note that even an infinitely long lens cannot form a perfect image because the rays going away from the lens will never arrive at the image point. Of course, the paraxial approximation will not hold in this case either.

The only optical systems which can form a perfect image are the mirrors having the shape of conical surfaces. These are shown in Fig. 1.3.2. For example, the source, S, placed at one of the focii of the elliptical mirror, will form a perfect image at the point, I, the other focus of the ellipsoidal surface as shown in Fig. 1.3.2(a). Similarly, the parabolic and hyperbolic mirrors can form the perfect image.

For practical purposes, we define the so-called "approximate image". If the rays from the source, S, within a very small solid angle converge to the image after passing through the optical system, I, then we say that a real image has formed. If the rays actually diverge but appear to diverge from the image point, I, we say that a virtual image has formed. Figure 1.3.3 shows the approximate image formation. Note the large number of rays emanating from S that are not even incident on the optical system.

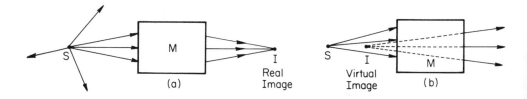

Fig. 1.3.3. Practical and approximate image formation: (a) real image and (b) virtual image.

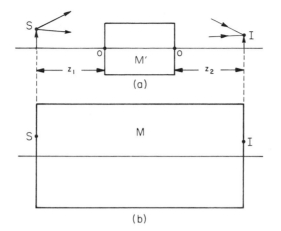

Fig. 1.3.4. Schematics and coordinates for approximate image formation calculations: (a) actual system and (b) equivalent system used for calculations.

In most practical cases, we will be interested in the image of an object, which is not a point source. However, summation of many points make an object and, thus, once we know how to find the image of a point, we can easily obtain the image of an extended object. However, before we proceed any further, it is worthwhile to find the mathematical equivalent of the approximate image formation. In Fig. 1.3.4, we note that the source, S, is situated at a distance z_1 from the edge of the optical system and at a distance x_1 from the optical axis. The optical system is M', so let us consider that the image I is formed at a distance z_2 from the other edge of M' and at a distance x_2 from the optical axis. Note that z_1 and x_1 are known. What we want to find out is z_2, called the image distance, and x_2, which is related to the size of the image.

Let us consider the bundle of rays $X_1(x_1, \theta_1), \ldots$, all starting from S within a very small solid angle. Note that all the rays have the same x value, x_1. However, the angles are different. These rays, first of all, go through a translation, of distance z_1, then through the optical system matrix, M', and finally through the image distance z_2, to form the image. If the rays arriving at the image point, I, are denoted by

$$X_2 = MX_1, \tag{1.3.1}$$

then

$$M = T(z_2)M'T(z_1) \tag{1.3.2}$$

is the equivalent system matrix for the image formation, which includes the optical system and the image and object distances. Thus, in terms of the M matrix elements,

$$x_2 = M_{11}x_1 + M_{12}\theta_1,$$
$$\theta_2 = M_{21}x_1 + M_{22}\theta_1. \tag{1.3.3}$$

Now, if $M_{12} = 0$, then we find that all the rays from x_1 will pass through x_2 irrespective of at what angle they originated. Thus the condition for image formation is

$$M_{12} = 0. \tag{1.3.4}$$

The solution of (1.3.4) gives the value of z_2, the image distance, in terms of the other constants of the system. The value of x_2 is obtained from the equation

$$x_2 = M_{11}x_1. \tag{1.3.5}$$

The lateral magnification is given by

$$m_x = \frac{x_2}{x_1} = M_{11}, \tag{1.3.6}$$

and the angular magnification is given by

$$\frac{\Delta\theta_2}{\Delta\theta_1} = m_\alpha = M_{22}. \tag{1.3.7}$$

If the object is located in a medium having a refractive index n_1, and the image is formed in a medium having a refractive index n_2, then we know that

$$\det M = \det T(z_1) \times \det M' \times \det T(z_2)$$

$$= 1 \times \frac{n_1}{n_2} \times 1 = \frac{n_1}{n_2}, \tag{1.3.8}$$

or

$$M_{11}M_{22} - M_{12}M_{21} = \frac{n_1}{n_2}. \tag{1.3.9}$$

As $M_{12} = 0$, for the image formation condition, we obtain an important relationship

$$M_{11}M_{22} = \frac{n_1}{n_2},$$

$$m_x m_\alpha = \frac{n_1}{n_2}. \tag{1.3.10}$$

In general, the object and image spaces are in air. In that case, $n_1 = n_2 = 1$, or

$$m_x = \frac{1}{m_\alpha}. \tag{1.3.11}$$

1.3.1. Image Formation by a Thin Lens in Air

As the object and the image are in air, this example of image formation simplifies significantly. The image distance, v, and the object distance, u, are

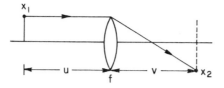

Fig. 1.3.5. Image formation through a lens. u = object distance, v = image distance, x_1 = size of the object, and x_2 = size of the image.

shown in Fig. 1.3.5. A ray is traced through the optical system, its different values at different points on the optical axis are denoted by X_1 and X_2, respectively,

$$
\begin{aligned}
X_2 &= T(v)M(f)T(u)X_1 \\
&= \begin{pmatrix} 1 & v \\ 0 & 1 \end{pmatrix}\begin{pmatrix} 1 & 0 \\ -1/f & 1 \end{pmatrix}\begin{pmatrix} 1 & u \\ 0 & 1 \end{pmatrix}\begin{pmatrix} x_1 \\ \theta_1 \end{pmatrix} \\
&= \begin{pmatrix} 1 - v/f & u - uv/f + v \\ -1/f & -u/f + 1 \end{pmatrix}\begin{pmatrix} x_1 \\ \theta_1 \end{pmatrix}.
\end{aligned}
\tag{1.3.12}
$$

Using the image formation condition (1.3.4), we obtain

$$
M_{12} = 0 = u - \frac{uv}{f} + v = 0,
$$

or

$$
\frac{1}{u} + \frac{1}{v} = \frac{1}{f}.
\tag{1.3.13}
$$

The lateral magnification, m_x (in this case) is given by

$$
m_x = 1 - \frac{v}{f} = -\frac{v}{u},
\tag{1.3.14}
$$

whereas the angular magnification, m_α, is given by

$$
m_\alpha = 1 - \frac{u}{f} = -\frac{u}{v}.
\tag{1.3.15}
$$

As expected, the condition

$$
m_x m_\alpha = 1
\tag{1.3.16}
$$

is satisfied.

For the thin lens, the following special cases are of great importance.

Case I. $u \to \infty$ or the object is far away from the lens. Then

$$
v = f.
$$

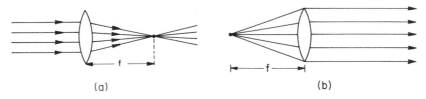

Fig. 1.3.6. (a) All parallel rays passing through the lens converge in the focal plane. (b) All rays from the sources situated at the focal plane become parallel rays after passing through the lens.

This also means that all the rays parallel to the optical axis and incident on the lens pass through the focus which is at a distance f from the lens (Fig. 1.3.6(a)).

Case II. $u = f$ or the object is at a distance f from the lens. Then

$$v = \infty.$$

This is shown in Fig. 1.3.6(b).

Case III. $u = 2f$. Then

$$v = 2f,$$

$$m_x = -1,$$

$$m_\alpha = -1.$$

Finally, in Fig. 1.3.7, the variation of the image distance and the lateral and angular magnification are plotted as a function of the object distance. Note that m_x and m_α are positive for the virtual image. This happens when the image distance is negative, or the virtual image is formed on the same side of the lens as that of the object.

It is also interesting to obtain the image formation through graphical construction. Remember that the image will be formed at the point when two different rays, emanating from the source and passing through the lens, cross each other. It is advantageous to consider the following four rays:

(i) $x_1 = 0$ and $\theta_1 = 0$.

Then from (1.3.12), or for any matrix,

$$x_2 = 0 \quad \text{and} \quad \theta_2 = 0.$$

This is a very important ray. The ray identical to the optical axis always passes undeviated through an optical system.

(ii) $x_1 =$ any value, $\theta_1 = 0$.

$$x_2 = 0, \quad \theta_2 = -\frac{x_1}{f}.$$

x_2 is at $z = f$ in the image space.

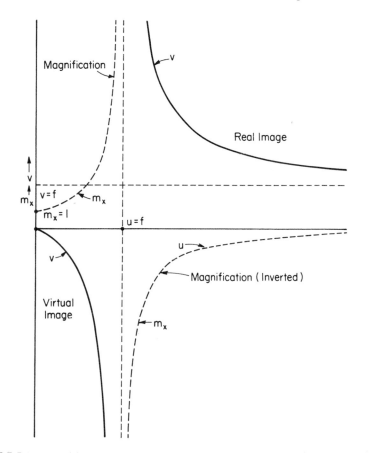

Fig. 1.3.7. Image position and magnification for a lens with focal length, f, as a function of object distance, u, positive only.

Any ray parallel to the optical axis passes through the focus in the image space.

(iii) $x_1 = 0$ at $z_1 = -f$ and $\theta_1 =$ any value. $\theta_2 = 0$.

Any ray passing through the focus in the object space emerges parallel to the optical axis in the image space.

(iv) $x_1 = 0$ at $z = 0$; $\theta_2 = \theta_1$.

Rays passing through the center of the lens emerge undeviated in the image space.

In Fig. 1.3.8 several cases of image formation are illustrated using the above-mentioned four types of ray tracing. Different types of lenses, such as biconvex, concavo-convex, biconcave, plano-concave, and plano-convex are also illustrated in Fig. 1.3.9.

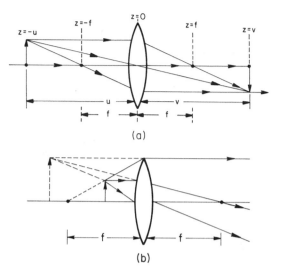

Fig. 1.3.8. Image formation through ray tracing: (a) real image and (b) virtual image.

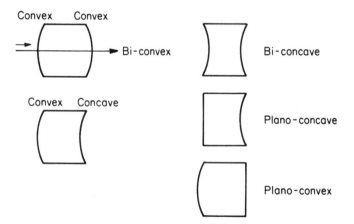

Fig. 1.3.9. Different types of lenses.

1.4. Complex Systems

A complex optical system is formed by two or more thin lenses or by different combinations of thick lenses, of which the equivalent system matrix is given by (1.2.35) and (1.2.40). First, let us consider one of the simplest complex systems, the two thin lenses separated by a distance "d", as shown in Fig. 1.4.1. The focal lengths of the lenses are f_1 and f_2, respectively. There are two ways we can discuss the image formation of this two-lens combination. The first

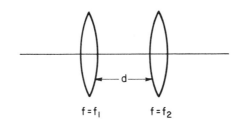

Fig. 1.4.1. Schematics for an equivalent thin-lens formulation.

method is to use the arguments given in the previous section to calculate the intermediate image position due to the first lens only. Then the calculations are repeated for the second lens to obtain the final image. Let the image of the object S, due to the first lens only, be I_1. The position and magnification of I_1 can be obtained using the equations developed in the previous section. Then, consider I_1 as the object for the image formation by the second lens. Again, using the same formulas, the final image I is obtained. It is more illustrative to take a numerical example. Let $f_1 = 5$ cm, $f_2 = 10$ cm, $d = 12$ cm, $u_1 = 15$ cm.

Then $v_1 = 7.5$ and $u_2 = 12 - 7.5 = 4.5$. (*Note:* u_2 is not 7.5 but 4.5 cm.) The final image distance $v_2 = -90/11$ cm. The total lateral magnification, m_x, is given by

$$m_x = m_{x_1} m_{x_2} = \left(-\frac{v_1}{u_1}\right)\left(-\frac{v_2}{u_2}\right) = -\frac{7.5}{15} \times \frac{90}{11 \times 4.5} = \frac{10}{11}.$$

The second method is more general and can also be applied to the combination of compound lenses or any optical system. In this method, we obtain the equivalent system matrix including the image and object distances, as discussed in Section 1.3, and we obtain the relevant quantities using the approximate image formation conditions. The student should work out the previous example by this method and obtain the same answers. The equivalent system matrix will be given by

$$(M) = T(v_2)M'T(u_1),$$

where

$$(M') = M(f_2)T(d)M(f_1). \tag{1.4.1}$$

However, a variation of this second method is more interesting and often used. This method can be called image formation, using the equivalent thin lens formulation.

The method is based on the fact that any optical system in conjunction with two additional empty spaces can be shown to be an equivalent thin lens. Consider the optical system, which can be very complex, consisting of many thin and thick lenses having different materials with different refractive indices.

Still, the equivalent matrix for this complex system can be written as

$$(M) = \begin{pmatrix} M_{11} & M_{12} \\ M_{21} & M_{22} \end{pmatrix},$$

where M_{11}, M_{12}, M_{21}, and M_{22} are arbitrary. However, if the complex system is in air, that is, if both the image and object space is air, then

$$\det M = 1. \tag{1.4.2}$$

Because $\det M = (\det M_1) \times \det(M_2) \times \det(M_3),\ldots$, where M_1 and M_2, etc., are the individual matrices. However, we know that the determinant of the translational matrices is 1, and that the determinant of the refraction or lens matrix is the ratio of the initial refractive index and the final refractive index. Thus,

$$\det(M) = \frac{1}{n_1} \cdot \frac{n_1}{n_2} \cdot \frac{n_2}{n_3} \cdots \frac{n_i}{1} = 1. \tag{1.4.3}$$

Thus, in any general optical system matrix, only three elements are arbitrary.

To obtain the equivalent thin-lens formulation, consider the total system shown in Fig. 1.4.2. This system consists of an empty space of length D in front of the complex system, and another empty space of length D' behind the complex system. The total system matrix is given by

$$\begin{aligned}
(M_{\text{sys}}) &= \begin{pmatrix} 1 & D' \\ 0 & 1 \end{pmatrix} \begin{pmatrix} M_{11} & M_{12} \\ M_{21} & M_{22} \end{pmatrix} \begin{pmatrix} 1 & D \\ 0 & 1 \end{pmatrix} \\
&= \begin{pmatrix} M_{11} + D'M_{21} & M_{11}D + M_{12} + D'(M_{21}D + M_{22}) \\ M_{21} & M_{21}D + M_{22} \end{pmatrix}.
\end{aligned} \tag{1.4.4}$$

Remember that D and D' have arbitrary values. Thus, we can choose the values of D and D' such that

$$(M_{\text{sys}})_{11} = M_{11} + D'M_{21} = 1,$$

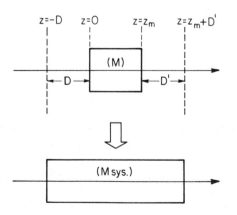

Fig. 1.4.2. An optical system and its equivalent thin-lens system.

and

$$(M_{sys})_{22} = M_{21}D + M_{22} = 1,$$

or

$$D' = \frac{1 - M_{11}}{M_{21}}, \tag{1.4.5}$$

and

$$D = \frac{1 - M_{22}}{M_{21}}.$$

Of course, the above equations are meaningless unless

$$M_{21} \neq 0 \tag{1.4.6}$$

(since $f = 1/M_{21}$).

Using the values of D and D' obtained in (1.4.6), we obtain

$$(M_{sys})_{12} = 0, \tag{1.4.7}$$

where use of (1.4.2) has been made. Thus, with these chosen values of D and D', we obtain

$$(M_{sys}) = \begin{pmatrix} 1 & 0 \\ -1/f & 1 \end{pmatrix}, \tag{1.4.8}$$

where

$$'f = -\frac{1}{M_{21}} \tag{1.4.9}$$

and is only valid where $M_{21} \neq 0$.

This is a startling result in the sense that any complex optical system can be made equivalent to a thin lens. The planes perpendicular to the optical axis at $z = D$ and $z = z_m + D'$ are known as the principal planes. z_m is the width of the optical system with matrix M. Note that for image formation using this equivalent thin lens, the object distance must be measured from the D principal plane and the image distance must be measured from the D' principal plane. For an actual thin lens

$$D = D' = 0$$

and the principal planes coincide with the physical position of the lens.

It is to be noted that we have derived equations (1.4.2)–(1.4.9) under the assumption that the initial object and image spaces are air or, rather, vacum. However, if they are n and n', respectively, then the equations are modified as follows:

$$\det M = \frac{n}{n'},$$

$$\tag{1.4.10}$$

$$M_{21}D + M_{22} = \frac{n}{n'},$$

or

$$D = \frac{n/n' - M_{22}}{M_{21}}, \tag{1.4.11}$$

$$M_{sys} = \begin{pmatrix} 1 & 0 \\ -1/fn' & n/n' \end{pmatrix}. \tag{1.4.12}$$

1.4.1. Image Formation Using an Equivalent Thin-Lens Formulation

Image formation calculations using this technique are easy to perform when we remember the formulas derived in Section 1.3.1, with the important reminder that the object and image distances are measured from the imaginary principal planes. Let us consider two examples: (i) the thick lens, and (ii) the two thin-lens combination.

Thick Lens

The equivalent matrix for the thick lens is given by (from (1.2.40))

$$M = \begin{pmatrix} 1 - P_1 d/n_2 & d/n_2 \\ -P_2(1 - P_1 d/n_2) - P_1 & 1 - P_2 d/n_2 \end{pmatrix}. \tag{1.4.13}$$

Using (1.2.10) and setting $(M_{sys})_{11} = 1$ and $(M_{sys})_{22} = 1$, we find the principal planes located at

$$D = \frac{-P_2 d}{n_2 P},$$
$$D' = \frac{-P_1 d}{n_2 P}, \tag{1.4.14}$$

where

$$P = \frac{1}{f_{eq}} = P_1 + P_2 - \frac{P_1 P_2 d}{n_2}. \tag{1.4.15}$$

The equivalent system is shown in Fig. 1.4.3. This figure also illustrates how the ray tracing can be performed using special rays.

(i) The parallel ray coming from infinity hits the D-plane at A. It emerges at B on the D'-plane and passes through the focus. It is interesting to point out that we need not know the exact path the ray takes between the D–D'-plane.

(ii) The ray passing through the focus in the object space comes out parallel to the optical axis from the D'-plane. Of course, it is at the same vertical distance from the optical axis, both on the D-plane and the D'-plane. This is true for all rays.

(iii) The ray incident on the intersection of the D-plane and the optical axis emerges at the same angle, with respect to the optical axis on the D'-plane.

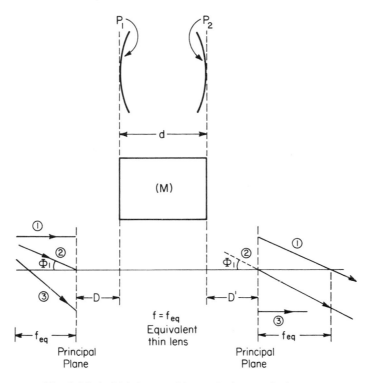

Fig. 1.4.3. A thick lens and its equivalent optical system.

(iv) The formula $1/S + 1/S' = 1/f_{eq}$ holds where S is the object distance measured from D, and S' is the image distance measured from D'.

$$m_x = -\frac{S'}{S}, \qquad m_\alpha = -\frac{S}{S'}, \qquad (1.4.16)$$

and

$$m_x m_\alpha = 1.$$

The above discussion is somewhat modified if the image and object spaces are not air. This can easily be derived by the reader.

Two Thin Lenses Separated by a Distance, d

For this case

$$
(M) = \begin{pmatrix} 1 & 0 \\ -1/f_2 & 1 \end{pmatrix} \begin{pmatrix} 1 & d \\ 0 & 1 \end{pmatrix} \begin{pmatrix} 1 & 0 \\ -1/f_1 & 1 \end{pmatrix}
$$
$$
= \begin{pmatrix} 1 - d/f_1 & d \\ -(1/f_2)(1 - d/f_1) - 1/f_1 & -d/f_2 + 1 \end{pmatrix} \qquad (1.4.17)
$$

$$\frac{1}{f_{eq}} = \frac{1}{f_1} + \frac{1}{f_2} - \frac{d}{f_1 f_2},$$

$$D = -\frac{df_{eq}}{f_2},$$

$$D' = -\frac{df_{eq}}{f_1}.$$

(1.4.18)

Using the previous example of Fig. 1.4.1

$$u_1 = 15 \text{ cm}, \qquad f_1 = 5 \text{ cm}, \qquad f_2 = 10 \text{ cm}, \qquad d = 12 \text{ cm},$$

we obtain

$$f_{eq} = 16\tfrac{2}{3} \text{ cm},$$

$$D = -20,$$

$$D' = -40.$$

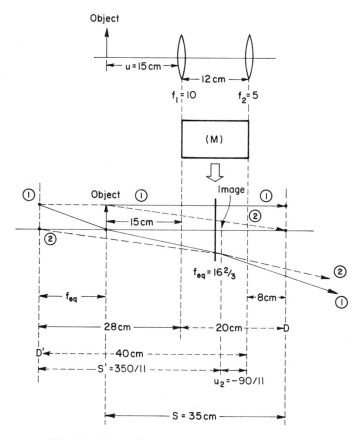

Fig. 1.4.4. Image formation due to two thin lenses.

Thus, $S = u_1 - D = 15 - (-20) = 35$,

$$\frac{1}{S'} = \frac{1}{f_{eq}} - \frac{1}{S},$$

$$S' = \frac{350}{11},$$

$$u_2 = S' + D' = \frac{350}{11} + (-40) = -\frac{90}{11},$$

$$m_x = -\frac{S'}{S} = -\frac{10}{11}.$$

As expected, the values agree with those derived previously. The image formation using ray tracing is also shown in Fig. 1.4.4.

1.5. The Telescoping System

If $M_{21} = 0$, then the equivalent thin-lens formulation that we have discussed earlier does not hold good. These systems are called telescopic systems. For image-formation calculations in the telescopic system, we have to use the method discussed in Section 1.3. As shown in Fig. 1.5.1, the total optical system matrix (including the object distance, z, the image distance, z', and the telescopic system matrix, M), is given by

$$(M)_{tot} = T(z')MT(z)$$

$$= \begin{pmatrix} 1 & z' \\ 0 & 1 \end{pmatrix} \begin{pmatrix} M_{11} & M_{12} \\ 0 & M_{22} \end{pmatrix} \begin{pmatrix} 1 & z \\ 0 & 1 \end{pmatrix}$$

$$= \begin{pmatrix} 1/P_\alpha & M_{12} + z'P_\alpha + z/P_\alpha \\ 0 & P_\alpha \end{pmatrix}, \tag{1.5.1}$$

where

$$P_\alpha = M_{22} = \frac{1}{M_{11}}. \tag{1.5.2}$$

The last equality is obtained by noting that the determinant of the M matrix is 1. For image formation

$$M_{12} + z'P_\alpha + \frac{z}{P_\alpha} = 0,$$

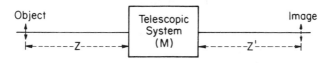

Fig. 1.5.1. A telescopic system.

or

$$z' = -\frac{z}{P_\alpha^2} - \frac{M_{12}}{P_\alpha}. \tag{1.5.3}$$

Also, for this case,

$$(M)_{tot} = \begin{pmatrix} 1/P_\alpha & 0 \\ 0 & P_\alpha \end{pmatrix}.$$

Thus,

$$x_2 = \frac{x_1}{P_\alpha},$$

$$\theta_2 = P_\alpha \theta_1.$$

From which we obtain

$$m_x = \frac{1}{P_\alpha}, \tag{1.5.4}$$

and

$$m_\alpha = P_\alpha. \tag{1.5.5}$$

An important quantity for the telescopic system is the longitudinal magnification. This is given by

$$\frac{\Delta z'}{\Delta z} = -\left(\frac{1}{P_\alpha}\right)^2. \tag{1.5.6}$$

For a two thin-lens combination, the condition for a telescopic system is obtained from (1.4.18), and is given by

$$d = f_1 + f_2. \tag{1.5.7}$$

For this important case, (1.4.17) becomes

$$(M) = \begin{pmatrix} -f_2/f_1 & f_1 + f_2 \\ 0 & -f_1/f_2 \end{pmatrix}, \tag{1.5.8}$$

or

$$P_\alpha = -\frac{f_1}{f_2}. \tag{1.5.9}$$

1.6. Some Comments About the Matrix Method

(1) Some of you have probably noticed that although any optical system is three dimensional, we have only been concerned with two-dimensional systems; or, we have only considered the plane containing the ray and the optical axis. This does not, however, restrict our calculations because, in general, the optical axis is the axis of symmetry. Also, the rays from an object can be always split up into rays having x and y components. Once they are resolved into x and y components, they can be treated independently and thus the two-

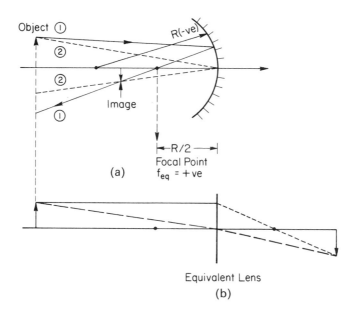

Fig. 1.6.1. An optical mirror and its equivalent system: (a) actual system and (b) equivalent system, unfolded.

dimensional calculations become one dimensional. The above statement can be proved mathematically. See, for example, Klein and Furtak [1].

(2) If the optical system contains a mirror, for example, a Galilean telescope, then the rays can also be calculated using the matrix approach. However, in this case, we consider that the image space has a refractive index $n = -1$, and that the optical axis direction is in $-z$ direction for the image space. For example, consider the imaging problem shown in Fig. 1.6.1. The convex mirror has a radius of curvature, R. Thus, the equivalent matrix for reflection is given by (1.2.28)

$$R(R) = \begin{pmatrix} 1 & 0 \\ -P & 1 \end{pmatrix},$$

where

$$P = \frac{(-1) - 1}{R} = -\frac{2}{R},$$

or

$$R(R) = \begin{pmatrix} 1 & 0 \\ -1/f & 1 \end{pmatrix},$$

where

$$f_{eq} = -\frac{R}{2}.$$

Image formation using ray tracing is also illustrated in Fig. 1.6.1.

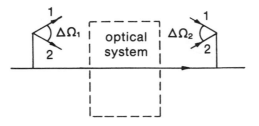

Fig. 1.6.2. Real and virtual objects and images.

(3) The physical meaning of lateral magnification is obvious; however, the meaning of angular magnification may not be so. To clarify it, consider Fig. 1.6.2 which shows a bundle of rays with a solid angle, $\Delta\Omega_2$ converging at the image point. These rays originated at the object point having a solid angle $\Delta\Omega$,

$$\Delta\Omega_2 = \Delta\Omega_1 \times m_\alpha^2$$

and the image will appear brighter if $m_\alpha > 1$.

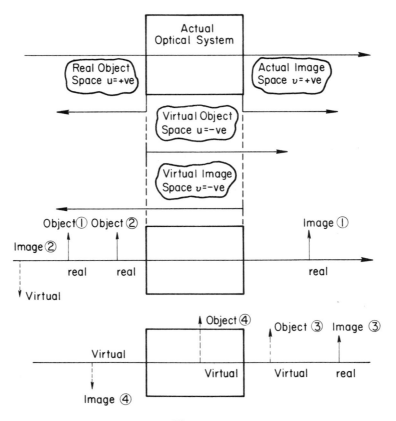

Fig. 1.6.3

(4) We have mentioned image space and object space. In general, the object space is to the left-hand side of the lens and the image space is to the right-hand side, if the rays are traveling from left to right. However, an image can be to the left and, in the same way, the object can be to the right. Of course, these are the virtual image and object, respectively. Actually, the object space is the whole space with the positive direction to the left of the lens, and the negative direction to the right of the lens. Similarly, the image space is the whole space where the positive direction is to the right side of the lens, whereas the negative side is to left side of the lens. Similar situations can arise with respect to the principal planes and the position of the focus. Some interesting cases are shown in Fig. 1.6.3.

1.7. Apertures and Stops

Up until now we have not discussed anything regarding the size of the lenses used in optical systems. Although we have used paraxial approximation, we have implied that the size of the lens is of infinite dimension normal to the optical axis. However, the lens diameter or lens aperture, as it is sometimes called, is finite. There are two points to be considered in connection with this finite lens size:

(i) the paraxial approximation may not hold well;
(ii) the rays emanating from an object may be partially or completely blocked due to finite lens size or other restrictions such as a mechanical lens holder or intentional aperture.

The effect of the first point is the so-called subject of aberration and will be discussed later. The latter point is discussed in this section. It is to be noted that there is a third effect, the diffraction effect, which will be discussed in Part II.

1.7.1. The Aperture Stop

To understand the aperture stop, let us first consider the simple case depicted in Fig. 1.7.1. The same object is imaged by different lenses with identical focal length, f, but having different diameters D_1 and D_2 ($D_1 < D_2$). The light within the solid angle, $(\Delta\alpha_1)^2$, will be imaged by the first lens where

$$\Delta\alpha_1 \approx \frac{D_1}{2u} \tag{1.7.1}$$

under paraxial approximation and for the object distance u. For the second case,

$$\Delta\alpha_2 = \frac{D_2}{2u}, \tag{1.7.2}$$

and as $D_2 > D_1$, the image in the second case will be brighter. Thus, we see

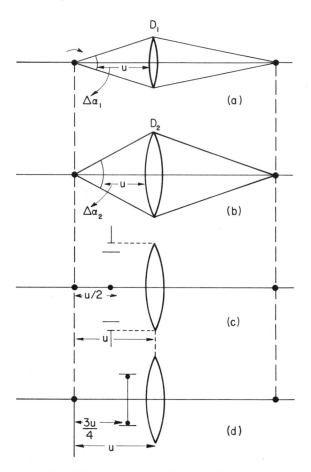

Fig. 1.7.1. Aperture stop for optical systems.

that in this simple case the size of the lenses determines the aperture stop—and the second lens has a larger aperture stop.

Now consider the situations in Fig. 1.7.1(c) and (d), where the same two lenses are used but both lenses have a mechanical aperture of diameter D_3, placed symmetrically on the optical axis as shown in the figure. For the first lens, the maximum angle the ray from the source can have, with respect to the optical axis and still passing through the aperture, is given by

$$\Delta\alpha_{S1} = \frac{D_3}{u}. \tag{1.7.3}$$

Comparing $\Delta\alpha_1$ and $\Delta\alpha_{S1}$, we see that if $D_3 > D_1/2$, then the brightness of the image (note that the position of the image is not altered at all) will still be determined by the angle $\Delta\alpha_1$, and the aperture stop will be the lens.

Similarly, for the second lens

$$\Delta\alpha_{S2} = \frac{4D_3}{2 \times 3u},$$ (1.7.4)

and the aperture stop will be the mechanical aperture unless

$$D_3 > \frac{3D_2}{4}.$$ (1.7.5)

If we assume that the condition defined by (1.7.5) is satisfied, then we find that the image brightness will be determined by the mechanical aperture and not by the lens diameter.

In Fig. 1.7.1 we have discussed two simple cases. However, an optical system can have many components. The particular component which physically limits the solid angle of rays passing through the system from an on-axis object is called the aperture stop. Thus to calculate the aperture stop, evaluate the "$\Delta\alpha$" angle for each component. Remember that when calculating $\Delta\alpha$ for a particular component, assume that the other components have infinite size. Then the component which makes the lowest "$\Delta\alpha$" is called the aperture stop.

It is of interest to define the $\Delta\alpha$ angle again. From the on-axis source, let us consider the rays which slowly make larger and larger angles with respect to the optical axis. Then, for the ray that has the largest angle and still passes through the optical component, this angle is called the "$\Delta\alpha$" angle.

Two other important quantities are generally considered. These are the exit pupil and the entrance pupil. The entrance pupil is the image of the aperture stop in object space, whereas the exit pupil is the image of the aperture stop in image space. The entrance pupil, in a sense, determines the amount of light which will pass through the optical system unobstructed, whereas the exit pupil determines where the light rays are expected to come out through the optical system.

These concepts concerning aperture stop, entrance pupil, and exit pupil are somewhat confusing. The two examples in the next section will help clarify some of this confusion. One important word of caution—the aperture stop is dependent on the position of the object. Thus, for some positions of the object, one component can be the aperture stop, whereas for a different position of the object a different element of the optical system can be the aperture stop. Of course, when the aperture stop changes, the exit and entrance pupils will change also.

Examples of Aperture Stop

Before we start discussing a new example, let us calculate the exit and entrance pupils in Fig. 1.7.1, where $f = 10$ cm, $D_1 = 5$ cm, $D_2 = 10$ cm, $D_3 = 10$ cm, and $u = 20$ cm.

First system: aperture stop = first lens,
      entrance pupil = first lens,
      exit pupil = first lens.

Second system: aperture stop = mechanical aperture,
entrance pupil = mechanical aperture,
exit pupil = located at $v = -10$ and of size 20 cm in diameter.

The Matrix Method for Finding the Aperture Stop

Instead of calculating the angle $\Delta\alpha$ by geometrical arguments, we can use the optical matrices developed previously. Consider the optical system shown in Fig. 1.7.2, where we are interested in finding the $\Delta\alpha$ angle for element 4 which is a mechanical aperture. The ray with the angle $\Delta\alpha$ will just touch the rim of element 4 which has a radius ρ. We can calculate the optical system matrix (M) from the source A to B which is the position of element 4. Then we know that

$$X_2 = (M)x_1,$$

where, for our case,

$$X_1 = \begin{pmatrix} 0 \\ \Delta\alpha \end{pmatrix}.$$

Also

$$X_2 = \begin{pmatrix} \rho \\ \theta_2 \end{pmatrix}.$$

Thus

$$\rho = M_{11} \times 0 + M_{12}\Delta\alpha,$$

$$\theta_2 = M_{21} \times 0 + M_{22}\Delta\alpha,$$

or

$$\Delta\alpha = \frac{\rho}{M_{12}}. \tag{1.7.6}$$

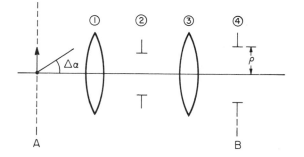

Fig. 1.7.2. Optical system to demonstrate the matrix method evaluation of aperture stop.

So to determine the aperture stop, calculate $\Delta\alpha$ for each element of the system by using the M_{12} element of the equivalent system matrix.

It should be mentioned that ordinary imaging by lenses can also be used in place of the matrix method to determine the aperture stop. Depending on the circumstances, this method may even give a better physical insight to the situation.

Examples

Consider the optical system in Fig. 1.7.3 which consists of two lenses, one having $f = 5$ cm and $D = 4$ cm and the other one having $f = 6$ cm and $D = 6$ cm. The object is located at a distance 10 cm from the first lens.

(i) Calculate α_{L1}.
   By inspection
$$\alpha_{L1} = \tfrac{2}{10} = \tfrac{1}{5}, \tag{1.7.7}$$

or we can use the matrix method. The total system matrix is simply $T(10)$. Thus,
$$M_{12} = 10,$$
$$\rho = 2,$$

therefore,
$$\alpha_{L1} = \frac{\rho}{M_{12}} = \frac{1}{5}.$$

(ii) Calculate α_{L2}.
   The image of the object through lens L_1 is at
$$v = 10 \text{ cm} \qquad \text{with} \quad m_\alpha = -1.$$

Thus, the equivalent source for lens L_2 is at $u = 12$ cm. Thus,
$$\alpha_{L2} = \frac{3}{12} \cdot \frac{1}{m_\alpha} = \frac{1}{4}. \tag{1.7.8}$$

(Note that in determining α for the aperture the sign is not important.) Note

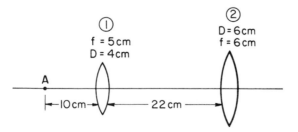

Fig. 1.7.3. A two lens optical system.

that we have to divide by the angular magnification factor. If we use the matrix method, then the total matrix is given by

$$M = T(22)M(f = 5)T(10)$$

$$= \begin{pmatrix} -\frac{17}{5} & -12 \\ -\frac{1}{5} & -1 \end{pmatrix}.$$

Thus,

$$\alpha_{L2} = \frac{-1}{4},$$

$$|\alpha_{L2}| = \frac{1}{4}.$$

Aperture stop: Comparing α_{L1} and α_{L2} we find $\alpha_{L1} < \alpha_{L2}$. Thus, lens L_1 is the aperture stop.

Entrance pupil: As lens L_1 is in the object space already, lens L_1 is again the entrance pupil.

Exit pupil: The image of lens L_1 through lens L_2 is located at $v = 8\frac{1}{4}$ cm and $m_x = \frac{3}{8}$. Thus, the exit pupil is located at a distance $8\frac{1}{4}$ cm from lens L_2 and has a diameter $\frac{3}{2}$ cm.

1.7.2. The Field Stop

The aperture stop limits the illumination of an on-axis point image, whereas the field stop does it for an off-axis source. As we shall see, the field stop is dependent on the aperture stop.

Before we define the field stop it is advantageous to define *chief* and *marginal* rays. Chief rays are those which emanate from off-axis sources and pass through the center of the aperture stop. Alternatively, we can think of the chief rays as all the rays which emanate from a source placed at the center of the aperture stop. Marginal rays are those passing through the edges of the aperture. The chief and marginal rays are illustrated in Fig. 1.7.4 for the

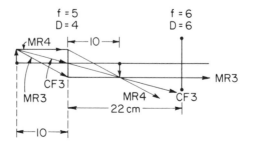

Fig. 1.7.4. Chief and marginal rays in an optical system.

optical system discussed in Fig. 1.7.3. It is obvious that all these rays do not pass through all the components of the optical system. The chief ray, CF3, and the marginal ray, MR4, fail to pass through the second lens.

The particular element in an optical system which limits the chief rays is called the field stop. Thus, if the object is placed further off-axis, the angle subtended by the chief rays at the center of the aperture stop will be larger, and the field stop determines how far the off-axis object can be situated from the optical axis and still have at least one ray which will pass through the system. If we think for a while it becomes clear that the field stop is nothing but the new aperture stop, when the object is placed at the center of the actual aperture stop. Thus, to determine the field stop of an optical system, first calculate the aperture stop and then recalculate this new aperture stop, the field stop.

Similar to the definition of entrance and exit pupils, the entrance window is the image of the field stop in object space. The exit window is defined as the image of field stop in image space. If the entrance window is near the object, then it determines how far the object can be off-axis and how an image can still be obtained. Thus, the entrance window is related to the field of view of the optical system. However, the image of the off-axis object may not be as bright as that for the on-axis object. It is quite possible that there is a gradual loss of light as the object is moved off-axis. This gradual loss of light, or the variation of brightness of the image point as the source is moved off-axis, is known as *vignetting*. To an observer in the image space the exit window tends to limit the area of the image, just as a window limits the view an observer can see when he looks through it. This concept will be clearer if we again consider the problem discussed in connection with the aperture stop in Fig. 1.7.3.

Example

To obtain the field stop, we place the object at the center of the aperture stop, which in this case is the lens L_1. Then

$$\alpha_{L2} = \frac{3}{22}.$$

As there are no other elements, we obtain the trivial answer that the field stop must be the lens L_2. The exit window is also the lens L_2. The entrance window is located at

$$v = \frac{110}{17} \quad \text{with} \quad m_x = \frac{5}{17}.$$

Thus the size of the window diameter is 30/17 cm and is located 110/17 cm in front of the lens L_1.

For this problem consider the off-axis object as shown in Fig. 1.7.4 which traces some of the rays emanating from the object through the optical system.

It is obvious that not all the rays incident on the aperture stop are incident on the second lens. Thus, vignetting will occur for this optical system. To avoid or minimize vignetting, one obvious solution is to increase the sizes of the lenses, which might be expensive or impossible. However, there is a clever solution to this problem, the use of a *field lens*.

The field lens is an extra lens which is only added to the optical system to stop or to minimize vignetting. However, this lens does not change the position of the original image. For example, in the previous problem, if we add a so-called field lens between the two lenses, at the position of the inter-mediate image plane, 10 cm from lens L_1, then the position of the final image is unchanged. However, if the diameter of this field lens is also 5 cm and its focal length is chosen properly, then it will stop vignetting. The best focal length for the field lens is when it images the aperture stop onto the field stop

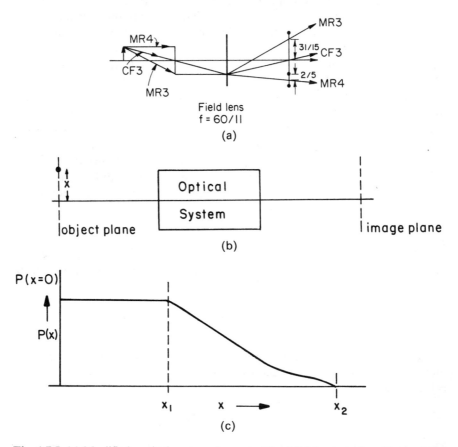

Fig. 1.7.5. (a) Modified optical system shown in Fig. 1.7.4 to stop vignetting using a field lens. (b) A general optical system with an object and image plane. (c) Plot of image intensity as a function of the object position.

which, in this case, is determined by the equation

$$\frac{1}{f} = \frac{1}{10} + \frac{1}{12} \quad \text{or} \quad f = \frac{60}{11} \text{ cm.}$$

Using this lens, the same rays are retraced through the optical system and, as shown in Fig. 1.7.5, they all pass through lens L_2.

1.7.3. Field of View

Let us consider an optical system, shown in Fig. 1.7.5(b), where the object and image planes are indicated. If a point source of 1 W power, radiating uniformly in all directions, is placed on the optical axis in the image plane, then the total power at the image point for this object, at $x = 0$, will be given by

$$P(x = 0) = \frac{\Omega}{4\pi},$$

where Ω is the solid angle subtended by the aperture stop and is given by

$$\Omega = \frac{\text{area of the entrance pupil}}{(\text{distance between the point source and the entrance pupil})^2}.$$

Now, if we move the point source away from the optical axis but still in the object plane, we obtain the total image power given by $P(x)$ where x is the position of the point source. Typically, $P(x)$ will be equal to or less than $P(x = 0)$. This is so because for x large, marginal rays will start missing the other elements in the optical system. A typical plot of $P(x)$ versus x is shown in Fig. 1.7.5(c). For a distance up to $x = x_1$, $P(x) = P(x_0 = 0)$. For $x_1 \le x \le x_2$, $P(x) < P(x = 0)$ and for $x \ge x_2$, $P(x) = 0$.

So we see that if the radius of the object is less than x_1, the image will be a true replica of the object in brightness. However, if the radius is between the values x_1 and x_2, vignetting occurs. We define x_2 as the radius of the field of view. The field of view is related to the entrance window.

1.8. Radiometry and Photometry

1.8.1. Radiometry

To discuss the brightness of an image quantitatively, we need to define the following quantities.

Radiant energy (E) is the total amount of energy radiated by an optical source, or transferred or collected in an optical system. It is measured in units of joule (J). Radiant energy density (D) is the optical energy per unit volume and its unit is joule per cubic meter (J/m^3). Radiant power (P) is the amount of radiant energy transferred per unit time. The unit of radiant power is the watt (W). Thus, if 1 W of laser power is incident on a material for 10 s, then the total incident energy is 10 J.

In general, optical energy is not monochromatic. It often contains distributions of wavelengths. To denote the energy or power contained by each wavelength we define the spectral version of each quantity. The spectral energy, $E(\lambda)\, d\lambda$, denotes the amount of optical energy in a wavelength range between λ and $\lambda + d\lambda$. Thus,

$$E = \int_0^\infty E(\lambda)\, d\lambda. \tag{1.8.1}$$

Similarly,

$$D = \int_0^\infty D(\lambda)\, d\lambda, \tag{1.8.2}$$

and

$$P = \int_0^\infty P(\lambda)\, d\lambda. \tag{1.8.3}$$

Optical energy is generally emitted from surfaces having some finite area. The radiant exitance (M) is the total power emitted per unit area. It has a unit of watt per square meter (W/m^2). Thus, for example, if a 1-W laser is emitted through a window with an area of 1 cm^2, the radiant exitance of the source will be 10^4 W/m^2. When the optical energy is incident on a surface we define the radiant incidence (N) to denote the incident power per unit area. Note that the element of area, dA, must be perpendicular to the direction of the light propagation. Otherwise, a $\cos\theta$ factor must be included if the normal to the area is not parallel to the direction of the light propagation. This is shown in Fig. 1.8.1. Using vector notation, where the element of area, $d\mathbf{A}$, is represented by a vector having magnitude dA, and direction along its normal, then we obtain the incident power to be given by

$$P = \int_s \mathbf{N} \cdot d\mathbf{A}. \tag{1.8.4}$$

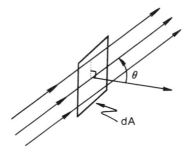

Fig. 1.8.1. Figure showing vector $d\mathbf{A}$.

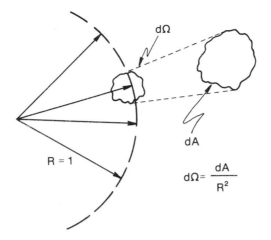

Fig. 1.8.2. Calculation of solid angle.

The spectral version is given by

$$P(\lambda) = \int_s N(\lambda) \cdot d\mathbf{A}. \tag{1.8.5}$$

The integration in the above two equations has to be performed over the whole surface on which light is incident. Irradiance (Q) is the total power incident per unit area. Its unit is joule per square meter (J/m^2).

Radiant intensity (I) is also a vector quantity and it represents the power emitted by an entire source per unit solid angle in a particular direction. Its unit is watt per steradian (W/sr). The solid angle is defined to be the ratio of area under consideration divided by the square of the radius, as shown in Fig. 1.8.2. This is also equivalent to the area subtended by the cone on a sphere of radius unity

$$d\Omega = \frac{dA}{R^2}. \tag{1.8.6}$$

For some sources, light will be independent of direction and position. This is generally known as the Lambert sources or Lambert emitters. For a spherical uniform source placed at the center of a sphere, having radius R_0, the total emitted power is

$$E_{tot} = \iint \mathbf{M} \cdot dA = 4\pi R_0^2 M. \tag{1.8.7}$$

Also

$$E_{tot} = \int d\Omega = 4\pi I.$$

Thus, for this case $I = R_0^2 M$.

Finally, radiance (L) is defined as the flux per unit solid angle per unit projected area. Its unit is watt per square meter steradian (W/(m² sr)). Thus we have the following relationships between radiance, intensity, and radiant exitance:

$$I = \iint \mathbf{L} \cdot \mathbf{dA},$$

$$M = \iint \mathbf{L} \cdot \mathbf{d\Omega}.$$

$$(1.8.8)$$

Blackbody Radiation

Let us consider a cavity or a completely enclosed source in an equilibrium condition. Within this cavity the light is completely randomized, i.e., it propagates uniformly in all directions. If a small hole is made in the cavity without disturbing the equilibrium, then the radiance of the emitted energy can be calculated.

If D is the energy density inside the cavity, then in a small volume, dV, the total amount of energy is $D\ dV$. Of this amount, only a portion is propagating within a small solid angle, $d\Omega$. This amount is given by

$$D\ dV \frac{d\Omega}{4\pi},$$

since the energy within the cavity is randomly propagating. As the light propagates with a velocity, c, the power flowing for time Δt through an area dA at an angle θ with respect to the direction of propagation, is given by

$$dP = \frac{1}{\Delta t} D\ dV \frac{d\Omega}{4\pi} = \frac{1}{\Delta t} D \cdot c\Delta t \cdot dA \cdot \cos\theta \frac{d\Omega}{4\pi}$$

$$= \frac{Dc}{4\pi} dA \cos\theta\ d\Omega.$$

Thus the radiance and radiant exitance in the cavity are given by

$$L = \frac{Dc}{4\pi} \quad \text{and} \quad M = \frac{Dc}{4}. \qquad (1.8.9)$$

If the cavity is a blackbody at an isothermal temperature T then from experimental results we obtain

$$D(f)\Delta f = \frac{8\pi n^3 hf^3}{c^3} \frac{\Delta f}{e^{hf/kT} - 1} \ \text{J/m}^3, \qquad (1.8.10)$$

where $D(f)$ is the energy density in the frequency range between f and $f + \Delta f$; h is the Planck constant, 6.626×10^{-3} J s; k is the Boltzmann constant, 1.381×10^{-23} J/K; and T is the temperature of the blackbody in degrees Kelvin.

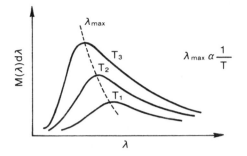

Fig. 1.8.3. Blackbody exitance $M(\lambda)$ as a function of wavelength.

If the blackbody has a small hole then the exitance of that radiation is given by

$$M(f)\Delta f = \frac{2\pi n^2 hf^3}{c^2} \frac{\Delta f}{e^{hf/kT} - 1} \text{ W/m}^2, \tag{1.8.11}$$

or

$$M(\lambda)\,d\lambda = \frac{2\pi hc^2 \lambda^{-5}}{e^{hf/kT} - 1}\,d\lambda \text{ W/m}^2.$$

The plot of $M(\lambda)$ is shown in Fig. 1.8.3 for a few temperatures. The total exitance can be obtained by integrating (1.8.11), obtaining

$$M = 5.672 \times 10^{-8}\,T^4 \text{ W/m}^2. \tag{1.8.12}$$

The peak exitance occurs at λ_m given by

$$\lambda_m T = 2897.8\,(\mu m \times \text{degrees Kelvin}), \tag{1.8.13}$$

where the dimensions of λ_m are measured in microns.

1.8.2. Photometric Unit

A human observer has different sensitivities to differing wavelengths incident on his eye. The actual sensitivity, of course, is dependent on the individual observer; however, a standard luminosity curve, shown in Fig. 1.8.4, shows the human eye response variation as a function of wavelength.

In photometry, which deals with human observers, a different set of units has been in vogue. These are shown in Table 1.8.1.

The primary standard of the photometric system is the radiant exitance of a blackbody radiator at 2043.5 K (the melting point of platinum). This radiant exitance is 60 cd/cm^2 = 60 lm/cm^2 sr. To convert the units from photometry to radiometry we note that at $\lambda = 5500\text{Å}$, 1 W is equivalent to 680 lm where y_λ is given by the standard luminosity curve shown in Fig. 1.8.4. Thus

$$P\,(W) = 680 \int P(\lambda)y_\lambda\,d\lambda \text{ lm}.$$

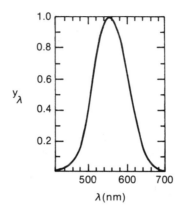

Fig. 1.8.4. Standard luminosity curve.

Some Examples

Typical examples of source and background brightness are shown in Tables 1.8.2 and 1.8.3.

Table 1.8.1. Radiometric and photometric units.

	Physical	Psychophysical
Energy	radiant energy (E), J	luminous energy, talbot
Energy density	radiant density (D), J/m^3	luminous density, talbot/m^3
Power	radiant flux (P), W	luminous flux, lm
Exitance	radiant exitance (M), W/m^2	luminous exitance, lm/m^2
Intensity	radiant intensity (I), W/Ω	luminous intensity, cd (lm/Ω)
	irradiance (Q), W/m^2	illuminance, lm/m^2
	radiance (L), W/m^2 Ω	luminance, lm/Ω m^2
Ω = unit solid angle (steradian), cd = caudle, lm = lumen.		

Table 1.8.2. Approximate luminance values of various sources.

Source	Luminance, cd/m²
Atomic fission bomb; 0.1 ms after firing 90-ft diameter ball	2×10^{12}
Blackbody; 6500 K	3×10^9
Sun; at surface	2.25×10^9
Sun; observed at zenity from the earth's surface	1.6×10^9
High intensity carbon arc; 13.6-mm rotation—positive carbon	0.75×10^9 to 1.50×10^9
Photoflash lamps	1.6×10^8 to 4.0×10^8
Blackbody; 4000 K	2.5×10^8
High intensity mercury short arc; type SAH1000A, 30 atoms	2.4×10^8
Zenon short arc; 900-W direct current	1.8×10^8
Zirconium concentrated arc; 300-W size	4.5×10^7
Tungsten filament incandescent lamp; 1200-W projection, 31.5 lm/W	3.3×10^7
Tungsten filament; 750-W, 26 lm/W	2.4×10^7
Tungsten filament; gas filled, 29 lm/W	1.2×10^7
Sun; observed from the earth's surface, sun near horizon	6.0×10^6
Blackbody; 2042 K	6.0×10^5
Inside-frosted bulb; 60 W	1.2×10^5
Acetylene flame; Mees burner	1.05×10^5
Welsback mantle; bright spot	6.2×10^4
Sodium arc lamp; 10,000-lm size	5.5×10^4
Low-pressure mercury arc; 50-in. rectifier tube	2.0×10^4
T-12 bulb-fluorescent lamp; 1500-mA extra high loading	1.7×10^4
Clear sky; average brightness	8.0×10^3
T-17 bulb-fluorescent; 420-mA low loading	4.3×10^3
Illuminating gas flame; fish-tail burner	4.0×10^3
Moon; observed from the earth's surface bright spot	2.5×10^3
Sky; overcast	2.0×10^3
Clear glass neon tube; 15 mm, 60 mA	1.6×10^3
Clear glass blue tube; 15 mm, 60 mA	8.0×10^2
Self-luminous paint	$0.0–0.17$

Table 1.8.3. Approximate luminances of backgrounds.

Source	Luminance, cd/m^2
Horizon sky	
overcast, no moon	3.4×10^{-5}
clear, no moon	3.4×10^{-4}
overcast, moon	3.4×10^{-3}
clear, moonlight	3.4×10^{-2}
deep twilight	3.4×10^{-1}
twilight	3.4
very dark day	34
overcast day	3.4×10^2
clear day	3.4×10^3
clouds, sun-lighted	3.4×10^4
Daylight fog	
dull	3.4×10^2 to 10×10^2
typical	10×10^2 to 34×10^2
bright	3.4×10^3 to 17×10^3
Ground	
on sunny day	3.4×10^2
on overcast day	34–100
snow, full sunlight	17×10^3

1.9. Exact Matrices and Aberration

1.9.1. Exact Matrices

Until now we have considered only paraxial rays which are valid for $\theta \leq 5°$. However, in practice, rays for which the paraxial condition does not hold are also incident on an optical system. In this section we shall formulate the exact translation and refraction matrices which are valid for all rays.

To obtain the exact matrices, it is convenient to redefine the rays as

$$X = \begin{pmatrix} x \\ \sin \theta \end{pmatrix}. \tag{1.9.1}$$

Of course, for the paraxial case, $\sin \theta \approx \theta$. Using this new definition we immediately obtain the translation matrix, $T(D)$, for free space propagation through a distance D along the z-axis

$$T(D) = \begin{pmatrix} 1 & D/\cos \theta \\ 0 & 1 \end{pmatrix}. \tag{1.9.2}$$

Or, as shown in Fig. 1.9.1,

$$x_2 = x_1 + \frac{D}{\cos \theta_1} \cdot \sin \theta_1,$$

$$\sin \theta_2 = 0 + \sin \theta_1.$$

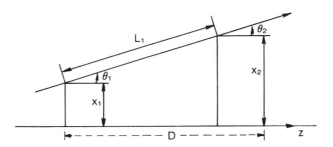

Fig. 1.9.1. Exact translation matrix formulation.

We also note that $D = L_1 \cos \theta_1$. Thus, in terms of L_1, the translation matrix becomes

$$T = \begin{pmatrix} 1 & L_1 \\ 0 & 1 \end{pmatrix}. \qquad (1.9.3)$$

To obtain the refraction matrix through a surface with radius of curvature R_1, separating two media with refractive indices n_1 and n_2, consider Fig. 1.9.2.

We note, as before, $x_2 = x_1$. Using Snell's law we obtain

$$n_2 \sin \alpha_2 = n_1 \sin \alpha_1, \qquad (1.9.4)$$

as

$$\alpha_2 = \psi + \theta_2,$$

and

$$\alpha_1 = \psi + \theta_1.$$

Equation (1.9.4) becomes

$$n_2 \sin(\psi_2 + \theta_2) = n_1 \sin(\psi + \theta_1),$$

or

$$\sin \theta_2 = \tan \psi \frac{n_1}{n_2} \cos \theta_1 - \cos \theta_2 \tan \Psi + \frac{n_1}{n_2} \sin \theta_1.$$

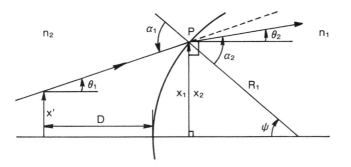

Fig. 1.9.2. Exact refraction matrix formulation.

As

$$\sin \psi = \frac{x_1}{R_1}, \tag{1.9.5}$$

$$\sin \theta_2 = -x_1 \frac{P}{n_2} + \frac{n_1}{n_2} \sin \theta_1, \tag{1.9.6}$$

where P is the bending power and is given by

$$P = \frac{n_2 \cos \theta_2 - n_1 \cos \theta_1}{R_1 \cos \psi} \tag{1.9.7}$$

$$= \frac{n_2 \cos(\alpha_2 - \psi) - n_1 \cos(\alpha_1 - \psi)}{R_1 \cos \psi}$$

$$= \frac{n_2 \cos \alpha_2 - n_1 \cos \alpha_1}{n_2 R_1}$$

$$= \frac{\sqrt{n_2^2 - n_1^2 \sin^2(\psi + \theta_1)} - \sqrt{n_1^2 - n_1^2 \sin^2(\psi + \theta_1)}}{R_1}.$$

Thus, we obtain

$$R(R_1) = \begin{pmatrix} 1 & 0 \\ -P/n_2 & n_1/n_2 \end{pmatrix}. \tag{1.9.8}$$

We note that P is more complex and of course, for the paraxial case, it reduces to the value

$$P \to \frac{n_2 - n_1}{R_1}.$$

For calculations using paraxial approximations, the translation matrix and the refraction matrix are all we need to calculate the equivalent matrix for any optical system. However, as shown in Fig. 1.9.2, just knowing the ray $(x_1, \sin \theta_1)$ and the distance D, we cannot locate the point P using the $T(D)$ matrix only. We will have to modify the translational matrix to obtain the point of incidence. Consider Fig. 1.9.3, where a ray $(x_1, \sin \theta_1)$ is incident on a refractive surface with radius R_1. We are interested in finding the distance $L_1 = AP$. As ED is parallel to AC, and OC is perpendicular to AC,

$$L_1 = AP = ED - AB - PC$$

$$= (d + R_1) \cos \theta_1 - x_1 \sin \theta_1 - \sqrt{R_1^2 - OC^2}.$$

As

$$OC = OD + DC = (d + R_1) \sin \theta_1 + BE$$

$$= (d + R_1) \sin \theta_1 + x_1 \cos \theta_1.$$

Thus

$$L_1 = (d + R_1) \cos \theta_1 - x_1 \sin \theta_1$$

$$- \sqrt{R_1^2 - \{(d + R_1) \sin \theta_1 + x_1 \cos \theta_1\}^2}. \tag{1.9.9a}$$

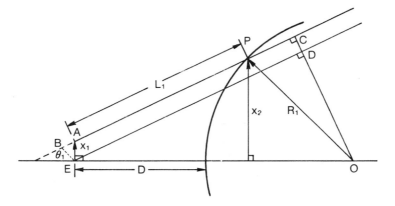

Fig. 1.9.3. Point of incidence needed for nonparaxial ray tracing.

Thus, this value of L_1 must be substituted in the translation matrix in (1.9.3). Also, note that for R_1, negative as shown in Fig. 1.9.4, L_1 is given by

$$L_1 = (D + R_1) \cos \theta_1 - x_1 \sin \theta_1$$
$$+ \sqrt{R_1^2 - \{(D + R_1) \sin \theta_1 + x_1 \cos \theta_1\}^2}$$
$$= (D - |R_1|) \cos \theta_1 - x_1 \sin \theta_1$$
$$+ \sqrt{|R_1|^2 - \{(D - |R_1|) \sin \theta_1 + x_1 \cos \theta_1\}^2}. \qquad (1.9.9b)$$

1.9.1.1. Example

We are interested in tracing the ray X_1 through a thick lens having radii of curvature R_1 and R_2. The rays $X_1 - X_6$, as shown in Fig. 1.9.5, can be obtained

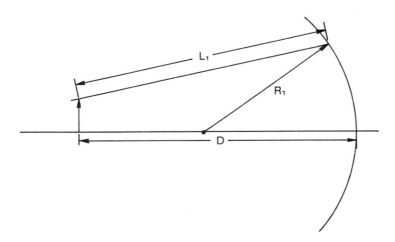

Fig. 1.9.4. The relationship between L_1, R_1, and D is shown for the case of negative R_1.

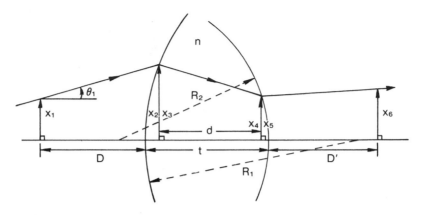

Fig. 1.9.5. Nonparaxial ray tracing for a lens.

as follows:

$$X_2 = T(L_1)X_1,$$

where

$$L_1 = (D + R_1)\cos\theta_1 + x_1 \sin\theta_1$$
$$- \sqrt{R_1^2 - \{(D + R_1)\sin\theta_1 + x_1 \cos\theta_1\}^2},$$

$$X_3 = R(R_1)X_2,$$

where P_1 in $R(R_1)$ is given by

$$P_1 = \frac{\sqrt{n^2 - \sin^2(\psi_1 + \theta_1)} - \cos(\psi_1 + \theta_1)}{nR_1},$$

and

$$\sin\psi_1 = \frac{x_2}{R_1},$$

$$X_4 = T(L_2)X_3,$$

where

$$L_2 = (d - |R_2|)\cos\theta_3 - x_3 \sin\theta_3$$
$$+ \sqrt{|R_1|^2 - \{(d - |R_2|)\sin\theta_3 + x_3 \cos\theta_3\}^2},$$

and d is related to the thickness of the lens, t, along the optical axis, by the following equation:

$$d = t - (L_1 \cos\theta_1 - D),$$

$$X_5 = R(R_2)X_4,$$

where

$$P_2 = \frac{\sqrt{1 - n^2 \sin^2(\psi_2 + \theta_4)} - n\cos(\psi_2 + \theta_4)}{R_2}$$

and

$$\sin \psi_2 = \frac{x_4}{R_2}.$$

Note that R_2 is negative. Finally,

$$X_6 = T(L_3)X_5,$$

where

$$L_3 = \frac{D' - (L_2 \cos \theta_4 - d)}{\cos \theta_4}.$$

The calculations are somewhat tedious and a calculator or a computer should be used.

1.9.2. Exact Matrices for Skew Rays

The meridional plane is defined as the plane which contains the point or line object and the optical axis. In the last section we restricted ourselves to the so-called meridional rays which lie in the meridional plane. All other rays, not lying in this plane, are called skew rays. In this section we develop the exact matrices for the skew rays.

A typical skew ray, SS_1 is shown in Fig. 1.9.6. The point S, has the coordinates x, y, and z, where z is the optical axis and xz is the meridional plane. x_1, y_1, and z_1 are the coordinates for S_1 and the directional cosines of the ray SS_1 are given by γ, δ, and ε. As $\gamma^2 + \delta^2 + \varepsilon^2 = 1$, only two of these quantities are independent. We note that

$$\gamma = \cos \theta_x = \sin \theta. \tag{1.9.10}$$

Thus, the equation for the translational matrix for the xz-plane can be written

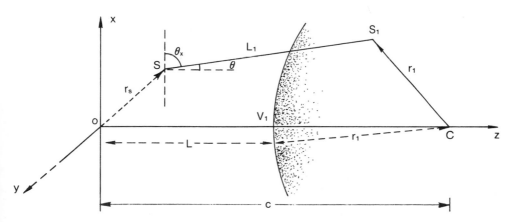

Fig. 1.9.6. Nonparaxial skew rays.

as

$$\begin{pmatrix} x_2 \\ \gamma \end{pmatrix} = \begin{pmatrix} 1 & L_1 \\ 0 & 1 \end{pmatrix}\begin{pmatrix} x_1 \\ \gamma \end{pmatrix}, \qquad (1.9.11)$$

or

$$X_2 = T_x(L_1)X_1.$$

Here, L_1 is the distance SS_1 along the ray.

Similar equations can be written for the y component, $T_y(L_1)$, and the z component, $T_z(L_1)$. $T_y(L_1)$ is given by

$$T_y(L_1) = \begin{pmatrix} 1 & L_1 \\ 0 & 1 \end{pmatrix}$$

and

$$\begin{pmatrix} z_1 + L \\ \varepsilon_1 \end{pmatrix} = \begin{pmatrix} 1 & L \\ 0 & 1 \end{pmatrix}\begin{pmatrix} z \\ \varepsilon \end{pmatrix}. \qquad (1.9.12)$$

where z_1 is measured with V_1 as the origin.

As x_1, y_1, and z_1 are on the surface of the sphere having radius r_1, we know that

$$S_1 C = V_1 C,$$

or

$$x_1^2 + y_1^2 + (z_1 - r_1)^2 = r_1^2. \qquad (1.9.13)$$

Thus the three quantities, x, y, and z, are not independent and only two of the three matrices are necessary to determine the ray uniquely. However, when performing numerical computations, it is best to determine x, y, and z independently through the matrices, and check the accuracy of the calculation through (1.9.13).

L_1 can be calculated by taking projections of OC along the ray and subtracting the projections of OS and CS_1 along the ray. Thus,

$$L_1 = (L + r_1)\varepsilon - r \cos \alpha - r_1 \cos \alpha_1, \qquad (1.9.14)$$

where r denotes the distances OS, α denotes the angle between OS and SS_1, and α_1 is the angle between $S_1 C$ and SS_1. We also note that

$$\cos \alpha = \frac{x}{r}\gamma + \frac{y}{r}\delta + \frac{z}{r}\varepsilon, \qquad (1.9.15)$$

or

$$r \cos \alpha = x\gamma + y\delta + z\varepsilon.$$

To obtain $\cos \alpha_1$ in terms of known quantities, we obtain the following equations by considering the triangles $SS_1 C$ and $SV_1 C$:

$$L_1^2 + 2L_1 r_1 \cos \alpha + r_1^2 = x^2 + y^2 + (L - z)^2 + 2(L_1 - z)r_1 + r_1^2. \quad (1.9.16)$$

As

$$L_1 = \{\varepsilon(L + r_1) - (x\gamma + y\delta + z\varepsilon)\} - r_1 \cos \alpha_1$$

$$= \beta - r_1 \cos \alpha_1, \tag{1.9.17}$$

where

$$\beta = \varepsilon(L + r_1) - (x\gamma + y\delta + z\varepsilon).$$

We obtain by substitution in (1.9.16)

$$\cos \alpha_1 = \pm\frac{1}{r_1}\sqrt{\beta^2 - x^2 - y^2 - (z - L)^2 + 2(z - L)r_1}. \tag{1.9.18}$$

Finally, we obtain

$$L_1 = \beta \pm \sqrt{\beta^2 - x^2 - y^2 - (z - L)^2 + 2(z - L)r_1}. \tag{1.9.19}$$

The three refraction matrices for the skew rays can also be written as

$$\begin{pmatrix} x_2 \\ \gamma_2 \end{pmatrix} = \begin{pmatrix} 1 & 0 \\ -P_1/n_2 & n_1/n_2 \end{pmatrix}\begin{pmatrix} x_1 \\ \gamma_1 \end{pmatrix}, \tag{1.9.20}$$

$$\begin{pmatrix} y_2 \\ \delta_2 \end{pmatrix} = \begin{pmatrix} 1 & 0 \\ -P_1/n_2 & n_1/n_2 \end{pmatrix}\begin{pmatrix} y_1 \\ \delta_1 \end{pmatrix}, \tag{1.9.21}$$

$$\begin{pmatrix} z_2 \\ \varepsilon_2 - P_1 r_1/n_2 \end{pmatrix} = \begin{pmatrix} 1 & 0 \\ -P_1/n_2 & 1 \end{pmatrix}\begin{pmatrix} z_1 \\ \varepsilon_1 \end{pmatrix}. \tag{1.9.22}$$

Again, only two matrices are needed as the third angle can be calculated from the equation given by

$$\gamma_2^2 + \delta_2^2 + \varepsilon_2^2 = 1. \tag{1.9.23}$$

1.9.3. Aberration

In Section 1.9.1 we mentioned that for a real lens, under practical circumstances, the paraxial approximation does not hold and this gives rise to aberrations. If the object light is not monochromatic, then due to the dispersion of the lens media, the focal length of the lens, even in the paraxial approximation, will be wavelength-dependent giving rise to chromatic dispersion.

The origin and different aspects of these aberrations can be better understood in the following example. Let us consider a lens on which monochromatic light is incident, parallel to the optical axis as shown in Fig. 1.9.7. The paraxial rays will converge to a focus at $z = f$. Theoretically, the size of the focus tends to zero or a true point. Mathematically, it is represented by a delta function, $\delta(z - f)$.* The value of f for a thin lens is given by the lens maker's

---

* Readers not familiar with the concept of the delta function are referred to the Appendix A.

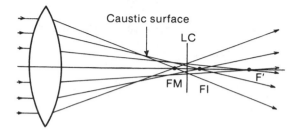

Fig. 1.9.7. Spherical aberration showing the caustic surface and the circle of least confusion (LC). F' is the paraxial focus and FM is the focus for the marginal rays.

formula which is repeated here, stressing the point that the refractive index is a function of incident wavelength

$$\frac{1}{f} = [n(\lambda) - 1]\left\{\frac{1}{|R_1|} + \frac{1}{|R_2|}\right\}.$$

Note that although all the incident rays, being parallel to the axis, satisfy the paraxial condition, the rays exiting from the lens may not be, especially the rays near the edge of the lens.

For these nonparaxial rays, we must use the exact matrices derived in the last section. Thus the discussion of aberration is rather a complex subject which is best analyzed for individual lens systems by actual numerical computation. However, there are certain important features of aberration which can be understood by analytical reasoning and these are discussed below.

First of all, as the lens is symmetrical around the optical axis, each small zone of the lens, defined by the distance from the optical axis, will bend the rays by the same angle, even if we consider exact matrices. Thus, we can define, for this example only, different effective focii of the lens zones, as shown in the Fig. 1.9.7. For a lens with positive focal length, $f_4 < f_3 < f_2 < f_1 < f$. The light cone from each zone focuses to a point corresponding to the zone focal length. We define the *caustic surface* as the envelope of the focii of these sets of cones. Thus the image, in place of being a point at $\delta(z - f)$, is a circle determined by the intersection of the caustic surface with plane $z = f'$ where $f_4 < f' < f$. The circle having the smallest radius is called the circle of least confusion and is shown in the Fig. 1.9.7.

Although we have discussed only parallel rays being focused by a lens, the discussion above remains the same as long as the object is on the optical axis. Because of the spherical symmetry, this aberration is called spherical aberration. However, if the point source is off-axis, or the parallel rays incident on the lens are not parallel to the optical axis, then we have a completely different situation. This is shown in Fig. 1.9.8. Again, the different cones of light focus to a point. However, these individual zone focii may not be in a straight line as shown in the figure. This form of aberration is known as a coma.

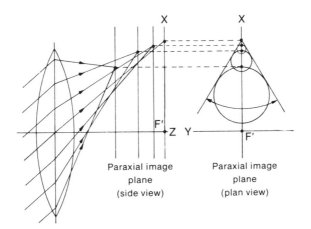

Fig. 1.9.8. Aberration "coma" due to rays not parallel to the optical axis.

The spherical aberrations, coma and astigmatism, are the aberration of a point source. The spherical aberration is related to a point source on the optical axis, whereas coma and astigmatism arise from an off-axis point source. For an object of extended size, two other aberrations can be distinguished. These are called curvature of field and distortion.

The subject of aberration can also be discussed from the following expansion of $\sin \theta$:

$$\sin = \theta - \frac{\theta^3}{3!} + \frac{\theta^5}{5!} - \cdots. \tag{1.9.24}$$

It can be shown that the presence of the second term, $\theta^3/3!$, leads to the five distinct types of aberrations mentioned before and are called third order or Seidel aberrations. The fifth-order aberration due to the third term, $\theta^5/5!$, is generally smaller in magnitude and is therefore negligible.

To discuss aberration quantitatively, it is convenient to define a quantity called optical path length. Optical path length along a ray from point A to point B, as shown in Fig. 1.9.9, is defined as

$$L(AB) = \int_{A\,Ray}^{B} n(\mathbf{r})\, dr. \tag{1.9.25}$$

For a straight line in a homogeneous medium, for a propagation distance of D, this becomes

$$L(AB = D) = n \cdot D = \frac{c}{v} \cdot D = c\Delta t, \tag{1.9.26}$$

whee c is the velocity of light in a vacuum, v is the velocity of light in the medium, and Δt is the time taken by the phasefront to move from A to B.

If we consider an imaging system as shown in Fig. 1.9.10 the aberration free phasefront, P_R, converging to the image point I_0 will be exactly spherical.

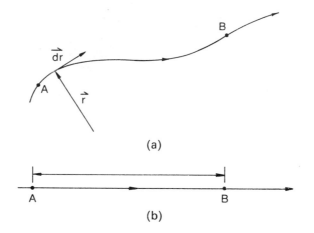

Fig. 1.9.9. Geometry for optical path-length definition: (a) arbitrary rays in a medium with $n(\mathbf{r})$ and (b) a homogeneous medium.

However, because of the aberration present the actual phasefront, P, is somewhat different. If we now trace a ray from the object, S_0, to the aberration free image, I_0, then the aberration function, $AB(\rho)$, is given by

$$AB(\rho) = L(S_0QABI_0) - L(S_0VTI_0) = -L(AB)$$
$$= -n'|AB|. \tag{1.9.27}$$

Here ρ is the radial distance of the point in the phasefront from the optical axis. For an on-axis object, as shown in Fig. 1.9.10, the third-order aberration

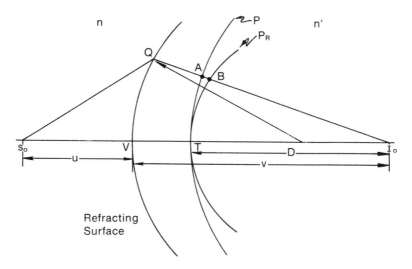

Fig. 1.9.10. The actual and ideal phasefront after passing through a refractive surface.

function can be shown to be given by

$$AB(\rho) = -\frac{c}{4}\left(\frac{v}{D}\right)^4 \rho^4, \tag{1.9.28}$$

where v is the image distance, D is the distance of the phasefront from the image point, and c is given by

$$c = \frac{1}{2}\left(\frac{n'}{n}\right)^2 \left(\frac{n'-n}{n'}\right)^2 \left(\frac{1}{v}-\frac{1}{R}\right)^2 \left(\frac{n'}{R}-\frac{n+n'}{v}\right),$$

where R is the radius of curvature of the refractive surface. For an off-axis point the third-order aberration function can be shown to be given by (Klein and Furtak [1])

$$AB(h', r', q) = -\frac{c}{4}\left(\frac{v}{D}\right)^4 [r'^4 + 4bh'r'^3 \cos\varphi + 4b^3h'^3r' \cos\varphi$$

$$+ 2b^2h'^2r'^2(2\cos^2\varphi + 1)], \tag{1.9.29}$$

where the quantities h', r', and φ are defined in Fig. 1.9.11 and

$$b = \frac{R+D-v}{v-R}.$$

Until now we have discussed only one refractive surface. However, for a complex optical system, we can define a composite aberration function, at the exit pupil, as the optical path difference between the true phasefront and the

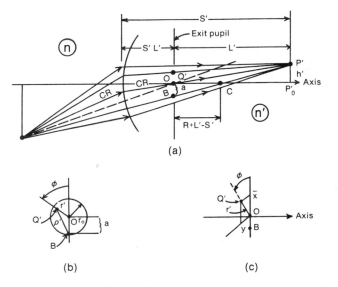

Fig. 1.9.11. Imaging of an off-axis point including third-order aberration, $L' = D$; $s' = v$.

Table 1.9.1. Components of aberration.

Spherical aberration	$c_{400}r'^4$
Coma	$c_{311}r'^3 \cos \varphi h'$.
Astigmatism	$c_{222}r'^2 h'^2 \cos^2 \varphi$
Curvature of field	$c_{202}r'^2 h'^2$
Distortion	$c_{113}r' \cos \varphi h'^3$

reference phasefront obtained for the paraxial or aberration-free situation. The contributions of the individual refractive surfaces to $AB(r')$ will be additive for $AB(r') \ll \rho$. Thus, in the most general case, the aberration function can be expanded as

$$AB = c_{400}r'^4 + c_{310}r'^3 \cos \varphi + c_{222}r'^2 \cos^2 \varphi h'^2$$
$$+ c_{202}h'^2 r'^2 + c_{113}r' \cos \varphi h'^3, \qquad (1.9.30)$$

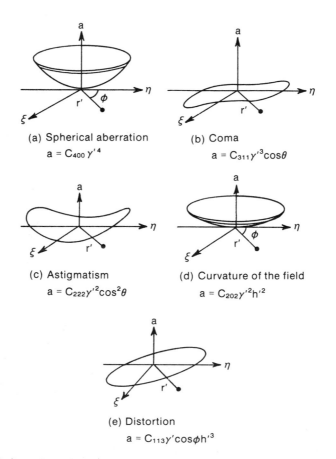

(a) Spherical aberration
$a = C_{400} \gamma'^4$

(b) Coma
$a = C_{311}\gamma'^3 \cos\theta$

(c) Astigmatism
$a = C_{222}\gamma'^2 \cos^2\theta$

(d) Curvature of the field
$a = C_{202}\gamma'^2 h'^2$

(e) Distortion
$a = C_{113}\gamma' \cos\phi h'^3$

Fig. 1.9.12. Wavefront distortions for the primary aberrations. (From F.G. Smith and J.H. Thomson, *Optics*, Wiley, New York, 1971.)

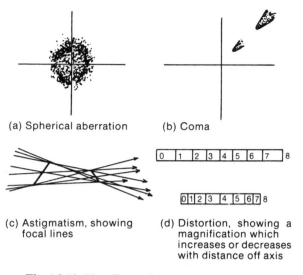

(a) Spherical aberration   (b) Coma

(c) Astigmatism, showing   (d) Distortion, showing a
    focal lines   magnification which
                                      increases or decreases
                                      with distance off axis

Fig. 1.9.13. The effects of the various aberrations.

where the subscripts of the c coefficients refer to the powers of r', $\cos \varphi$, and h', respectively. The form of the above equation can also be understood from the symmetry arguments, since the aberration, which does not depend on φ, must be independent of the signs of h' and r'. They must occur in the form h'^2, r'^2 or $h'^2 r'^2$ or r'^4. Also as φ is the angle measured from the meridional plane the aberration function must be symmetric with respect to φ, and thus only $\cos \varphi$ can occur.

From (1.9.30), we can identify the different components of aberration as given in Table 1.9.1.

Figures 1.9.12 and 1.9.13 show the effect of these so-called primary aberrations in connection with the propagation of a plane wave through the lens and the ray aberrations around ideal point images. Figure 1.9.12(a) shows the effect of the spherical aberration only and is plotted as a deviation from an idealized plane wave. Actually, it is the plot of the equation given by

$$\sigma(r', \varphi, h') = c_{400} r'^4.$$

The constant σ surface is plotted as a function of r', φ, and h'. Similarly, Fig. 1.9.12(b), (c), (d) and (e) present the situations for coma, astigmatism, curvature of the field, and distortion. Figure 1.9.13(a) shows the effect of spherical aberration on a point image. The imaging rays form a halo around the image point.

In the following we shall discuss different aberrations separately.

1.9.4. Spherical Aberration

Spherical aberration is generally referred to either as longitudinal spherical aberration (LSA) or transverse spherical aberration (TSA). For a single lens,

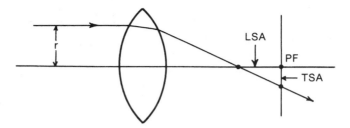

Fig. 1.9.14. Representation of the spherical aberration of a single lens. PF = paraxial focus; LSA = longitudinal spherical aberration; TSA = transverse spherical aberration.

these quantities are defined in Fig. 1.9.14. Thus LSA is the distance between the paraxial image and the marginal image, whereas TSA is the distance between the paraxial image and the point in the paraxial image plane where the marginal rays meet. We define a quantity, δ, called angular aberration, given by

$$\delta = \frac{1}{h'}\frac{d}{dr'}[AB(r')]. \tag{1.9.31}$$

Then TSA for a single refractive surface is given by

$$\text{TSA} = -D\delta = \frac{4D}{n'}c_{400}r'^3, \tag{1.9.32}$$

and

$$\text{LSA} = \frac{4D^2}{n'}c_{400}r'^2. \tag{1.9.33}$$

where D is the distance between the image plane and the refractive surface plane. For a thin lens with radii of curvature, R_1 and R_2, and a refractive index, n, it can be shown that

$$c_{400} = \frac{1}{32f^3n(n-1)}\left[\frac{n+1}{n-1}S^2 + 4(n+1)PS\right.$$
$$\left. + (3n+2)(n-1)P^2 + \frac{n^3}{n-1}\right], \tag{1.9.34}$$

where S (the slope factor) and P (the position factor) are given by

$$S = \frac{R_2 + R_1}{R_2 - R_1}, \tag{1.9.35}$$

and

$$P = 1 - \frac{2f}{v}.$$

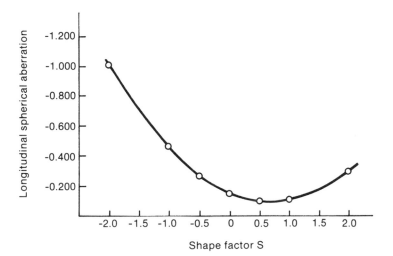

Fig. 1.9.15. LSA versus S for the seven lens combination of the same thickness and refractive index with the following radii, and for $r' = 1$. (From A. Nussbaum and R.A. Phillips, *Contemporary Optics for Scientists and Engineers*, Prentice-Hall, Engelwood Cliffs, NJ, 1976.)

r_1	r_2	S
-10.00	-3.33	-2.00
∞	-5.00	-1.00
20.00	-6.67	-0.50
10.00	-10.00	0.00
6.67	-20.00	0.50
5.00	∞	1.00
3.33	10.00	2.00

For a doubly convex symmetric lens, $R_2 = -R_1$ and $S = 0$. Thus, the shape factor measures the deviation from symmetry. A plot of LSA versus S for $P = 1$ (i.e., for parallel rays) is shown in Fig. 1.9.15. It is found that, for the values chosen, LSA is a minimum for $\delta \approx 0.7$; thus, properly choosing the values S, and still keeping the focal length constant, we can minimize the spherical aberration. This is known as "bending the lens".

Thus far, we have considered only third-order theory. However, if we extend the theory to fifth order, we find

$$\text{LSA} = ar'^2 + br'^4, \tag{1.9.36}$$

where a and b are third- and fifth-order constants, respectively. By Choosing the values of a and b properly we can also minimize the spherical aberration. As the spherical aberration is dependent on the sign of the focal length of the lens, we can reduce LSA by using a doublet, a positive and a negative lens combination. If we wish to make LSA go to zero at $r = r_{max}$, then from (1.9.36)

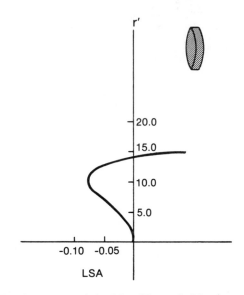

Fig. 1.9.16. LSA for the corrected doublet. (From A. Nussbaum and R.A. Phillips, *Contemporary Optics for Scientists and Engineers*, Prentice-Hall, Engelwood Cliffs, NJ, 1976.)

we obtain

$$\text{LSA} = (-r_{max}^2 r^2 + r^4),$$

or

$$\frac{\text{LSA}}{r_{max}^4} = \left(\frac{r}{r_{max}}\right)^2 \left[\left(\frac{r}{r_{max}}\right)^2 - 1\right].$$

This is plotted in Fig. 1.9.16 for the doublet with the following specifications:

r	t'	n'
61.070		
	4.044	1.56178
-47.107	2.022	1.70100
-127.098		

It is observed that although LSA is zero at $r = 0$ and $r = r_{max}$, it has a maximum at $r = r_{max}/\sqrt{2}$.

1.9.5. Coma

Coma is an aberration for point objects off-axis. The rays converging to the image point intersect the paraxial image plane in a cometlike spread image whose length increases as the square of the distance off-axis. Rays coming

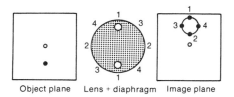

Object plane   Lens + diaphragm   Image plane

Fig. 1.9.17. Coma and comatic circle.

from a line across the aperture at $\varphi = 90°$ do not contribute to coma. The rays along the line $\varphi = 90°$ do not focus at a point and spread around the image point. The comatic image, in general, consists of many circular images superposed. These circles are shifted successively further from the axis and focused less sharply (see Fig. 1.9.13(b)).

To understand how the circles are produced out of a point source, due to the off-axis rays passing through different zones of the lens, consider Fig. 1.9.17. In the object plane, we show the off-axis point object and the optical axis. We also consider the lens to have an opaque diaphragm with two holes. If the holes are at position 1–1, we obtain image 1 at the image plane; similarly for 2–2, 3–3, and so on. Larger zones in the lens through which the light passes produce larger comatic circles. The radius of the comatic circle is proportional to the square of the radius of the lens zone. The distance from the center of the comatic circle of the optic axis is proportional to the square of the radius of the zone. Combining all these we obtain the comatic flare. Note that the flare may point towards the axis or away from it depending on the type of the lens.

It is customary to specify the magnitude of the coma as shown in Fig. 1.9.18. The length of the comatic pattern along the meridional or tangential direction

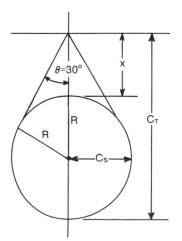

Fig. 1.9.18. Comatic circle showing tangential coma, C_T, and sagital coma, C_S.

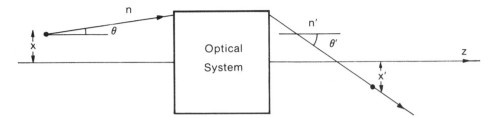

Fig. 1.9.19. Aplanatic optical system satisfying Abbé's sine condition.

is called tangential coma, C_T, and its half width (equal to R) is the sagital coma, C_s. The area of the comatic aberration is then defined as

$$A = C_s C_T. \tag{1.9.37}$$

Reducing spherical aberration automatically reduces coma. We can show that coma is absent for a lens when the shape factor S (given by (1.9.35)) is

$$S = \left(\frac{2n^2 - n - 1}{n - 1}\right)\left(\frac{u - v}{u + v}\right). \tag{1.9.38}$$

A lens or optical system which does not have any spherical aberration and coma is called aplanatic. It can be shown that these systems obey Abbe's sine condition. For any refracting surface, the Abbe sine condition is given by

$$xn \sin \alpha = x'n' \sin \alpha', \tag{1.9.39}$$

where the quantities x, n, α and x', n', α' are defined in Fig. 1.9.19.

1.9.6. Astigmatism

Astigmatism is due to cylindrical wavefront aberration. Astigmatism increases as the square of the distance off-axis and the square of the aperture readius, r'. To understand astigmatism, imagine a narrow bundle of rays having a circular cross section incident on the lens away from the optical axis. On the lens surface, the ray boundaries will form an ellipse with the major axis pointing towards the vertex of the refracting surface. The rays lying in the major axis will come to a focus at point f_T, called the tangential focus (see Fig. 1.9.13(c)). Rays in the minor axis come to the sagital or radial focus denoted by f_R. Thus, if an off-axis object is imaged, two focused image planes will result. Any radial line in the object will be focused as a radial in the f_R plane, and any tangential line will be focused as a line in the f_T plane. The separation between these planes is called astigmatism. Note that a point object is imaged as a line due to astigmatism. The distance between f_T and f_R is called the astigmatic interval. Note that the optimum lies between f_T and f_R where a point object produces a circle of least confusion.

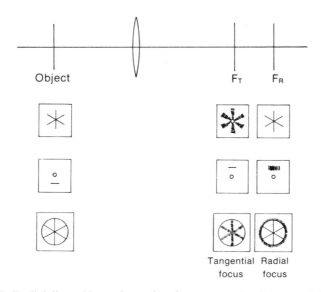

Fig. 1.9.20. Radial line object of rotational symmetry (top), tangential line object (center), and astigmatic images of a spoked wheel (bottom).

Figure 1.9.20 further illustrates the effect of astigmatic aberration. An object, like the spoke of a wheel containing radial lines only, will be sharply imaged at f_R. The circular object, just the wheel, will be imaged properly at f_T. However, a spoked wheel will be distorted in any plane.

Elimination of astigmatism requires that the tangential and sagital surfaces be made to coincide. If this can be done, then the common surface is defined by the Petzeval equation given by

$$\frac{n'}{r} + \frac{n}{r'} = \frac{n' - n}{R}, \tag{1.9.40}$$

where the quantities, n, n', r, r', are shown in Fig. 1.9.21 for the single refracting surface.

R = radius of the refractive surface;
r' = radius of the image curvature;
r = radius of the object curvature.

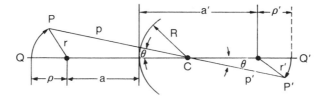

Fig. 1.9.21. Petzeval surface for a single refracting surface.

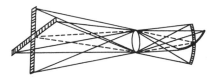

Fig. 1.9.22. Curvature of a field.

1.9.7. Curvature of Field

Curvature of field results because of the failure of a lens to transform a plane object into a plane image. Thus a flat object will give a curved image surface. Curvature of field and astigmatism are closely related. Curvature of field is symmetrical about the optical axis. However, both aberrations increase with the off-axis distance of the object and with the aperture of the refracting surface.

In many cases, the Petzeval surface is curved when astigmatism is removed. To record sharp images under these conditions, the film must be curved to fit the Petzeval surface. Figure 1.9.22 illustrates the curvature of field aberration for an object shaped like a cross.

1.9.8. Distortion

In distortion, the transverse linear magnification in the image varies with the distance from the optic axis. Note that a point object is imaged as a point image. However, an object shaped like a rectangular grid will look like Fig. 1.9.23(b) which illustrates pincushion distortion. Figure 1.9.23(c) shows barrel distortion. The image in either case is sharp but distorted. Distortion often results due to the limitation of ray bundles by stops or optical elements acting as stops. This is illustrated in Fig. 1.9.24. Due to the placement of the stop near the lens to reduce astigmatism and curvature of field, distortion is introduced because the rays for large values of x and y are limited to an off-center portion of the lens. The situation shown in the figure results in barrel distortion. If we place a stop at an equal distance on the other side of the lens,

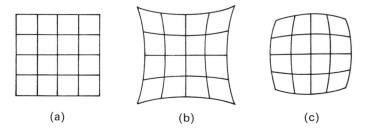

(a)   (b)   (c)

Fig. 1.9.23. Images of a square grid showing: (a) pincushion distortion, (b) barrel distortion, and (c) is due to nonuniform magnifications.

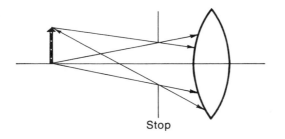

Fig. 1.9.24. Distortion resulting from mechanical stops.

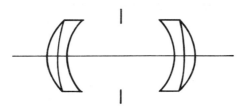

Fig. 1.9.25. An optical system to correct distortion.

it will result in an equal amount of pincushion distortion. This gives us a clue of how to correct for distortion—place symmetrical stops on both sides of the lens. Another possibility is to use two identical lens groupings with an iris diaphragm in the center—this is shown in Fig. 1.9.25. As will be discussed in Section 4.2, many highly corrected camera lenses use this trick.

1.9.9. Chromatic Aberration

We have already discussed the fact that the refractive index of a material is a function of the light wavelength. A demonstration of this is shown in Fig. 1.9.26, where argon laser is dispersed to a multicolored beam by a prism.

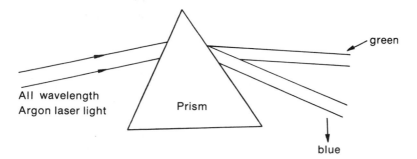

Fig. 1.9.26. Incident all wavelength argon light is dispersed by the prism.

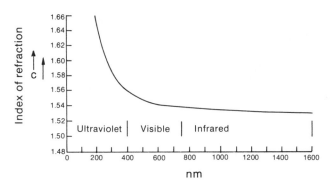

Fig. 1.9.27. Refractive index of quartz versus light wavelength showing dispersion in quartz.

This is similar to the famous experiment performed by Newton to demonstrate the multicolor nature of white light. A typical variation of the refractive index with wavelength for glass is shown in Fig. 1.9.27. Table 1.9.2 lists the important parameters for other materials. The table also lists a quantity called the Abbé number or the dispersive power defined as

$$V = \left[\frac{n(F) - n(C)}{n(D) - 1}\right]^{-1} = \left[\frac{\Delta n}{n - 1}\right]^{-1}, \tag{1.9.41}$$

where F and C represent the blue and red lines of hydrogen at $\lambda = 0.4861\ \mu m$ and $\lambda = 0.6563\ \mu m$, respectively, and D is the sodium yellow line with $\lambda = 0.5893\ \mu m$.

In the paraxial approximation, chromatic aberration can be corrected using compound lenses consisting of two, three, or four lenses. Two lenses can correct the chromatic aberration at two wavelengths, or if the refractive indices of the lenses are linear functions of the wavelength, then for all colors. These are called achromatic lens-pair. Three lenses can correct for three colors and this combination is known as apochromatic. Lenses corrected for four colors using four lenses are referred to as superachromats.

To correct for two colors, a doublet is used consisting of two lenses, A and B, in contact, having focal lengths f_A and f_B. The lenses A and B use materials

Table 1.9.2

Fraunhofer line	Color	Wavelength (nm)	Index (crown glass)	Index (flint glass)
C	red	656.28	1.51418	1.69427
D	yellow	589.59	1.51666	1.70100
F	blue	486.13	1.52225	1.71748
V			187.94	73.29

of refractive indices index n_A and n_B and Abbé numbers V_A and V_B, respectively. Thus, we have

$$P_A = \frac{1}{f_A} = (n_A - 1)\left(\frac{1}{R_{A1}} - \frac{1}{R_{A2}}\right) = (n_A - 1)A_A, \qquad (1.9.42)$$

$$P_B = \frac{1}{f_B} = (n_B - 1)\left(\frac{1}{R_{B1}} - \frac{1}{R_{B2}}\right) = (n_B - 1)B_B, \qquad (1.9.43)$$

$$P = P_A + P_B = \frac{1}{f_A} + \frac{1}{f_B} = \frac{1}{f}. \qquad (1.9.44)$$

R_{A1}, R_{A2} and R_{B1}, R_{B2} are radii of curvature for the lenses A and B, respectively. Note that A_A and B_B are constants and are independent of the wavelength. Also, for a doublet $R_{A2} = -R_{B1}$. Denoting the two colors to be corrected as 1 and 2 (these are generally red and blue), for chromatic correction we have

$$f_A - f_B = 0, \qquad (1.9.45)$$

or

$$P_A - P_B = 0,$$

or

$$\Delta n_A A_A + \Delta n_B B_B = 0,$$

or

$$\frac{\Delta n_A A_A (n_A - 1)}{(n_A - 1)} + \frac{\Delta n_B B_B (n_B - 1)}{(n_B - 1)} = 0,$$

or

$$\frac{1}{f_A V_A} + \frac{1}{f_B V_B} = 0. \qquad (1.9.46)$$

Equation (1.9.46) is the condition for the achromatic lens.

For the case of a spaced doublet, with a separation, d, between lenses, we have

$$\frac{1}{f} = P = P_A + P_B - P_A P_B d. \qquad (1.9.47)$$

Differentiating (1.9.47) we obtain

$$\Delta P = \Delta P_A + \Delta P_B - d(P_B \Delta P_A + P_A \Delta P_B). \qquad (1.9.48)$$

The condition for the correction of the chromatic correction is given by

$$\Delta P = 0, \qquad (1.9.49)$$

or

$$\Delta n_A A_A + \Delta n_B B_B - d\left(\frac{\Delta n_A A_A}{f_B} + \frac{\Delta n_B B_B}{f_A}\right) = 0,$$

or

$$\frac{1}{f_A V_A} + \frac{1}{f_B V_B} - \frac{d}{f_A f_B}\left(\frac{1}{V_A} + \frac{1}{V_B}\right) = 0. \tag{1.9.50}$$

Equation (1.9.50) is the condition for the spaced doublet. For $V_A = V_B$, this simplifies to

$$f_A + f_B = 2d. \tag{1.9.51}$$

Numerical Example

(a) Design a doublet with the following specifications:

$$f = 10 \text{ cm}.$$

lens A—borasilicate glass

$$V_A = 64.5, \qquad R_{A1} = R_{A2} \qquad \text{and} \qquad n_A = 1.517.$$

lens B—dense flint

$$V_A = 36.6, \qquad R_{B1} = -R_{A1} \qquad \text{and} \qquad n_B = 1.617.$$

Using (1.9.46)

$$\frac{1}{f_B V_A} = -\frac{1}{f_B V_B},$$

or

$$\frac{1}{f_B} = \frac{V_B}{f}(V_B - V_A) = 13.12 \text{ m}^{-1},$$

$$\frac{1}{f_A} = \frac{1}{f} - \frac{1}{f_B} = 23.12 \text{ m}^{-1},$$

$$\frac{1}{f_A} = (n_A - 1)\frac{2}{R_A},$$

or

$$R_A = 4.47 \text{ cm},$$

$$\frac{1}{f_B} = (n_B - 1)\left(-\frac{1}{4.47} - \frac{1}{R_{B2}}\right),$$

or

$$R_{B2} = 91.2 \text{ cm}.$$

(b) Design a separated doublet using two identical lenses with the following specifications:

$$f = 10 \text{ cm},$$

$$f_A = f_B,$$

$$n_A = n_B$$

and

$$V_A = V_B.$$

Using (1.9.47) we have

$$\frac{2}{f_A} - \frac{d}{f_A^2} = 10 \text{ m}^{-1}.$$

Using (1.9.51), we finally obtain

$$d = 10 \text{ cm}.$$

References

[1] M.V. Klein and T.E. Furtak, *Optics*, 2nd ed. Wiley, 1986.
[2] A. Nussbaum and R.A. Phillips, *Contemporary Optics for Scientiest and Engineers*, Prentice-Hall, 1976.
[3] F.A. Jenkins and H.E. White, *Fundamentals of Optics*, 3rd ed., McGraw-Hill, 1957.
[4] W. Brouwer, *Matrix Methods in Optical Instrument Design*, Benjamin, 1964.
[5] F.G. Smith and J.H. Thomson, *Optics*, Wiley, 1971.
[6] J.R. Meyer-Arendt, *Introduction to Classical and Modern Optics*, Prentice-Hall, 1972.
[7] F.L. Pedrotti and L.S. Pedrotti, *Introduction to Optics*, Prentice-Hall, 1987.

PART II

Physical Optics, Wave Optics, and Fourier Optics

2.1. Fundamentals of Diffraction

When any of the dimensions or sizes of the components in an optical system is on the order of wavelength, then the methods discussed under geometrical optics do not give the proper solution to the problem. For this case, we must start from Maxwell's equations, and derive the wave equation from which the proper solution to this problem can be obtained.

It is of interest to point out the wide range of wavelengths in the electromagnetic spectrum as discussed in Table 1.1.1. To observe the dramatic effects of diffraction, for light we need dimensions on the order of 1 μm, whereas for microwaves a few centimeters is all we need.

2.1.1. Maxwell's Equations

$$\nabla \times \mathbf{E} = -\frac{\partial \mathbf{B}}{\partial t},$$

$$\nabla \times \mathbf{H} = +\frac{\partial \mathbf{D}}{\partial t} + \mathbf{J}, \qquad (2.1.1)$$

$$\nabla \cdot \mathbf{D} = \rho,$$

$$\nabla \cdot \mathbf{B} = 0,$$

where t represents time, $\mathbf{E}$ is the electric field vector and has units of V/m, $\mathbf{H}$ is the magnetic field vector and has units of A/m, $\mathbf{B}$ is the magnetic induction and has units of Wb/m^2, $\mathbf{D}$ is the electric induction and has units of C/m^2, ρ is the free charge density with units of C/m^3, and $\mathbf{J}$ is the current density with units of A/m^2.

In conjunction with the above four equations, we need the so-called constitutive equations. These are

$$\mathbf{D} = \varepsilon \mathbf{E},$$

$$\mathbf{B} = \mu \mathbf{H}, \qquad (2.1.2)$$

$$\mathbf{J} = \sigma \mathbf{E},$$

where ε is the permittivity with units of F/m, μ is the permeability with units of H/m, and σ is the conductivity with units of mho/m. To simplify the discussion we will assume that the medium is linear, isotropic, and insulating. Then (2.1.2) can be rewritten as

$$\mathbf{D} = \varepsilon_0 \varepsilon_r \mathbf{E},$$

$$\mathbf{B} = \mu_0 \mu_r \mathbf{H}, \qquad (2.1.3)$$

$$\mathbf{J} = 0.$$

Here ε_r is the dielectric constant, ε_0 is the permittivity of the vacuum $= 8.8542 \times 10^{-12}$ F/m, μ_r is the relative permeability of the medium, and $\mu_0 = 4\pi \times 10^{-7}$ H/m.

Substituting (2.1.3) into (2.1.1) we obtain the wave equations in $\mathbf{E}$ and $\mathbf{H}$

$$\nabla^2 \mathbf{E} - \frac{1}{v^2} \frac{\partial^2 \mathbf{E}}{\partial t^2} = 0,$$

$$ \qquad (2.1.4)$$

$$\nabla^2 \mathbf{H} - \frac{1}{v^2} \frac{\partial^2 \mathbf{H}}{\partial t^2} = 0,$$

where it is assumed that no sources are present and

$$v = \frac{1}{\sqrt{\varepsilon_0 \mu_0}} \times \frac{1}{\sqrt{\varepsilon_r \mu_r}} = \frac{c}{\sqrt{\varepsilon_r \mu_r}},$$

$c = 3 \times 10^8$ m/s $=$ the velocity of electromagnetic waves in a vacuum.

To study the electromagnetic wave propagation in this media, we assume time dependence in the form $e^{j\omega t}$. This modifies (2.1.4) as

$$\nabla^2 \mathbf{E} + \frac{\omega^2}{v^2} \mathbf{E} = 0, \qquad (2.1.5)$$

$$\nabla^2 \mathbf{H} + \frac{\omega^2}{v^2} \mathbf{H} = 0. \qquad (2.1.6)$$

If we look for a plane wave solution, then

$$\mathbf{E} \propto e^{j(\omega t - \mathbf{k} \cdot \mathbf{r})},$$

$$ \qquad (2.1.7)$$

$$\mathbf{H} \propto e^{j(\omega t - \mathbf{k} \cdot \mathbf{r})},$$

where $\mathbf{k}$ is the propagation vector and denotes the direction in which the plane wave is propagating. In the Cartesian coordinate system

$$\mathbf{k} \cdot \mathbf{r} = (\mathbf{i}_x k_x + \mathbf{i}_y k_y + \mathbf{i}_z k_z) \cdot (\mathbf{i}_x x + \mathbf{i}_y y + \mathbf{i}_z z) = k_x x + k_y y + k_z z. \quad (2.1.8)$$

For the case $k_y = k_z = 0$, we see that (2.1.7) represents the one-dimensional plane waves discussed in Section 1.1.1.

Substituting (2.1.7) into (2.1.5) we obtain the relationship between ω and $\mathbf{k}$

$$\frac{\omega^2}{|k|^2} = v^2,$$

or

$$k = \pm \frac{\omega}{v}. \tag{2.1.9}$$

The $+$ sign is associated with the forward-going wave and the $-$ sign with the backward-traveling wave.

Substituting (2.1.7) into (2.1.1), we also obtain

$$\mathbf{E} = \frac{Z}{k}(\mathbf{k} \times \mathbf{H}),$$

$$\mathbf{H} = \frac{\mathbf{k} \times \mathbf{E}}{kZ}, \tag{2.1.10}$$

and

$$\mathbf{P} = \frac{1}{2} R_e(\mathbf{E} \times \mathbf{H^*}) = \frac{1}{2} \frac{|E|^2}{Z} \left(\frac{\mathbf{k}}{k} \right),$$

where $\mathbf{P}$ is known as the Poynting vector which denotes the direction in which the energy is propagating and Z is called the characteristic impedance, given by $Z = \sqrt{\mu/\varepsilon}$, and has units of ohms. Thus the electric field, the magnetic field, and the propagation directions are mutually perpendicular to each other. For propagation in the x direction with the E field in the y direction, the H field must be in the z direction

$$\mathbf{E} = \mathbf{i}_y E_0 e^{j(\omega t - kx)},$$

$$\mathbf{H} = \mathbf{i}_z H_0 e^{j(\omega t - kx)}, \tag{2.1.11}$$

and

$$\frac{E_0}{H_0} = Z = \sqrt{\frac{\mu}{\varepsilon}} = Z_0 \sqrt{\frac{\mu_r}{\varepsilon_r}}, \tag{2.1.12}$$

where $Z_0 = 377\ \Omega$ and E_0 is the magnitude of the electric field.

The power density of this wave propagating in the x direction is given by

$$P\ (\text{W/m}^2) = \frac{1}{2} \frac{E_0^2}{Z} = \frac{1}{2} H_0^2 Z. \tag{2.1.13}$$

If we solve the wave equation in the spherical coordinate system, the solution can be shown to be given by

$$E_0 = A \frac{1}{j\lambda r} e^{j(\omega t - kr)} \cdot \mathbf{i}_E, \tag{2.1.14}$$

and

$$\mathbf{H} = \frac{\mathbf{k} \times \mathbf{E}}{kZ},$$

where $\mathbf{i}_E$ is the unit vector, which can have only a θ or a φ component, and A is a constant.

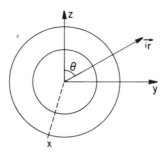

Fig. 2.1.1. Spherical waves.

The dependence on the amplitude, as $1/r$, can be understood from intuitive arguments. As shown in Fig. 2.1.1, let us consider the waves starting from the point source located at the origin. Then the wavefronts will be spheres, the radius of the sphere being larger for larger values of r. However, since all the wavefronts emanate from the same source at the origin, the power density must decrease proportional to $1/r^2$ to keep the total power constant. Thus, the electric field vector must be inversely proportional to r.

Using a similar argument, we can also obtain the cylindrical electromagnetic waves, shown in Fig. 2.1.2, to be given by

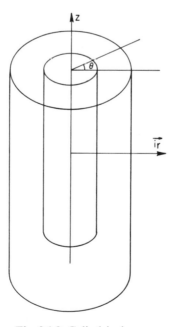

Fig. 2.1.2. Cylindrical waves.

$$E_z \propto \frac{2}{\pi\sqrt{r}} e^{j(\omega t - kr)},$$

$$H_\varphi \propto \frac{2}{\pi\sqrt{r}} e^{j(\omega t - kr)}.$$

(2.1.15)

It must be obvious by this time that the refractive index n is related to ε_r and μ_r by the relationship

$$n = \frac{1}{\sqrt{\varepsilon_r \mu_r}}.$$

(2.1.16)

Usually, at optical frequencies,

$$\mu_r \approx 1.$$

Thus

$$n^2 \approx \frac{1}{\varepsilon_r}.$$

(2.1.17)

2.2. Radiation from a Source

We have found that a particular component of the electric field, due to the electromagnetic radiation from a unit source situated at the origin, can be written as

$$E(\mathbf{r}) = \frac{1}{j\lambda r} e^{j(\omega t - kr)},$$

(2.2.1)

where $r = \sqrt{x^2 + y^2 + z^2}$ and $\mathbf{r} = \mathbf{i}_x x + \mathbf{i}_y y + \mathbf{i}_z z$. However, if the source is located at (x', y', z') or at $\mathbf{r}'$, then the electric field at the point $\mathbf{r}$ will be given by

$$E(\mathbf{r}) = \frac{1}{j\lambda |\mathbf{r} - \mathbf{r}'|} e^{j(\omega t - k|\mathbf{r} - \mathbf{r}'|)}.$$

(2.2.2)

As shown in Fig. 2.2.1, we call the point r the position of the detector. Now we ask ourselves what happens to the electric field at the detector point if we have not one source but many sources, which can be discrete and/or con-

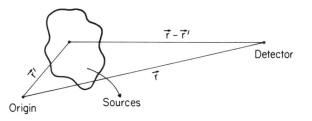

Fig. 2.2.1. Schematics for the general radiation problem from a multitude of sources distributed over a volume.

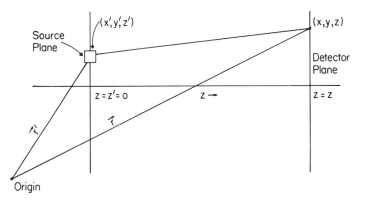

Fig. 2.2.2. Schematics for the radiation problem showing the source plane and the detector plane.

tinuous and distributed over some volume. Remembering that Maxwell's equations are linear, and that superposition should hold well for the solutions to Maxwell's equations we immediately obtain, for this case,

$$E(\mathbf{r}) \propto \int\int\int \frac{E(x', y', z')}{j\lambda|\mathbf{r} - \mathbf{r}'|} e^{j(\omega t - k|\mathbf{r} - \mathbf{r}'|)} d^3r', \qquad (2.2.3)$$

where integration has to be performed over all the sources. The source of the point (x', y', z') has an electric field strength which is proportional to $E(x', y', z')$. In most optics problems, we simplify the above equation by noting that most of the time our sources will all be situated in a single plane which is perpendicular to the optical axis, the z-axis. Thus, for Fig. 2.2.2, the equation becomes

$$E(x, y, z) = \int\int \frac{E(x', y')}{j\lambda|\mathbf{r} - \mathbf{r}'|} e^{j(\omega t - k|\mathbf{r} - \mathbf{r}'|)} dx' \, dy', \qquad (2.2.4)$$

where we have chosen the source plane as the plane $z = 0$.

2.3. The Diffraction Problem

In many problems in optics, not only do we need to consider the radiations from the sources, but we need to know what happens to these radiations as they pass through an obstacle like an aperture—this is shown in Fig. 2.3.1. To solve this problem, we need to consider the effect of boundaries. The technique for solving this boundary-value problem is rather involved. In place of deriving the result we shall state the result, which is also known as Huygen's approximation.

In Fig. 2.3.1 we calculate the electric field incident on the boundary from (2.2.4). To calculate the electric field at the detector plane, we assume that the

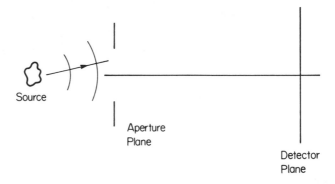

Fig. 2.3.1. Schematics for the diffraction problem showing the source, the diffracting aperture, and the detector.

boundary region is replaced by sources of the electric field having the same magnitude as the incident electric field. That is, the electric field at the detector plane z is given by

$$E(x, y, z) = \int\int E_{\text{inc}} T(x', y', 0) \frac{1}{j\lambda |\mathbf{r} - \mathbf{r}'|} e^{j(\omega t - k|\mathbf{r} - \mathbf{r}'|)} \, dx' \, dy'$$

$$= \int\int E_{\text{trans}}(x', y') \frac{1}{j\lambda |\mathbf{r} - \mathbf{r}'|} e^{j(\omega t - k|\mathbf{r} - \mathbf{r}'|)} \, dx' \, dy'. \qquad (2.3.1)$$

Remember that if there is an obstruction on the boundary (which we have called the source plane), or any variation in amplitude or phase, due to a different transparent material placed at the source plane, it is included in the transmission function $T(x', y', 0)$. The transmission function is defined to be

$$T(x', y') = \frac{E_{\text{trans}}(x', y', 0^+)}{E_{\text{inc}}(x', y', 0^-)}, \qquad (2.3.2)$$

where $E_{\text{trans}}(x', y', 0^+)$ is the transmitted electric field.

Our fundamental diffraction formula is (2.3.1), and the rest of this section is based on the application of this diffraction equation. However, before we proceed any further it is of interest to note the different cases. For example, if the transmission function is such that we have only discrete sources (see Fig. 2.3.2), then the problem is generally known as an interference problem. If the transmission function is such that the equivalent sources are distributed, then the problem is called a diffraction problem.

The diffraction integral, for most purposes, cannot be evaluated simply. However, for most cases of practical importance, certain approximations can be performed. These are generally known as the Fresnel and Fraunhofer diffraction approximations and are discussed in the next section. It should be mentioned that (2.2.3), (2.2.4), and (2.3.1) should include an obliquity factor to be more precise; however, this factor is negligible for most of the applications considered here.

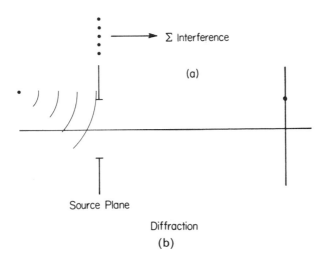

Fig. 2.3.2. Diffraction due to: (a) discrete apertures or sources and (b) continuous apertures or sources.

2.4. Different Regions of Diffraction

In the last section we derived the diffraction integral. As far as the use of this integral is considered, we think of the equivalent system model shown in Fig. 2.4.1. The incident electric field is $E_{inc}(x', y', 0)$ and the output electric field is $E(x, y, z)$, where they are related by (2.3.1). It is of interest to point out that the diffraction formula can be derived from this system concept considering that the optical system is linear as it represents a solution of the wave equation. This derivation is given in Reference 7.

To simplify (2.3.1), we first consider the far-zone approximation. That is, we assume that the detector plane distance, D, on the optical axis, z, is much greater than any value of x', y', x, or y which we shall be interested in. It turns out that, for the most practical optical systems, this is a valid assumption

$$z \gg x, y, x', \text{ or } y'.$$

In that case

$$|\mathbf{r} - \mathbf{r}'| = \{z^2 + (x - x')^2 + (y - y')'\}^{1/2}$$

$$= \left\{1 + \frac{1}{2}\left(\frac{x - x'}{z}\right)^2 + \frac{1}{2}\left(\frac{y - y'}{z}\right)^2\right\} + \dots$$

$$\simeq z,$$

or

$$\frac{1}{|\mathbf{r} - \mathbf{r}'|} \simeq \frac{1}{z}. \tag{2.4.1}$$

Thus, the term $1/(|\mathbf{r} - \mathbf{r}'|)$ in the diffraction integral can be approximated simply by $1/z$. We are tempted to replace the term $k|\mathbf{r} - \mathbf{r}'|$ in the exponential

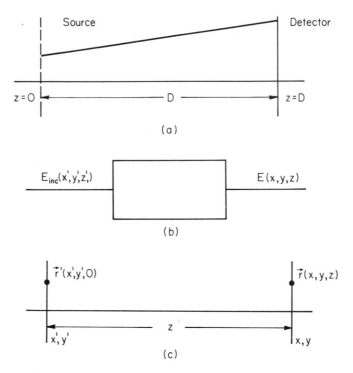

Fig. 2.4.1. Equivalent system model for a diffraction problem: (a) actual problem, (b) equivalent system, and (c) schematics and coordinate system for the Fresnel approximation.

factor by kz also. However, this is a gross mistake because

$$k|\mathbf{r} - \mathbf{r}'| = \frac{2\pi}{\lambda}|\mathbf{r} - \mathbf{r}'| = \frac{2\pi \times 10^6}{0.5}|\mathbf{r} - \mathbf{r}'|, \qquad (2.4.2)$$

when $\lambda = 0.5 \ \mu$m, e.g., $|\mathbf{r} - \mathbf{r}'|$ is multiplied by such a large factor that even a minute error in computing $|\mathbf{r} - \mathbf{r}'|$ will be disastrous. Furthermore, it is in the phase term and cannot be neglected. Thus, the far field approximation can be written as

$$E(x, y, z) = \frac{e^{j\omega t}}{j\lambda z} \iint E(x', y')_{\text{trans}} e^{-jk|\mathbf{r} - \mathbf{r}'|} \, dx' \, dy'. \qquad (2.4.3)$$

2.4.1. The Fresnel Approximation

The far-zone diffraction formula given by (2.4.3) is still a formidable integral for most practical purposes. However, some further simplification can be obtained for the phase term in the exponential. Using the notation in Fig.

2.4.1(c) we obtain

$$|\mathbf{r} - \mathbf{r}'| = \{z^2 + (x - x')^2 + (y - y')^2\}^{1/2}$$

$$= z\left\{1 + \frac{1}{2}\left(\frac{(x - x')^2 + (y - y')^2}{z^2}\right)\right\}$$

$$- \frac{1}{8}\left\{\frac{(x - x')^2 + (y - y')^2}{z^2}\right\}^2 + \dots. \qquad (2.4.4)$$

The Fresnel approximation is valid when (2.4.4) is approximated by the following equation:

$$|\mathbf{r} - \mathbf{r}'| \simeq z\left\{1 + \frac{1}{2}\left(\frac{(x - x')^2 + (y - y')^2}{z^2}\right)\right\}. \qquad (2.4.5)$$

Thus, in the Fresnel region, the diffraction integral in (2.4.3) simplifies to

$$E(x, y, z) = \frac{e^{j(\omega t - kz)}}{j\lambda z} \iint E_{\text{trans}}(x', y')e^{-j(k/2z)\{(x-x')^2+(y-y')^2\}} \, dx' \, dy'$$

$$= \frac{e^{j(\omega t\ kz)}}{j\lambda z} \iint E_{\text{trans}}(x', y')e^{-j(\pi/\lambda z)\{(x-x')^2+(y-y')^2\}} \, dx' \, dy'. \quad (2.4.6)$$

This is only valid when

$$\exp\left\{\frac{-jk}{8z^3}[(x - x')^2 + (y - y')^2]^2\right\} \approx 1,$$

or

$$\frac{\pi}{4\lambda z^3}\{(x - x')^2 + (y - y')^2\}^2 \ll 1,$$

or

$$z^3 \gg \frac{\pi}{4\lambda}\{(x - x')^2 + (y - y')^2\}^2_{\text{max}}. \qquad (2.4.7)$$

In (2.4.7) we must consider the maximum possible value of $\{(x - x')^2 + (y - y')^2\}^2$, including all the nonzero source points on the source plane and the region of interest in the detector plane.

For the reader who is familiar with linear system theory, it might be obvious that (2.4.6) can be rewritten as

$$E(x, y, z) = \frac{e^{j(\omega t - kz)}}{j\lambda z} \cdot E(x', y')_{\text{trans}} * e^{-j(\pi/\lambda z)(x'^2+y'^2)}, \qquad (2.4.8)$$

where $*$ means two-dimensional convolution.

2.4.2. The Fraunhofer Approximation

The diffraction integral in (2.4.6) can be further simplified under a more restrictive condition on the distance of the detector plane. Again expanding the phase term,

$$\frac{\pi}{\lambda z}\{(x-x')^2 + (y-y')^2\} = \frac{\pi}{\lambda z}(x^2 + y^2) + \frac{\pi}{\lambda z}(x'^2 + y'^2)$$

$$- \frac{2\pi}{\lambda z}(xx' + yy'), \qquad (2.4.9)$$

we note that the first term is a constant as far as the integration variables x' and y' are concerned. In the Fraunhofer approximation, the second term is considered negligible. Thus, under the Fraunhofer approximation, (2.4.6) can

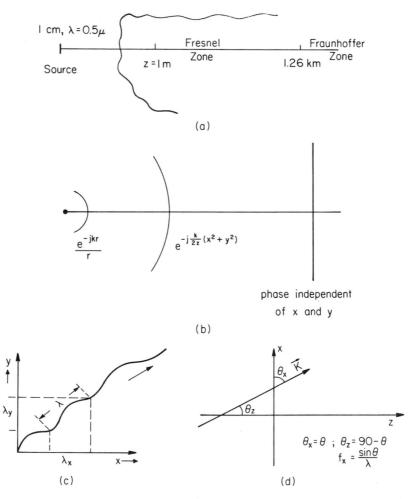

Fig. 2.4.2. (a) Regions of validity for different approximations, (b) wavefronts radiating from a print source for different approximations, (c) x and y components of wavelength, and (d) coordinates for the spatial frequency, f_x.

be rewritten as

$$E(x, y, z) = \frac{e^{j(\omega t - kz)}}{j\lambda z} e^{-j(\pi/\lambda z)(x^2 + y^2)} \cdot \iint E_{trans}(x', y') e^{+j(2\pi/\lambda z)(xx' + yy')} \, dx' \, dy'.$$

(2.4.10)

The above equation is valid only when

$$e^{-j(k/2z)(x'^2 + y'^2)} \simeq 1,$$

or

$$\frac{\pi}{\lambda z}(x'^2 + y'^2) \ll 1,$$

or

$$z \gg \frac{\pi}{\lambda}(x'^2 + y'^2)_{max}.$$

(2.4.11)

It is of interest to compare (2.4.7) and (2.4.11) numerically, to get a clearer idea about the limits of validity of different approximations. Consider $\lambda = 0.5 \ \mu m$ and the maximum value of x, x', y, and y' is on the order of 1 cm. Then

$$z_{Fraunhofer} \gg 1.26 \ km,$$

$$z_{Fresnel} \gg 1 \ m,$$

$$z_{far-zone} \gg 0.1 \ m.$$

Figure 2.4.2 depicts the different regions of validity for this numerical example. In nearly all practical cases dealing with optics, the Fresnel diffraction is quite good, even though (2.4.7) is not strictly satisfied. In this book, we shall be concerned only with Fresnel and Fraunhofer approximations.

2.4.3. The Spatial Frequency

The Fraunhofer diffraction formula given by (2.4.10) can be rewritten as

$$E(x, y, z) = \frac{e^{j(\omega t - Kz)}}{j\lambda z} e^{-j(k/2z)(x^2 + y^2)} \iint E_{trans}(x', y') e^{-j2\pi(f_x x' + f_y y')} \, dx' \, dy',$$

(2.4.12)

where we have defined two new variables, f_x and f_y. These are given by

$$f_x = \frac{x}{\lambda z},$$

(2.4.13)

and

$$f_y = \frac{y}{\lambda z}.$$

The dimensions of these new variables are $(meter)^{-1}$ and they are called spatial frequencies. These spatial frequencies in optics play a role very similar to the frequency (time) in electrical engineering, as will soon become evident.

However, we note that (1.1.9), which describes the many different forms of an expression for a one-dimensional wave, can also be written as

$$E(x, t) = Ae^{j2\pi(ft - f_x x)}, \tag{2.4.14}$$

where we have written

$$f_x = \frac{1}{\lambda} = \frac{k}{2\pi}. \tag{2.4.15}$$

Thus, we see that the spatial frequency of a one-dimensional wave is simply the inverse of λ. Going to a three-dimensional wavefront we obtain, from (2.1.7),

$$E(x, y, z, t) = Ae^{j2\pi(ft - f_x x - f_y y)} \cdot e^{-jK_z z}, \tag{2.4.16}$$

when

$$f_x = 2\pi k_x; \qquad f_y = 2\pi k_y. \tag{2.4.17}$$

We also know that

$$k_x^2 + k_y^2 + k_z^2 = k^2 = \left(\frac{2\pi}{\lambda}\right)^2.$$

Sometimes it is customary to define the quantities λ_x, λ_y, and λ_z as

$$\lambda_i = \frac{2\pi}{k_i}, \qquad i = x, y, z, \tag{2.4.18}$$

where, for example, λ_x is the projection or component of λ along the x direction, as shown in Fig. 2.4.2(c). Thus (2.4.17) becomes

$$f_x = \frac{1}{\lambda_x} \qquad \text{and} \qquad f_y = \frac{1}{\lambda_y}. \tag{2.4.19}$$

If the k vector makes an angle $(90 - \theta_x)$ with the x-axis, an angle $(90 - \theta_y)$ with the y-axis, and an angle θ_z with the z-axis, then

$$f_x = \frac{\sin \theta_x}{\lambda} \qquad \text{and} \qquad f_y = \frac{\sin \theta_y}{\lambda}, \tag{2.4.20}$$

and

$$\sin^2 \theta_x + \sin^2 \theta_y + \cos^2 \theta_z = 1. \tag{2.4.21}$$

The situation where $\theta_y = 0$ is shown in Fig. 2.4.2(d) and is a very important case and will be used often, later in this section.

Physically, spatial frequency means how many wavelengths can be in 1 m for a wavefront. However, for (2.4.13) it is rather complex, and we note that it depends on the position of the detector with respect to the source.

The reader familiar with Fourier transform theory immediately recognizes that (2.4.12) describes the electric field at the detector plane, as the two-dimensional Fourier transform of the transmitted electric field when spatial frequencies f_x and f_y, defined by (2.4.13), are used

$$E(x, y, z) = \frac{e^{j(\omega t - Kz)}}{j\lambda z} e^{-j(k/2x)(x^2+y^2)} \{E_{\text{trans}}(x', y')\} \Big|_{\substack{f_x = x/\lambda z \\ f_y = y/\lambda z}}$$

where the symbol $\mathscr{F}\{\ \}$ means the Fourier transform.

Because of (2.4.21) and also the fact that the Fresnel transformation becomes equivalent to the Fraunhofer approximation (in conjunction with a lens), the Fourier transform plays a very important role in the understanding and applications of wave optics. As we shall see later, the concept of holography also becomes easier to comprehend using this concept. That is one reason this part of wave optics is also known as Fourier optics.

A point of historical note is worth mentioning. Although optics has been a subject of scientific interest for a long time, this analogy between the frequency in electrical engineering and spatial frequency in optics has only been utilized since the 1960s. So you can see that Fourier optics is rather a new subject.

2.4.4. Summary of Formulas

Realizing the importance of the Fourier transform we will review it in the next section. However, before we do that, it is of convenience to rewrite all the diffraction formulas and approximations in one place.

The General Formula

$$E(\mathbf{r}) \propto \frac{e^{j\omega t}}{j\lambda} \int_r E(\mathbf{r}') \frac{e^{-jk|\mathbf{r}-\mathbf{r}'|}}{|\mathbf{r} - \mathbf{r}|} d^3 v'.$$

The Far-Field Approximation

$$E(x, y, z) = \frac{e^{j\omega t}}{j\lambda z} \iint E(x', y', 0)_{\text{trans}} e^{-jk|\mathbf{r}-\mathbf{r}'|} \, dx' \, dy',$$

where $z \gg x, x', y$ and y'.

The Fresnel Approximation

$$E(x, y, z) = \frac{e^{j(\omega t - kz)}}{j\lambda z} \iint E(x', y', 0)_{\text{trans}} e^{-j(k/2z)\{(x-x')^2+(y-y')^2\}} \, dx' \, dy'$$

$$= \frac{e^{j(\omega t - kz)}}{j\lambda z} E(x', y', 0)_{\text{trans}} * e^{-j(k/2z)(x'^2+y'^2)}$$

$$= \frac{e^{j(\omega t - kz)}}{j\lambda z} e^{-j(\pi/\lambda z)(x^2+y^2)} \iint E(x', y', 0)_{\text{trans}} e^{-j(\pi/\lambda z)(x'^2+y'^2)}$$

$$\cdot e^{+j2\pi(f_x x'+f_y y')} \, dx' \, dy'$$

$$= \frac{e^{j(\omega t - kz)}}{j\lambda z} e^{-j(\pi/\lambda z)(x^2 + y^2)}$$

$$\cdot \mathscr{F}\{E(x', y', 0)_{\text{trans}} e^{-j(\pi/\lambda z)(x'^2 + y'^2)}\}_{\substack{f_x = x/\lambda z \\ f_y = y/\lambda z}}. \tag{2.4.22}$$

The Fraunhofer Approximation

$$E(x, y, z) = \frac{e^{j(\omega t - kz)}}{j\lambda z} e^{-j(\pi/\lambda z)(x^2 + y^2)} \iint E(x', y', 0) e^{j(k/2z)(xx' + yy')} \, dx' \, dy'$$

$$= \frac{e^{j(\omega t - kz)}}{j\lambda z} e^{-j(\pi/\lambda z)(x^2 + y^2)} \mathscr{F}\{E(x', y', 0)\}_{\substack{f_x = x/\lambda z \\ f_y = y/\lambda z}}$$

$$= \alpha_{\text{con}}(z) \mathscr{F}\{E(x', y', 0)\}_{\substack{f_x = x/\lambda z \\ f_y = y/\lambda z}}. \tag{2.4.23}$$

Note that in (2.4.22) and (2.4.23) there are some new forms of the diffraction formulas which were not given previously and

$$\alpha_{\text{con}} = \frac{e^{j(\omega t - kz)}}{jz\lambda} e^{-j(\pi/\lambda z)(x^2 + y^2)}.$$

2.5. The Fourier Transform

For a function, $\varphi(t)$, the Fourier transform* is defined as

$$\mathscr{F}(f) = \int \varphi(t) e^{+j2\pi ft} \, dt = \mathscr{F}\{\varphi(t)\}, \tag{2.5.1}$$

where $\varphi(t)$ is a square integrable function and goes to zero as $t \to \pm \infty$. It is actually a mapping of the φ function from the t-plane to the f-plane according to the prescription defined by (2.5.1). In electrical engineering, t is the time variable and f is called frequency. However, as we have discussed before, for optics, t will be replaced by either x or y, the space variable, and f by f_x or f_y, the spatial frequencies. An inverse Fourier transform formula, which can be proven, is given by

$$\varphi(t) = \int \mathscr{F}(f) e^{-j2\pi ft} \, df = \mathscr{F}^{-1}\{\mathscr{F}(\varphi(t))\}. \tag{2.5.2}$$

Let us consider the Fourier transform of the function

$$\varphi(t) = e^{-j2\pi f_0 t}, \tag{2.5.3}$$

which is a monochromatic or single-frequency time function. Using (2.5.1) we

---

* Note that for convenience we have interchanged the conventional definition of Fourier and its inverse.

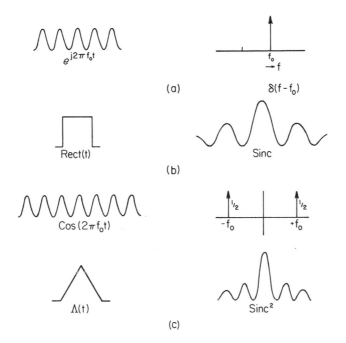

Fig. 2.5.1. Some functions and their Fourier transforms: (a) exponential function and its transform which is a delta function; (b) rectangular and sinc functions; and (c) cosine and triangular functions.

obtain

$$F(f) = \mathscr{F}\{e^{-j2\pi f_0 t}\} = \int_{-\infty}^{+\infty} e^{j2\pi(f-f_0)t}\, dt$$

$$= \delta(f - f_0). \tag{2.5.4}$$

The last equality resulting in a delta function is discussed further in the Appendix. It is assumed that the reader is familiar with the concept of delta functions. If not, please read the appendix discussing the delta function.

Figure 2.5.1(a) graphically illustrates (2.5.4). It is of interest to derive a few other Fourier transforms of different functions.

(i) $$\varphi(t) = \begin{cases} 1 & \text{for } -T/2 \le t \le T/2, \\ 0 & \text{otherwise.} \end{cases} \tag{2.5.5}$$

This particular function is also known as a rectangular function, rect(T), and is shown in Fig. 2.5.1(b). The Fourier transform of rect(T) can be evaluated as follows:

$$\mathscr{F}(f) = \int_{-T/2}^{+T/2} e^{+j2\pi ft}\, dt$$

$$= \frac{1}{j2\pi f}[e^{+j\pi fT} - e^{-j\pi fT}]$$

$$= \frac{1}{\pi f} \frac{e^{j\pi f T} - e^{-j\pi f T}}{2j} = T \frac{\sin \pi f T}{\pi f T}$$

$$= T \operatorname{sinc}(f T), \qquad (2.5.6)$$

where the "sinc" function is defined as

$$\operatorname{sinc} x = \frac{\sin \pi x}{\pi x}. \qquad (2.5.7)$$

It is obvious that the sinc function goes to zero for

$$\sin \pi x = 0 \quad \text{or} \quad x = \pm m \quad \text{when} \quad m = 1, 2, \dots.$$

For

$$x = 0, \quad \operatorname{sinc}(x) = 1. \qquad (2.5.8)$$

The sinc function has maxima or minima at x values given by

$$\tan \pi x = \pi x. \qquad (2.5.9)$$

The first maximum value is 1 at $x = 0$.

The function $T \operatorname{sinc}(f T)$ is plotted for different values of T in Fig. 2.5.2. It is observed that as $T \to \infty$ the sinc function approaches the delta function.

(ii) $\qquad \qquad \varphi(t) = e^{-j2\pi f_0 t} \cdot \varphi_1(t),$

$$\mathscr{F}\{\varphi(t)\} = \int \varphi_1(t) e^{+j2\pi(f - f_0)t} \, dt$$

$$= \mathscr{F}_1(f - f_0), \qquad (2.5.10)$$

where

$$\mathscr{F}_1(f) = \mathscr{F}\{\varphi_1(t)\}.$$

Thus multiplying by $e^{-j2\pi f_0 t}$, simply shifts the frequencies.

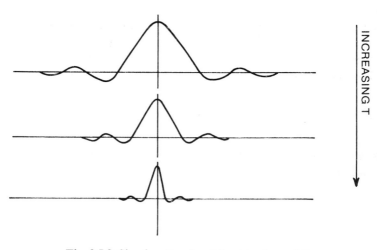

INCREASING T

Fig. 2.5.2. Sinc function for different values of T.

Table 2.5.1. Fourier transform theorems.

1. *Linearity theorem*

   $\mathscr{F}\{ag + \beta h\} = \alpha\mathscr{F}\{g\} + \beta\mathscr{F}\{h\}$; that is, the transform of a sum of two functions is simply the sum of their individual transforms.

2. *Similarity theorem*

   If $\mathscr{F}\{g(x, y)\} = G(f_x, f_y)$, then

   $$\mathscr{F}\{g(ax, by)\} = \frac{1}{|ab|}G\left(\frac{f_x}{a}, \frac{f_y}{b}\right),$$

   that is, a "stretching" of the coordinates in the space domain (x, y) results in a contraction of the coordinates in the frequency domain (f_x, f_y), plus a change in the overall amplitude of the spectrum.

3. *Shift theorem*

   If $\mathscr{F}\{g(x, y)\} = G(f_x, f_y)$, then

   $$\mathscr{F}\{g(x - a, y - b)\} = G(f_x, f_y)\exp[-j2\pi(f_x a + f_y b)],$$

   that is, translation of a function in the space domain introduces a linear phase shift in the frequency domain.

4. *Parseval's theorem*

   If $\mathscr{F}\{g(x, y)\} = G(f_x, f_y)$, then

   $$\int_{-\infty}^{\infty}\int_{-\infty}^{\infty}|g(x, y)|^2\,dx\,dy = \int_{-\infty}^{\infty}\int_{-\infty}^{\infty}|G(f_x, f_y)|^2\,df_x\,df_y.$$

   This theorem is generally interpretable as a statement of conservation energy.

5. *Convolution theorem*

   If $\mathscr{F}\{g(x, y)\} = G(f_x, f_y)$ and $\mathscr{F}\{h(x, y)\} = H(f_x, f_y)$, then

   $$\left\{\int_{-\infty}^{\infty}\int_{-\infty}^{\infty} g(\xi, \eta)h(x - \xi, y - \eta)\,d\xi\,d\eta\right\} = G(f_x, f_y)H(f_x, f_y).$$

   The convolution of two functions in the space domain (an operation that will be found to arise frequently in the theory of linear systems) is entirely equivalent to the more simple operation of multiplying their individual transforms.

The Fourier transform of other important functions are given in Tables 2.5.1 and 2.5.2 without proof. This also includes other important formulas which are useful.

2.5.1. Physical Interpretation of the Fourier Transform

Consider the Fourier transform of $\cos 2\pi f_0 t$, shown in Fig. 2.5.1(c). It has two components f_0 and $-f_0$ which are the only frequency components of this waveform with infinite-time duration. However, consider the Fourier transform of $\text{rect}(\Delta T) \cdot \cos 2\pi f_0 t$

$$\mathscr{F}[\text{rect}(\Delta T) \times \cos 2\pi f_0 t] = \Delta T\tfrac{1}{2}[\text{sinc}\{(f - f_0)\Delta T\} + \text{sinc}\{(f + f_0)\Delta T\}].$$

For a finite duration, a so-called "single-frequency wave" is found to have many frequencies centered around f_0 which is sometimes called the carrier frequency. The approximate relationship between the time duration ΔT and

Table 2.5.2

Function	Fourier transform
$\delta(t)$	1
$\mathrm{sgn}(t)$	$2/j\omega$
$u(t) = \frac{1}{2} + \frac{1}{2}\,\mathrm{sgn}(t)$	$\pi\delta(\omega) + 1/j\omega$
$P(t) = \mathrm{Rect}(1)$	$\mathrm{sinc}\ f$
$\cos\omega_0 t$	$\pi\delta(\omega - \omega_0) + \pi\delta(\omega + \omega_0)$
$e^{-\pi t^2}$	$e^{-\pi f^2}$
$\Lambda(t)$	$\mathrm{sinc}^2 (f)$

$$\mathrm{sinc}\ f = \frac{\sin \pi f}{\pi f}$$

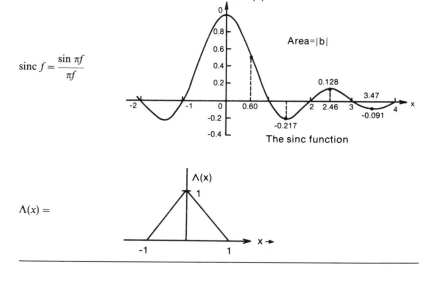

The sinc function

$$\Lambda(x) =$$

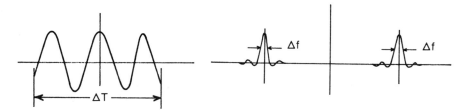

Fig. 2.5.3. Finite-time duration cosine function and its transform.

the spread in frequency, Δf (up to the half-power point shown in Fig. 2.5.3) is given by

$$\Delta f \cdot \Delta T = 1, \qquad (2.5.11)$$

a very important relationship.

2.5.2. The Two-Dimensional Fourier Transform

The two-dimensional Fourier transform of a function φ with x and y as variables is defined by

$$\mathcal{F}[\varphi(x, y)] = F(f_x, f_y) = \iint \varphi(x, y)e^{-j2\pi(f_x x + f_y y)}\, dx\, dy. \qquad (2.5.12)$$

where f_x and f_y are the frequencies associated with the x and y components, respectively.

In many examples, the function $\varphi(x, y)$ can be written as a product of two functions

$$\varphi(x, y) = \varphi_x(x)\varphi_y(y), \qquad (2.5.13)$$

where $\varphi_x(x)$ is a function of x only and $\varphi_y(y)$ is a function of y only. In that case

$$F\{\varphi(x, y)\} = F_x(f_x)F_y(f_y), \qquad (2.5.14)$$

where $F_x(f_x)$ and $F_y(f_y)$ are one-dimensional Fourier transforms defined by

$$F\{\varphi_x(x)\} = F_x(f_x),$$

and

$$F\{\varphi_y(y)\} = F_y(f_y).$$

However, in many cases this separation of variables is not possible. An example is the "circle" function shown in Fig. 2.5.4

$$\varphi(r) = \begin{cases} 1, & r \leq r_0, \\ 0, & r > r_0. \end{cases} \qquad (2.5.15)$$

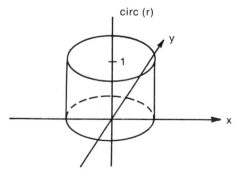

Fig. 2.5.4. Circle function and its transform.

The Fourier transform of this function, and other functions which possess circular symmetry, can be evaluated using Fourier–Bessel transforms.

By defining the following polar coordinate variables, r and θ:

$$
\begin{aligned}
r &= \sqrt{x^2 + y^2}, & x &= r \cos \theta, \\
\theta &= \tan^{-1}(y/x), & y &= r \sin \theta, \\
\rho &= \sqrt{f_x^2 + f_y^2}, & f_x &= \rho \cos \psi, \\
\psi &= \tan^{-1}(f_x, f_y), & f_y &= \rho \sin \psi,
\end{aligned}
\tag{2.5.16}
$$

in (2.5.12), we obtain

$$
F(\rho, \psi) = \int_0^{2\pi} d\theta \int_0^\infty dr \, r \cdot \varphi(r) e^{-j2\pi r \rho \cos(\theta - \psi)}.
\tag{2.5.17}
$$

The above equation can be written as

$$
F(\rho, \psi) = 2\pi \int_0^\infty r\varphi(r) J_0(2\pi r\rho) \, dr,
\tag{2.5.18}
$$

where J_0 is the zeroth-order Bessel function of the first kind. It is found that the Fourier transform is a function of ρ only. For the function $\varphi(r)$ defined in (2.5.15) we obtain

$$
F(\rho) = 2\pi \int_0^1 r J_0(2\pi r\rho) \, dr.
\tag{2.5.19}
$$

Using the integral formulas for Bessel functions, it can be shown that

$$
F(\rho) = \frac{J_1(2\pi\rho)}{\rho},
\tag{2.5.20}
$$

where J_1 is the first-order Bessel function of the first kind. A plot of the function

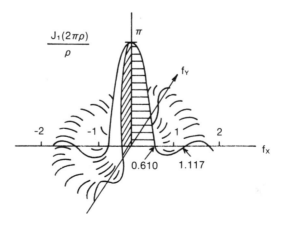

Fig. 2.5.5. Plot of eqn. (2.5.20).

Table 2.5.3

ρ	$J_1(2\pi\rho)/\pi\rho$	$[J_1(2\pi\rho)/\pi\rho]^2$
0	1	1
0.8175	0.132	0.0175
1.3395	0.065	0.0042
1.8495	0.04	0.0016

$F(f)$ is shown in Fig. 2.5.5. It is of interest to note that this function behaves similarly to the sinc function. Its zeros are located at

$$2\rho = 1.220, 2.233, 3.238\ldots . \tag{2.5.21}$$

The maximas and minimas are located at

$$2\rho = 0 \text{ (max)}, 1.635 \text{ (min)}, 2.679 \text{ (max)}, 3.699 \text{ (min)}\ldots . \tag{2.5.22}$$

These maximum and minimum values are tabulated in Table 2.5.3.

2.6. Some Examples of Fraunhofer Diffraction

2.6.1. The One-Dimensional Rectangular Aperture*

Consider the Fraunhofer diffraction of an aperture (in one dimension only) as shown in Fig. 2.6.1. The aperture is illuminated with a uniform light propagating parallel to the z-axis and having an amplitude E_0 at $z = 0$. The

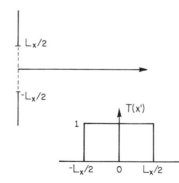

Fig. 2.6.1. One-dimensional rectangular aperture.

---

* One-dimensional problems are easier to handle mathematically. Unfortunately, one cannot just neglect the other dimension in the diffraction formulas. If the second dimension is just neglected, as we will often do for the sake of simplicity, the results obtained will be dimensionally incorrect (compare (2.6.3) and (2.6.9)).

transmission function $T(x')$ in this case is defined as

$$T(x') = \begin{cases} 1 & \text{for } -L_x/2 \le x' < L_x/2, \\ 0 & \text{otherwise,} \end{cases} \qquad (2.6.1)$$

or

$$T(x') = \text{rect}(L_x).$$

The diffracted electric field in the Fraunhofer approximation at $z = D$ and $x = x$ is given by

$$E(x) = \alpha_{\text{con}(D)} \mathscr{F}[E_0 \cdot T(x')]\}_{f_x = x/\lambda z}, \qquad (2.6.2)$$

where

$$\alpha_{\text{con}(D)} = \frac{e^{j(\omega t - kD)}}{j\lambda D} e^{-j(\pi/\lambda D)(x^2 + y^2)}.$$

Using the formulas discussed in the last section we obtain

$$\begin{aligned} E(x) &= \alpha_{\text{con}(D)} E_0 L_x \, \text{sinc}(f_x \cdot L_x) \\ &= \alpha_{\text{con}(D)} E_0 L_x \, \text{sinc}\left(\frac{x L_x}{\lambda D}\right). \end{aligned} \qquad (2.6.3)$$

We shall see later that, for most practical purposes, the important quantity is intensity due to the electric field. This intensity is defined as

$$I(x, y, z) = \tfrac{1}{2}\{E(x, y, z)E^*(x, y, z)\}, \qquad (2.6.4)$$

where the * means complex conjugate. Thus, for the case of a rectangular aperture

$$I(x) = \frac{E_0^2}{\lambda^2 D^2} L_x^2 \, \text{sinc}^2\left(\frac{x L_x}{\lambda D}\right), \qquad (2.6.5)$$

as

$$\alpha_{\text{con}(D)} \alpha_{\text{con}(D)}^* = \frac{1}{\lambda^2 D^2}.$$

$I(x)$ are plotted as functions of x and f_x in Fig. 2.6.2(a). The spot size Δx defined by the half-power points is given by

$$\Delta x \sim \frac{1}{L_x}(\lambda z). \qquad (2.6.6)$$

So as L_x tends to infinity, the spot size tends to zero. If we ponder over this result, we realize that this is really startling! Because, intuitively, as L_x increases we expect a larger spot, not a smaller spot. This contradictory result can easily be understood once we remember that the value of z in (2.6.3) must satisfy the Fraunhofer approximation condition given by (2.4.11), which is repeated here for the one-dimensional case

$$z \gg \frac{\pi}{\lambda}(x - x')_{\text{max}}^2 = z_f = \frac{\pi}{\lambda} L_x^2. \qquad (2.6.7)$$

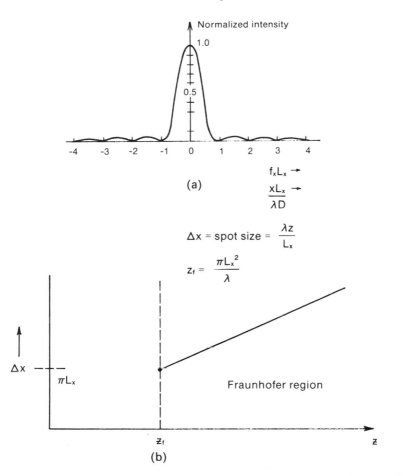

Fig. 2.6.2. (a) Diffracted electric field and intensity due to an aperture. (b) Spot size versus z.

If we put the value of $z = z_f$ in the equation for Δx we obtain

$$\Delta x|_{z=z_f} = \pi L_x.$$

Thus we see that the spot size really does not decrease but rather increases with the increasing value of L_x, as expected. This fact is shown in Fig. 2.6.2(b) which plots the spot size versus z.

2.6.2. The Two-Dimensional Rectangular Aperture

Now consider the Fraunhofer diffraction of a two-dimensional aperture defined by the following $T(x', y')$,

$$T(x', y') = \begin{cases} 1, & -L_x/2 \le x, L_x/2, \quad -L_y/2 \le y > L_y/2, \\ 0, & \text{otherwise.} \end{cases} \tag{2.6.8}$$

It is obvious that $T(x', y')$ is separable in the x' and y' functions

$$T(x', y') = \text{rect}(L_x)\,\text{rect}(L_y).$$

Thus the diffracted E field at $z = D$ is given by

$$
\begin{aligned}
E(x, y, D) &= \alpha_{\text{con}(D)}\mathscr{F}\{E_0\,T(x', y')\}_{\substack{fx=x/\lambda D \\ fy=y/\lambda D}} \\
&= \alpha_{\text{con}(D)}E_0\mathscr{F}[T(x')]\mathscr{F}[T(y')] \qquad (2.6.9) \\
&= \alpha_{\text{con}(D)}E_0[L_x\,\text{sinc}(f_x L_x)][L_y\,\text{sinc}(f_y L_y)].
\end{aligned}
$$

The intensity is given by

$$I(x, y) = \frac{E_0^2}{\lambda^2 D^2}\left[L_x\,\text{sinc}\left(\frac{xL_x}{\lambda D}\right)\cdot L_y\,\sin\left(\frac{yL_y}{\lambda D}\right)\right]^2. \qquad (2.6.10)$$

A picture of the intensity as a function of x, y and f_x and f_y is shown in Fig.

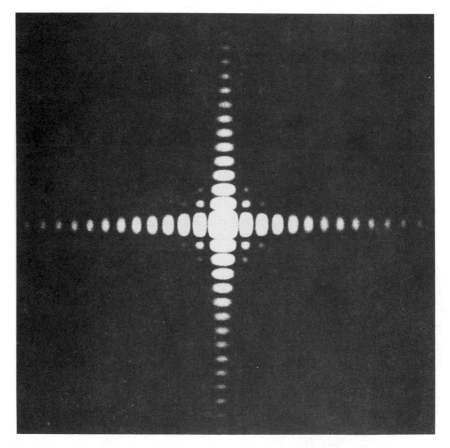

Fig. 2.6.3. Diffraction due to the two-dimensional aperture. (From M. Cagnet et al., *Atlas of Optical Phenomenon*, Springer-Verlag, New York, 1962).

2.6.3. It is interesting to note that along the x- and y-axes the intensity is strongest since one of the sinc functions has a value equal to 1.

2.6.3. One-Dimensional Aperture Centered at $x = x_0$

Consider the problem shown in Fig. 2.6.4, where the origin of the aperture is shifted from the origin of the coordinate system by a distance, x_0. Again the aperture is illuminated by a uniform electric field whose propagation vector is parallel to the optical axis. The transmission function $T(x')$ for this case is given by

$$T(x') = \begin{cases} 1, & x_0 - L_x/2 \leq x' < x_0 + L_x/2, \\ 0, & \text{otherwise.} \end{cases} \tag{2.6.11}$$

Thus

$$T(x') = \text{rect}(L_x - x_0).$$

For this case the diffracted electric field is given by

$$E(x) = \alpha_{\text{con}(D)} E_0 \mathscr{F} \{\text{rect}(L_x - x_0)\}_{f_x = x/\lambda z}$$

$$= \alpha_{\text{con}(D)} E_0 \int_{x_0 - L_x/2}^{x_0 + L_x/2} e^{+j2\pi f_x x'} \, dx$$

$$= \alpha_{\text{con}(D)} E_0 \int_{-L_x/2}^{L_x/2} e^{+j2\pi f_x (x'' + x_0)} \, dx'', \tag{2.6.12}$$

where we have substituted $x' = x'' + x_0$.

$$E(x) = \alpha_{\text{con}(D)} E_0 e^{-j2\pi f_x x_0} L_x(\text{sinc } f_x L_x). \tag{2.6.13}$$

(The above equation can also be obtained directly through the Fourier trans-

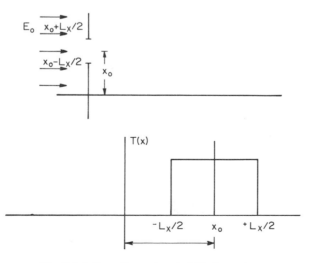

Fig. 2.6.4. One-dimensional shifted aperture.

form formulas.) Thus we obtain the important result that by shifting the aperture, only an extra term of the form $e^{-j2\pi f_x x_0}$ shows up which is just a phase term. Otherwise, the electric field is the same as if the aperture had not been moved. The intensity is given by

$$I(x) = \frac{E_0^2}{\lambda^2 D^2}\left[L_x \,\mathrm{sinc}\left(\frac{xL_x}{\lambda D}\right)\right]^2, \qquad (2.6.14)$$

which is exactly the same expression as that obtained for the case when the aperture was centered. Thus, if we measure intensity, the aperture shift is not detectable. Again, this somewhat startling result is easily understood by remembering that the expression is valid only in the Fraunhofer approximation.

2.6.4. One-Dimensional Rectangular Aperture with Uniform Light Shining at an Angle θ with Respect to the Optical Axis

Consider the problem shown in Fig. 2.6.5, when the incident light makes an angle θ with respect to the z-axis. In this case, the E field at $z = 0$ is given by*

$$E(x', 0) = E_0 e^{-j2\pi(\sin\theta/\lambda)x'}, \qquad (2.6.15)$$

---

* Note that the E field is actually given by

$$E(x', z) = E_0^{-j2\pi(\sin\theta/\lambda)x'} \cdot e^{j(\omega t - k_z z)},$$

where $k_z = k \cos\theta = (2\pi/\lambda)\cos\theta$.

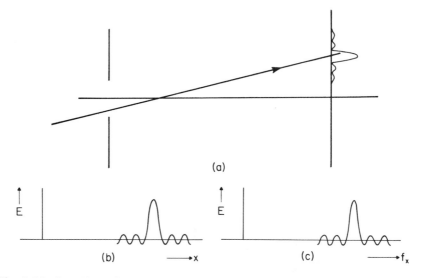

Fig. 2.6.5. One-dimensional rectangular aperture with uniform light shining at an angle.

as discussed in Section 2.4.3. This arises because the plane of constant phase is not in the x-plane but at an angle $(90^\circ - \theta)$ to the x-axis. Thus the wavefront along the x-axis has a phase variation given by the above equation. The limiting case of $\theta = 0$ is the case which has been considered so far. Equation (2.6.15) can also be written as

$$E(x', 0) = E_0 e^{-j2\pi f_{x0} x'}, \tag{2.6.16}$$

where

$$f_{x0} = \sin \theta / \lambda. \tag{2.6.17}$$

Thus, the diffracted E field is given by*

$$E(x) = \alpha_{\mathrm{con}(D)} \mathscr{F}[E_0 e^{-j2\pi f_{x0} x'} \cdot \mathrm{rect}(L_x)]$$

$$= \alpha_{\mathrm{con}(D)} E_0 \int \mathrm{rect}(L_x) e^{j2\pi (f_x - f_{x0}) x'}\, dx'$$

$$= \alpha_{\mathrm{con}(D)} E_0 \cdot L_x \, \mathrm{sinc}[(f_x - f_{x0}) L_x]. \tag{2.6.18}$$

Thus the sinc function is only shifted by f_{x0} in the frequency domain due to the presence of the incident light at an angle. This frequency, f_{x0}, can be called the spatial carrier frequency, in analogy with the carrier frequency in radio engineering terminology. Another important fact is that this spatial carrier frequency can easily be changed in value by simply shining light at different angles. This is a very important fact to understand, and as we shall see later in Section 2.8, the concept of holography is based on this fact.

   Again, the intensity is given by

$$I(x) = \frac{E_0^2}{\lambda^2 D^2} L_x^2 \, \mathrm{sinc}^2\{(f_x - f_{x0}) L_x\}. \tag{2.6.19}$$

The electric field and intensity are plotted as functions of x and f_x in Fig. 2.6.5(b) and (c). As expected, if the light is incident at an angle θ, the diffraction pattern is centered at

$$x = (f_{x0}) \times (\lambda z) = z \sin \theta.$$

Sometimes it is customary to write

$$f_x = \frac{\sin \alpha}{\lambda}, \tag{2.6.20}$$

where α is the angle subtended by the straight line joining the origin and the detector. This is illustrated in Fig. 2.6.5(a). For this case (2.6.18) can be rewritten as

$$E(x) = \alpha_{\mathrm{con}(D)} E_0 L_x \, \mathrm{sinc}\left[\left(\frac{\sin \theta + \sin \alpha}{\lambda}\right) L_x\right], \tag{2.6.21}$$

---

* Note that the value of "k" in α_{con} should be $k_z = k \cos \theta$.

which in the paraxial approximation becomes

$$E(x) = \alpha_{con(D)} E_0 L_x \text{ sinc} \left[\left(\frac{\theta + \alpha}{\lambda} \right) L_x \right]. \tag{2.6.22}$$

Sign Convention for Fourier Optics

$$f_{i0} = \frac{\sin \theta_i}{\lambda} \qquad (i = y \text{ or } x).$$

f_{i0} is negative if θ_i is measured clockwise with respect to the optical axis.

2.6.5. Some Discussion About the Free Space Propagation of Waves

Consider spherical waves emanating from a point source situated at the center of the coordinate axes, as shown in Fig. 2.1.1. We know that the amplitude of the wave at the point $\mathbf{r}$ is given by

$$E(\mathbf{r}) = e^{j(\omega t - \mathbf{k} \cdot \mathbf{r})}$$
$$= e^{j\{\omega t - k(x^2 + y^2 + z^2)^{1/2}\}}. \tag{2.6.23}$$

For large values of z, the quantity $|\mathbf{r}|$ can be expanded into a series as follows:

$$|r| = (x^2 + y^2 + z^2)^{1/2} = z \left(1 + \frac{x^2 + y^2}{z^2} \right)^{1/2}$$

$$= z \left(1 + \frac{x^2 + y^2}{2z^2} + \cdots \right)$$

$$\simeq z + \frac{x^2 + y^2}{2z} \qquad \text{for} \quad z \gg x \text{ and } y. \tag{2.6.24}$$

Thus the electric field at (x, y, z) becomes

$$E(x, y, z) = e^{j(\omega t - kz)} \cdot e^{-j(k/2z)(x^2 + y^2)} \tag{2.6.25}$$

under the Fresnel approximation. Of course, the above equation can be derived by inspection from the Fresnel diffraction formula given by (2.2.3) by substituting $E_{\text{trans}}(x', y', z') = \delta(x', y')$.

Similarly, in the Fraunhofer approximation, for this case

$$E(x, y, z) = e^{j(\omega t - kz)}. \tag{2.6.26}$$

This wave propagation in free space is illustrated in Fig. 2.4.2(b). It is observed that in the Fraunhofer region the spherical wave behaves like a plane wave whose phasefront is perpendicular to the z-axis. However, in the Fresnel region, the wavefront has a curvature with radius z, and phase in the z-plane varies in a fashion which is parabolic in x and y. Thus the major difference between the two wavefronts in the two regions is the presence of this curvature.

It is also now obvious that if we compare (2.4.22) and (2.4.23), it is found that the difference between the diffraction integrals is the presence of the factor $e^{-j(k/2z)(x'^2+y'^2)}$. This gives us a clue as to how we can eliminate this curvature using a lens. In the next section we shall prove the important fact that a lens changes the curvature of the wavefront or introduces a parabolic phase shift on the z-plane. Thus, it will be shown that a lens can convert the Fresnel approximation to a Fraunhofer-like diffraction integral.

2.7. Phase Transmission Functions and Lens

Up until now we have considered only the amplitude transmission function for diffraction problems. For example, in all the cases considered before

$$T(x', y') = |T(x', y')|.$$

However, in general,

$$T(x', y') = |T(x', y')|e^{j\varphi(x',y')}. \tag{2.7.1}$$

That is, the transmission function cannot only change the amplitude, but also the phase of the incident electric field. A special case is the phase transmission function where

$$T(x', y') = e^{j\varphi(x',y')},$$

and

$$|T(x', y')| = 1.$$

The best example of a phase transmission function is a transparent piece of glass of thickness $t(x', y')$, and having a fixed retractive index n. This is illustrated in Fig. 2.7.1.

If the glass was not present, the phase of the wavefront would have changed by

$$\theta_0 = kh = 2\pi\frac{h}{\lambda}$$

(by traveling a distance "h"). This is because at the point A the wave can be

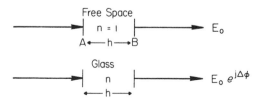

Fig. 2.7.1. Transmission through a transparent glass.

written as
$$e^{j(\omega t - kz)}$$

and at point B it is
$$e^{j(\omega t - kz - kh)}.$$

Thus the difference between the two phases is given by
$$\varphi_B - \varphi_A = kh.$$

However, when the glass is present in the path, the phase difference is
$$\varphi_{B'} - \varphi_{A'} = k'h,$$
where

$$k' = \frac{\omega}{v} = \frac{\omega}{c} n = kn,$$

where k is the free space propagation constant and k' is the propagation constant associated with the medium having a refractive index n. Thus, the phase difference introduced by the presence of the glass is

$$\Delta\varphi = (\varphi_{B'} - \varphi_{A'}) - (\varphi_B - \varphi_A)$$

$$= h(k' - k) = + \frac{(n-1)2\pi}{\lambda} h, \qquad (2.7.2)$$

where λ is the free space wavelength. Thus, we derive the important result that the transmission function of the transparent glass is given by

$$T(x', y') = e^{j\Delta\varphi} = e^{-j(2\pi/\lambda)(n-1)h(x,y)}, \qquad (2.7.3)$$

where the thickness of the glass is a function of x and y, $h(x, y)$.

In a sense a lens is nothing but a transparent piece of glass. However, the thickness of this glass is a function of x, y. If we can derive an expression for functional dependence on x and y, we can use (2.7.3) to obtain the transmission function of the lens.

A typical lens is shown in Fig. 2.7.2 having radii of curvature R_1 and R_2. The thickness along the optical axis (the center of the lens) is denoted by t_0. Then

$$h(x, y) = t_0 - t_1(x, y) - t_2(x, y), \qquad (2.7.4)$$

where $t_1(x, y)$ and $t_2(x, y)$ are defined in the figure. From geometrical considerations

$$t_1(x, y) = R_1 - (R_1^2 - x^2 - y^2)^{1/2}$$

$$= R_1 - R_1 \left(1 - \frac{x^2 + y^2}{R_1^2} \right)^{1/2}$$

$$= R_1 - R_1 \left[1 - \frac{x^2 + y^2}{2R_1^2} + \cdots \right]$$

$$= \frac{x^2 + y^2}{2R_1} + \text{higher-order terms.}$$

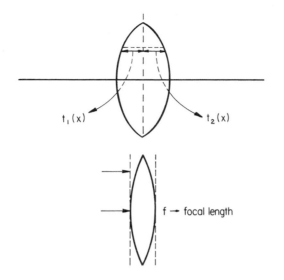

Fig. 2.7.2. Phase transmission function of a lens.

Paraxial approximation has been used in the above derivation. Similarly,

$$t_2(x, y) \simeq \frac{x^2 + y^2}{2|R_2|}.$$

(Note that R_2 is negative using our convention of geometrical optics.) Thus, we obtain

$$h(x, y) = t_0 - \frac{(x^2 + y^2)}{2}\left(\frac{1}{|R_1|} + \frac{1}{|R_2|}\right), \tag{2.7.5}$$

or the transmission function is given by

$$T(x', y') = e^{-j\frac{(n-1)2\pi}{\lambda}} h(x', y')$$

$$= e^{-j(n-1)\frac{2\pi}{\lambda}} t_0 e^{j(\pi/\lambda)\{(n-1)((1/|R_1|)+(1/|R_2|))(x^2+y^2)\}}$$

$$= e^{-j\varphi_0} e^{+j(\pi/\lambda f)(x^2+y^2)}, \tag{2.7.6}$$

where f is the focal length of the lens, and the lens designer's formula (1.2.39) has been used to derive the last expression. Thus we see that the lens has both a constant phase term and a phase term which varies parabolically as a function of x and y in the z-plane. This is illustrated in Fig. 2.7.3(a).

A constant phase wavefront, after passing through a lens, is converted to a spherical wave which converges to the focus. Whereas a spherical wave (see Fig. 2.7.3(b)) emanating from the focus becomes a plane wave after passing through the lens.

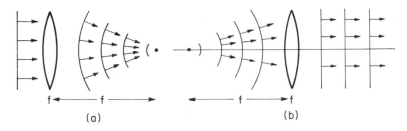

Fig. 2.7.3. Wavefronts after passing through lenses: (a) parallel incident wavefront and (b) diverging incident wavefront.

2.8. Fresnel Diffraction

In Fig. 2.4.2 we discussed the validity of different approximations in different regions. For $\lambda = 0.5$ μm and $(x')_{\max} = 1$ cm it was found that for Fraunhofer diffraction to be valid the detector must be placed at

$$z \gg 1.26 \text{ km}. \tag{2.8.1}$$

I am sure the reader will have wondered if z has to be that large; unless we are shining a laser on a moon to do some optics experiment, in most of the cases in the laboratory, Fraunhofer approximation does not hold in the usual laboratory experiments. We might then wonder why we have spent so much time in the last few sections on this approximation. Well, the reason will be clear in this section where we first show that in conjunction with a lens, the Fresnel diffraction formula becomes like that of Fraunhofer.

2.8.1. Fresnel Diffraction and Lens

Let us consider a situation where we have placed a lens with a focal length f at the aperture, as shown in Fig. 2.8.1. Let us consider that the aperture has a transmission function $T(x', y')$ by itself. Then the total transmission function, $T'(x, y)$, to be used in the formulas for the Fresnel diffraction formula, is given by

$$T'(x, y) = T(x', y')e^{-j\varphi_0}e^{+j[\pi(x'^2+y'^2)/\lambda f]}. \tag{2.8.2}$$

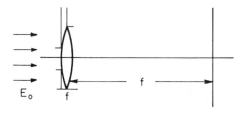

Fig. 2.8.1. Fresnel diffraction with a lens.

Substituting this transmission function in the Fresnel diffraction integral, (2.4.22), we obtain

$$E(x, y, z) = (\alpha_1) \iint E_{inc}(x', y', 0) T(x', y')$$

$$\times e^{-j\varphi_0} \cdot e^{-j(\pi/\lambda)(x^2+y^2)(1/z-1/f)} \cdot e^{-(j\pi/\lambda z)(x^2+y^2)} \qquad (2.8.3)$$

$$\times e^{+j2\pi(f_x x' + f_y y')} \, dx' \, dy',$$

where

$$\alpha_1 = \frac{e^{j(\omega t - kz)}}{jz\lambda}.$$

For $z = f$, we obtain

$$E(x, y, f) = (\alpha_{con}) e^{-j\varphi_0} \iint E_{inc}(x', y') T(x', y') e^{+j2\pi(f_x x' + f_y y')} \, dx' \, dy'$$

$$= (\alpha_{con}) e^{-j\varphi_0} \mathscr{F}\{E_{inc}(x', y') T(x', y')\} \qquad (2.8.4)$$

$$= \alpha_2 \mathscr{F}\{E_{inc}(x', y') T(x', y')\},$$

where

$$\alpha_2 = \alpha_{con} e^{-j\varphi_0} = \alpha_1 e^{-j(\pi/\lambda f)(x^2+y^2)} \cdot e^{-j\varphi_0}.$$

The last expression is identical to the Fraunhofer diffraction formula, except for the constant phase factor, $e^{-j\varphi_0}$.

Thus we see the important implication of the above result, in the sense that we can obtain the Fourier transform of a known $T(x', y')$, by using a lens and without sacrificing the Fresnel approximation.

Let us apply this result in a very important case—to understand the limitation of geometrical optics. To simplify the mathematics let us consider a one-dimensional case. Consider the finite lens of size L_x, as shown in Fig. 2.8.1. As parallel rays are incident on the lens, according to geometrical optics, all the rays pass through the focus. Or the focus is really a delta function on the z-axis of zero width.

However, if we use the diffraction integral equation (2.8.4) with

$$T(x') = rect(L_x) \qquad and \qquad E_{inc}(x', y') = E_0,$$

for this case, we find that

$$E(x, f) = (E_0 \alpha_2) L_x \, \text{sinc}\left(\frac{x}{\lambda f} L_x\right), \qquad (2.8.5)$$

or the intensity is given by

$$I(x, f) = \frac{E_0^2}{\lambda^2 f^2} L_x^2 \, \text{sinc}^2(f_x L_x), \qquad (2.8.6)$$

where

$$f_x = \frac{x}{\lambda f}.$$

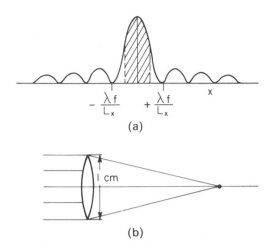

(a)

(b)

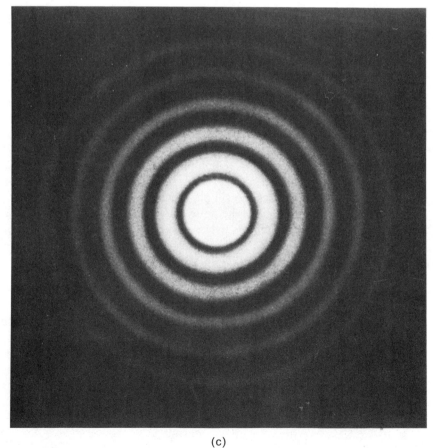

(c)

Fig. 2.8.2. Diffraction limitation of lenses: (a) square lens; (b) picture of diffraction due to a circular obstruction; and (c) a circular lens. (From M. Cagnet et al., *Atlas of Optical Phenomenon*, Springer-Verlag, New York, 1962.)

Remembering the discussions of the sinc function in Sections 2.5 and 2.6, it is obvious that the so-called focus has diffraction bands, as shown in Fig. 2.8.2(a). We can approximately define the size of the focus as spot width, given by

$$\text{size of focus} = \frac{\lambda f}{L_x}, \tag{2.8.7}$$

which is zero only when $L_x \to \infty$ or the size of the lens is much much larger than the wavelength. In other words, this lens can only resolve a dimension on the order of spot size when the resolving power is related to

$$\frac{L_x}{\lambda f} = \Delta f_x. \tag{2.8.8}$$

In most practical cases the lens is not square but circular. For this case, we should use the Bessel–Fourier transform discussed in Section 2.5.2. The result is

$$E(r, f) = E_0 e^{j(\omega t - kf)} e^{-jkr^2/2f} \cdot \frac{k\rho^2}{j8f} \left[2 \frac{J_1(k\rho r/2f)}{k\rho r/2f} \right], \tag{2.8.9}$$

where r is measured in the xy-plane and ρ is the radius of the lens. The intensity distribution is given by

$$I(r, f) = E_0^2 \frac{k\rho^2}{8f} \left[2 \frac{J_1(k\rho r/2f)}{k\rho r/2f} \right]^2.$$

For this case the spot size is given approximately by

$$\text{size of focus} = 1.22 \left(\frac{\lambda f}{\rho} \right). \tag{2.8.10}$$

Also, for a uniform light we obtain the circular ring pattern, as shown in Fig. 2.8.2, which follows the square of the $J_1(r)$ plot shown in Fig. 2.5.5.

2.8.2. Diffraction Grating

When we have not one aperture but many of them, in some periodic manner, then we form what is known as a grating. The transmission function of a one-dimensional amplitude grating is shown in Fig. 2.8.3(a). Each aperture in the grating is of size L_x, and there are N of them. The separation between the center of each is given by x_0. We shall consider only the Fraunhofer diffraction with the understanding that in laboratory experiments a lens is used, and the intensity pattern studied is at the focal plane. The problem of this diffraction grating can easily be formulated as a summation problem. For example, if $f(x')$ is the transmission function of one element of the diffraction grating, as shown in Fig. 2.8.3(a), then the total transmission function, $T(x')$, can be written as

$$T(x') = f(x') + f(x' - x_0) + f(x' - 2x_0) + \cdots + f(x' - (N-1)x_0). \tag{2.8.11}$$

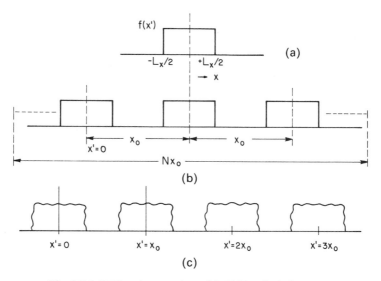

Fig. 2.8.3. Diffraction grating with N identical elements.

The diffracted field due to each slit or each element of the diffraction grating is obtained and then summed to obtain the total electric field. Thus,

$$
\begin{aligned}
E_{\text{tot}} &= \sum E(\text{due to each slit}) \\
&= E_{\text{slit}}(x' = 0) + E_{\text{slit}}(x' = x_0) \\
&\quad + E_{\text{slit}}(x' = 2x_0) + \cdots + E_{\text{slit}}(x' = (N-1)x_0). \quad (2.8.12)
\end{aligned}
$$

Let us consider, for a specific case, that the slit is of size L_x, as shown in Fig. 2.8.3(b). Then we know that

$$
\begin{aligned}
E_{\text{slit}}(x' = 0) &= (\alpha_2) E_0 \mathscr{F}\{f(x')\} \\
&= (\alpha_2) E_0 L_x \,\text{sinc}(f_x L_x). \quad (2.8.13)
\end{aligned}
$$

However, all other terms in (2.8.12) can be written in terms of $E_{\text{slit}}(x' = 0)$ as follows:

$$
E_{\text{slit}}(x' = x_0) = e^{j2\pi f_x x_0} E_{\text{slit}}(x' = 0). \quad (2.8.14)
$$

This result is obtained from (2.5.10). Similarly,

$$
E_{\text{slit}}(x' = 2x_0) = e^{j2 \cdot 2\pi f_x x_0} E_{\text{slit}}(x' = 0),
$$

$$
E_{\text{slit}}(x' = mx_0) = e^{jm2\pi f_x x_0} E_{\text{slit}}(x' = 0),
$$

$$
E_{\text{slit}}(x' = (N-1)x_0) = e^{j(N-1)2\pi f_x x_0} E_{\text{slit}}(x' = 0).
$$

So the total diffracted field is given by

$$
E_{\text{tot}} = E_{\text{slit}}(x' = 0) \times \{1 + e^{j2\pi f_x x_0} + \cdots + e^{j2\pi f_x x_0(N-1)}\}, \quad (2.8.15)
$$

where the first term denotes the diffraction due to a single slit and the term under the bracket is known as the interference term. Interference will be discussed in later sections with further detail. However, it should be pointed out that these arise because of the interference of different diffracted wavefronts with phase differences. The interference term can be written as

$$\text{interference term} = 1 + e^{j \cdot 2\pi f_x x_0} + e^{j2 \cdot 2\pi f_x x_0} + \ldots, e^{j2 f_x x_0 (N-1)}$$

$$= 1 + \gamma + \gamma^2 + \cdots + \gamma^{N-1}$$

$$= \frac{1 - \gamma^N}{1 - \gamma} \qquad \text{where} \quad \gamma = e^{j2\pi f_x x_0}$$

$$= \frac{1 - e^{j2\pi N f_x x_0}}{1 - e^{j2\pi f_x x_0}}$$

$$= e^{j\pi(N-1)f_x x_0} \left\{ \frac{\sin N\pi f_x x_0}{\sin \pi f_x x_0} \right\}. \tag{2.8.16}$$

So the total electric field at the detector plane due to the grating is given by

$$E_{\text{tot}}(x, y, z) = \left\{ \begin{array}{l} \text{Diffraction due to} \\ \text{single element} \end{array} \right\} \times e^{j\pi(N-1)f_x x_0} \times \left\{ \frac{\sin N\pi f_x x_0}{\sin \pi f_x x_0} \right\}. \tag{2.8.17}$$

It is of interest to study the properties of the most important term in the interference part

$$I_N = \frac{\sin N\pi f_x x_0}{\sin \pi f_x x_0}. \tag{2.8.18}$$

(i) For $N = 1$, of course, $I_1 = 1$.
(ii) For $N = 2$, $I_2 = 2 \cos \pi f_x x_0$.

This is the well-known interference between two sources and will be discussed in detail later. A plot of I_2 and I_2^2 is shown in Fig. 2.8.4. Remember that I_2^2 will be related to the intensity of the light and can be rewritten as

$$I_2^2 = 4 \cos^2 \left(\frac{\pi f_x x_0}{\lambda z} \right) = 4 \cos^2 \left(\frac{\pi x x_0}{\lambda z} \right)$$

$$= 2 \left[1 + \cos \left(\frac{2\pi x x_0}{\lambda z} \right) \right]. \tag{2.8.19}$$

(iii) For $N \to$ very large.

We note that for N very large, only time I_N has a large value when the denominator goes to zero. That is,

$$\sin \pi f_x x_0 = 0,$$

or

$$f_x = \frac{p}{x_0} = f_p \tag{2.8.20}$$

where p is an integer.

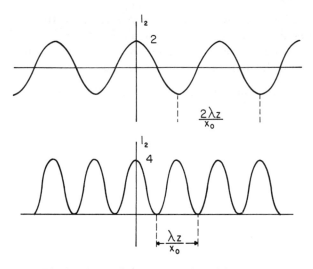

Fig. 2.8.4. Interference term plotted for $N = 2$.

Near a maximum we can write

$$I_N(\text{near } f_p) = \frac{\sin \pi N f_x x_0}{\sin \pi f_x x_0},$$

but

$$f_x - f_p = \text{very small} = f_x'.$$

Thus

$$I_N(\text{near } f_p) = \frac{\sin N\pi(f_x' + f_p)x_0}{\sin \pi(f_x' + f_p)}$$

$$= \frac{\sin N\pi f_x' x_0}{\sin \pi f_x' x_0}$$

$$\approx N \frac{\sin N\pi f_x' x_0}{N\pi f_x' x_0} \qquad \text{as } f_x' \text{ is small}$$

$$\approx N \operatorname{sinc}(N f_x' x_0). \tag{2.8.21}$$

Thus near a maxima, the interference term behaves like a sinc function. The maxima occurs for values of x given by

$$x = p\frac{\lambda z}{x_0}. \tag{2.8.22}$$

A plot of I_N for large N is shown in Fig. 2.8.5 both as a function of f_x and x.

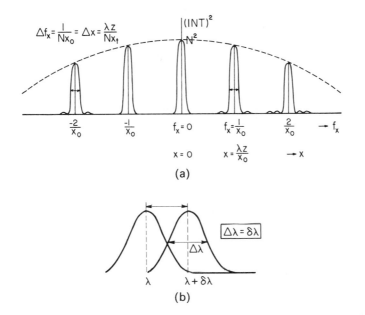

Fig. 2.8.5. (a) The interference term when N is large. (b) The Rayleigh criterion.

It is observed that the diffraction consists of a set of infinite waveform peaks. The spot size is independent of the diffraction order p, and is given by

$$\text{spot size} = \frac{\lambda z}{N x_0},$$

or in the spatial frequency domain it is

$$\Delta f_x = \frac{1}{N x_0}.$$

Thus, the spot size in this case is the same as if the whole aperture of size $(N x_0)$ is illuminated. Diffraction gratings can be used to determine different wavelengths of the composite incident light. The power to resolve this wavelength is measured by the resolving power, R, given by

$$R = \frac{\lambda}{\delta\lambda}, \tag{2.8.23}$$

where a wavelength, λ, and an adjacent one, $\lambda + \delta\lambda$, are called resolved if the diffraction spots by these two wavelengths obey the Rayleigh criterion. This is shown in Fig. 2.8.5(b). Thus

$$\frac{\lambda z}{N x_0} = \frac{p(\lambda + \delta\lambda)z}{x_0} - \frac{p\lambda z}{x_0} = \frac{p\delta\lambda z}{x_0}$$

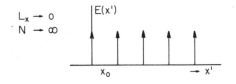

Fig. 2.8.6. The interference term when $N \to \infty$.

for the pth-order spot, or

$$R = \frac{\lambda}{\delta\lambda} = Np \qquad (2.8.24)$$

and the resolution, $\delta\lambda$, is given by

$$\delta\lambda = \frac{\lambda}{pN}. \qquad (2.8.25)$$

Thus for larger values of N and at higher diffraction orders, we obtain higher resolution.

(iv) For $N \to \infty$.

$$I_N = \underset{N\to\infty}{\mathrm{Lt}} \frac{\sin N\pi f_x x_0}{\sin \pi f_x x_0}. \qquad (2.8.26)$$

Then near a maxima

$$I_N = \underset{N\to\infty}{\mathrm{Lt}} \sum_p N \, \mathrm{sinc}(Nf'_x x_0)$$

$$= \sum_p \delta\left(f_x - \frac{N}{x_0}\right)$$

as the sinc function behaves like the delta function for $N \to \infty$. This is an interesting situation and is illustrated in Fig. 2.8.6.

One word of caution in the discussion of the interference term—it is not the diffracted field. To obtain the diffracted field, we must multiply it with the diffraction due to the single element of the grating. Thus for $L_x = x_0/5$, $N = 50$, the diffracted field is shown in Fig. 2.8.7. In the figure the interference

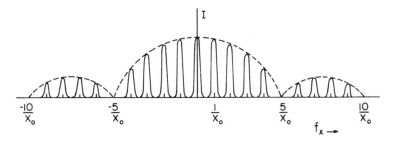

Fig. 2.8.7. Complete diffracted intensity for $N = 50$ and $L_x = x_0/5$.

and the diffraction due to single element terms are shown separately for clarification. It is obvious that for $x_0 = qLx$, the qth interference term will disappear.

A Typical Example Worked Out

It is of interest at this stage to work out a typical example that illustrates some other points which have not so far been stressed.

A photographic plate, 5 cm × 5 cm, is illuminated with uniform light ($\lambda = 8000$ Å) arriving at an angle of 5° with respect to the optical axis in the y'-plane. The transmission function of the plate is uniform in the y' direction and its dependence in the x' direction is shown in the following diagram:

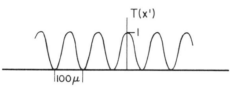

(a) Write an expression for $T(x')$ and its Fourier transform.
(b) If a 50-cm focal length lens is used, write an expression for the electric field at the focal plane showing the x and y dependence.
(c) Plot the intensity at the focal plane as a function of y.
(d) Plot the electric field amplitude at the focal plane as a function of the x component of spatial frequency.
(e) What change in the light intensity pattern is expected if the photographic plate breaks and its new size is 1 cm × 1 cm.

(a) $$T(x') = \tfrac{1}{2} + \tfrac{1}{2} \cos 2\pi \frac{x'}{100} \text{ (where } x' \text{ is in microns)}$$

$$= \tfrac{1}{2}(1 + \cos 2\pi f_{x0} x').$$

Changing x' from microns to meters where

$$f_{x0} = \frac{1}{10^2 \times 10^{-6}} \frac{1}{m} = 10^4 \text{ m}^{-1}.$$

$$\mathscr{F}\{T(x')\} = \tfrac{1}{2}\delta(f_x) + \tfrac{1}{4}[\delta(f_x - f_{x0}) + \delta(f_x + f_{x0})].$$

(b) $$E(x, y) = \frac{e^{j(\omega t - kf)}}{j\lambda f} e^{-j(\pi/\lambda f)(x^2 + y^2)} \mathscr{F}\{E(x', y', 0)\}_{\substack{f_x = x/\lambda f \\ f_y = y/\lambda f}}$$

$$= \alpha_{\text{con}}(f)\mathscr{F}\{E(x', y', 0)\}_{\substack{f_x = x/\lambda f, \\ f_y = y/\lambda f}}$$

$$E(x', y', 0) = E_0 E(x')E(y'),$$

where E_0 is the electric field amplitude of the light incident on the photographic plate.

$$E(x') = T(x') \operatorname{rect}(L_x),$$

$$E(y') = e^{-j2\pi f_0 y'} \cdot \operatorname{rect}(L_y),$$

where

$$L_x = L_y = 5 \times 10^{-2} \text{ m}; \qquad f_0 = \frac{\sin\theta}{\lambda} = \frac{\sin 5°}{0.8 \times 10^{-6}} = 1.0895 \times 10^5 \text{ m},$$

$$\mathcal{F}\{E(x')\}$$

$$= \tfrac{1}{2} L_x\{\operatorname{sinc}(f_x L_x) + \tfrac{1}{2}\operatorname{sinc}[(f_x - f_{x0})L_x] + \tfrac{1}{2}\operatorname{sinc}[(f_x + f_{x0})L_x]\},$$

$$\mathcal{F}\{E(y')\}$$

$$= L_y\{\operatorname{sinc}(f_y - f_0)L_y\},$$

$$\mathcal{F}\{E(x', y', 0)\}$$

$$= E_0 \mathcal{F}\{E(x')\}\mathcal{F}\{E(y')\},$$

$$E(x, y) = E_0 \alpha_{\text{con}(f)} \cdot \frac{L_y L_x}{2} \operatorname{sinc}\left\{\frac{yL_y}{\lambda f} - f_0 L_y\right\}$$

$$\times \frac{1}{2}\left[\operatorname{sinc}\left(\frac{xL_x}{\lambda f}\right) + \operatorname{sinc}\left\{\frac{xL_x}{\lambda f} - f_{x0}L_x\right\} + \operatorname{sinc}\left\{\frac{xL_x}{\lambda f} + f_{x0}L_x\right\}\right].$$

(c) $$I(y) \propto L_y^2 \operatorname{sinc}^2(f_y L_y - f_0 L_y)$$

$$= (5 \times 10^{-2})^2 \operatorname{sinc}^2\left\{\frac{y}{8 \times 10^{-6}} - 1.0894 \times 10^5 \text{ m}\right\}.$$

The intensity is plotted in the following figure.

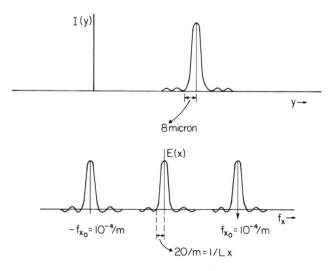

(d) $E(x) \propto L_x^2 [\text{sinc}(f_x L_x) + \text{sinc}\{f_x - f_{x0})L_x\} + \text{sinc}\{(f_x + f_{x0})\}L_x]^2.$

A plot of the electric amplitude is shown in the figure.

(e) The new electric field is the same as in section (b) except $L_x = L_y = 1 \times 10^{-2}$.

2.8.3. Sinusoidal Gratings

2.8.3.1. Phase Gratings

Consider the transmission function whose amplitude is shown in Fig. 2.8.8. It introduces sinusoidal phase variations. For example, if we make a sinusoidal ruling on a piece of glass such that the thickness of the glass, $t(x)$, is given by

$$t(x) = t_0 - t_1 \sin 2\pi f_{x0} x, \tag{2.8.27}$$

then the transmission function will be given by

$$T(x') = e^{-j(2\pi/\lambda)t_0 n} \cdot e^{j(2\pi/\lambda)t_1 n \sin 2\pi f_{x0} x'}$$

$$= e^{-j\varphi_0} e^{j(\rho/2) \sin f_{x0} x'}, \tag{2.8.28}$$

where ρ is defined as the index of modulation. As shown in Fig. 2.8.8, $f_{x0} = 1/x_0$ where x_0 is the period. If the transmission function is limited to width L_x, then the transmission function to be used in the diffraction integral is given by

$$T(x') = e^{-j\varphi_0} e^{j(\rho/2) \sin f_{x0} x'} \, \text{rect}(L_x). \tag{2.8.29}$$

To obtain the Fourier transform of $T(x')$ we note that

$$e^{-j(\rho/2) \sin 2\pi f_{x0} x'} = \sum_{q=-\infty}^{+\infty} J_q\left(\frac{\rho}{2}\right) e^{j2\pi q f_{x0} x'}, \tag{2.8.30}$$

where J_q represents the Bessel function of order q, and q is an integer. Using the above expression, the diffracted electric field in the Fraunhofer approximation is given by

$$E(x, y) = \alpha_{\text{con}(D)} L_x \, \text{sinc}(f_x L_x) * \left[\sum_{q=-\infty}^{+\infty} J_q\left(\frac{m}{2}\right) \delta(f_x - q f_{x0}) \right], \tag{2.8.31}$$

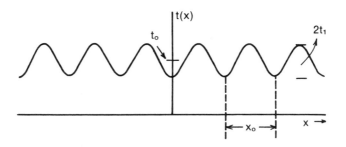

Fig. 2.8.8. Sinusoidal phase grating.

where the convolution formula form Table 2.5.2 has been used. Thus,

$$E(x) = \alpha_{con(D)} \sum_{q=-\infty}^{+\infty} J_q\left(\frac{\rho}{2}\right) \text{sinc}\left[\left(\frac{L_x}{\lambda z}\right)(x - qf_{xo}\lambda z)\right], \qquad (2.8.32)$$

or the intensity is given by

$$I(x) = \frac{1}{\lambda^2 D^2}\left\{\sum_{q=-\infty}^{+\infty} J_q\left(\frac{\rho}{2}\right) \text{sinc}\left[\left(\frac{L_x}{\lambda z}\right)(x - qf_{xo}\lambda z)\right]\right\}^2 \qquad (2.8.33)$$

A plot of the intensity for a phase grating is shown in Fig. 2.8.9(a). Readers familiar with the frequency modulation of radio waves recognize the similarity

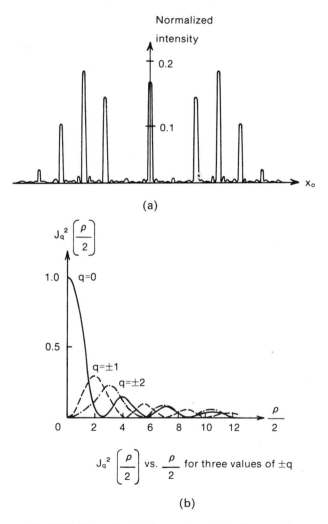

(a)

$$J_q^2\left[\frac{\rho}{2}\right] \text{ vs. } \frac{\rho}{2} \text{ for three values of } \pm q$$

(b)

Fig. 2.8.9. Intensity plot for a sinusoidal phase grating.

between that case and this phase grating in the spatial frequency domain. Figure 2.8.9(b) shows the plot of $J_q^2(\rho/2)$ for three values of q.

2.8.3.2. Amplitude Grating

Of course, a similar amplitude modulation case is found if we consider the transmission function as

$$T(x, y) = \left(\frac{1}{2} + \frac{\rho}{2} \cos 2\pi f_{x0}x\right) \text{rect}(L_x).\tag{2.8.34}$$

The diffracted field in this case is given by

$$E(x) = E_0 \alpha_{\text{con}(D)} \left\{ \text{sinc}\left(\frac{L_x x}{\lambda z}\right) + \frac{\rho}{2} \text{sinc}\left[\left(\frac{L_x}{\lambda z}\right)(x + f_{x0}\lambda z)\right]\right.$$

$$\left. + \frac{\rho}{2} \text{sinc}\left[\left(\frac{L}{\lambda z}\right)(x - f_{x0}\lambda z)\right]\right\},\tag{2.8.35}$$

$$I(x) = E_0^2 \left(\frac{1}{\lambda^2 D^2}\right)\left\{ \text{sinc}^2\left(\frac{L_x x}{\lambda z}\right) + \frac{\rho^2}{4} \text{sinc}^2\left[\left(\frac{L_x}{\lambda z}\right)(x + f_{x0}\lambda z)\right]\right.$$

$$\left. + \frac{\rho^2}{4} \text{sinc}^2\left[\left(\frac{L_x}{\lambda z}\right)(x - f_{x0}\lambda z)\right]\right\},\tag{2.8.36}$$

where it has been assumed that the grating frequency f_{x0} is much greater than $2/L_x$, so there is negligible overlap between the three sinc functions in (2.8.36). A plot of the diffracted intensity is shown in Fig. 2.8.10.

2.8.4. Fresnel Diffraction Without Lens

Until now we have considered the problems which are in the Fraunhofer regime, or using a lens, can be made to appear like a Fraunhofer diffraction. The problem of Fresnel diffraction is quite complex and here we shall consider

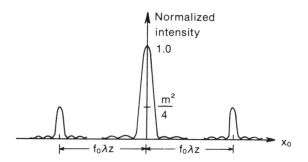

Fig. 2.8.10. Cross section of the Fraunhofer diffraction pattern of a sinusoidal amplitude grating.

only one problem, the rectangular aperture. The diffracted light for this case (see (2.6.8) for dimensions) is given by

$$E(x, y) = E_0 \frac{e^{j(\omega t - kz)}}{j\lambda z} \int_{-L_x/2}^{+L_x/2} \int_{-L_y/2}^{L_y/2} e^{j(k/2z)\{(x-x')^2 + (y-y')^2\}} \, dx' \, dy'. \quad (2.8.37)$$

Writing the above integral as the product of two integrals, we have

$$E(x, y) = E_0 \frac{e^{j(\omega t - kz)}}{j\lambda z} \int_{-L_x/2}^{L_x/2} e^{j(k/2z)(x-x')^2} \, dx' \int_{-L_y/2}^{L_y/2} e^{j(k/2z)(y-y')^2} \, dy'$$

$$= E_0 \frac{e^{j(\omega t - kz)}}{j\lambda z} \mathscr{I}(x)\mathscr{I}(y), \quad (2.8.38)$$

where

$$\mathscr{I}(x) = \int_{-L_x/2}^{L_x/2} e^{j(k/2z)(x-x')^2} \, dx'. \quad (2.8.39)$$

Substituting

$$\eta = \sqrt{\frac{k}{\pi z}}(x - x'), \quad (2.8.40)$$

we obtain

$$\mathscr{I}(x) = \sqrt{\frac{\pi z}{k}} \int_{\eta_1}^{\eta_2} e^{j(\pi/2)\eta^2} \, d\eta, \quad (2.8.41)$$

where

$$\eta_1 = -\sqrt{\frac{k}{\pi z}}\left(\frac{Lx}{2} + x\right),$$

$$\eta_2 = \sqrt{\frac{k}{\pi z}}\left(\frac{Lx}{2} - x\right).$$

The integral $\mathscr{I}(x)$ can be written in terms of the Fresnel integrals $C(x)$ and $S(x)$ as follows:

$$\mathscr{I}(x) = \sqrt{\frac{\pi z}{k}}\{[C(\eta_2) - C(\eta_1)] + j[S(\eta_2) - S(\eta_1)]\}, \quad (2.8.42)$$

$$C(x) = \int_0^x \cos\frac{\pi t^2}{2} \, dt, \quad (2.8.43)$$

$$S(x) = \int_0^x \sin\frac{\pi t^2}{2} \, dt. \quad (2.8.44)$$

Defining

$$\xi_1 = -\sqrt{\frac{k}{\pi z}}\left(\frac{L_y}{2} + y\right),$$

$$\xi_2 = \sqrt{\frac{k}{\pi z}}\left(\frac{L_y}{2} - y\right), \quad (2.8.45)$$

we finally obtain

$$E(x, y) = E_0 \frac{e^{j(\omega t - kz)}}{2j} \{[C(\eta_2) - C(\eta_1)] + j[S(\eta_2) - S(\eta_1)]\}$$

$$\times \{[C(\xi_2) - C(\xi_1)] + j[S(\xi_2) - S(\xi_1)]\}. \tag{2.8.46}$$

The intensity, $I(x, y)$, is given by

$$I(x, y) = \frac{E_0^2}{4} \{[C(\eta_2) - C(\eta_1)]^2 + [S(\eta_2) - S(\eta_1)]^2\}$$

$$\times \{[C(\xi_2) - C(\xi_1)]^2 + [S(\eta_2) - S(\eta_1)]^2\}. \tag{2.8.47}$$

The Fresnel integrals have a graphical interpretation in terms of the Cornu spiral as shown in Fig. 2.8.11. Here $C(x)$ and $S(x)$ is plotted in the complex plane. Note that

$$C(\alpha) = -C(-\alpha) = \tfrac{1}{2}, \tag{2.8.48}$$

and

$$S(\alpha) = -S(-\alpha) = \tfrac{1}{2}.$$

If we define the complex Fresnel integral as $A(x)$, given by

$$A(x) = C(x) + jS(x), \tag{2.8.49}$$

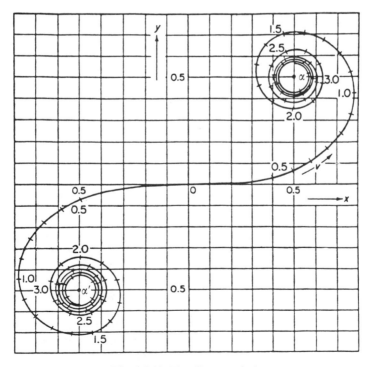

Fig. 2.8.11. The Cornu spiral.

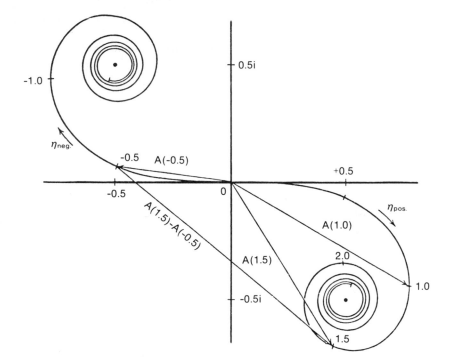

Fig. 2.8.12. Examples of the application of the Cornu spiral to determine the solution to the complex Fresnel integral.

we see that

$$E(x, y) = E_0 \frac{e^{j(\omega t - kz)}}{2j} \{A(\xi_2) - A(\xi_1)\} \{A(\eta_2) - A(\eta_1)\}. \quad (2.8.50)$$

In terms of the Cornu spiral, $A(\xi_2)$ is the distance from the origin to the point ξ_2 in the Cornu spiral. Thus $A(\xi_2) - A(\xi_1)$ can easily be obtained as shown in Fig. 2.8.12. Using this graphical construction we easily obtain the results for the straight edge and the rectangular aperture as shown in Fig. 2.8.13. Figure 2.8.14 also shows the transition from Fresnel to Fraunhofer diffraction as z increases.

For z very small, $\xi_2, \xi_1, \eta_2, \eta_1$ are very large. For this we easily obtain

$$E(x, y) = E_0 e^{j(\omega t - kz)} \text{rect}\left(\frac{x}{L_x}\right) \text{rect}\left(\frac{y}{L_y}\right), \quad (2.8.51)$$

which is the geometrical projection of the aperture. Using the principle of stationary phase, it can be shown that this result is not special for the rectangular aperture but valid for any arbitrary aperture.

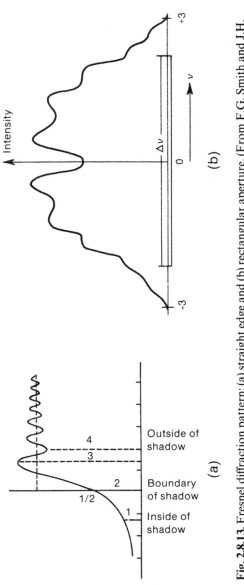

Fig. 2.8.13. Fresnel diffraction pattern: (a) straight edge and (b) rectangular aperture. (From F.G. Smith and J.H. Thomson, *Optics*, Wiley, New York, 1971.)

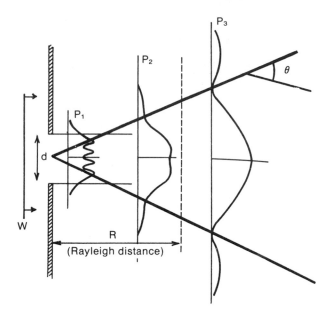

Fig. 2.8.14. Transition from Fresnel to Fraunhofer diffraction. A portion of a wave, W, passes through a slit of width, d. Intensity distributions across the wave are shown for planes P_1 (close to the slit), P_2 (just inside the Fresnel distance), and P_3 (beyond the Fresnel distance).

2.9. Detection and Coherence

Two topics of great interest, which we have avoided until now, are the subjects of the detection and coherence properties of light. Both these subjects need to be discussed before we can understand the subject of the next topic, interference.

2.9.1. Detection

All through this part of the book we have mentioned the detector plane. The discussion on detectors for the whole electromagnetic spectrum is a subject in itself. However, we will consider the relevant material for this book here. For low frequencies (up to a few gigahertz), it is possible to have detectors of electromagnetic energy which measure the electric field—the oscilloscope is one example. However, as we approach the optical region, nothing can respond quickly enough, and what we measure is the intensity. That is the reason, in the last few sections, that we have derived expressions for the intensity where many of the phase terms are lost.

The following is a partial list of the detectors which are used in the optical region:

Eye,
Photographic film,
Photomultiplier tube,
Photodetector diodes and transistors,
Bolometers.

Each of these detectors has a bandwidth, Δf, over which it can respond. This bandwidth is related to the response time, T, of the detector by the relation

$$\Delta f \simeq \frac{1}{T}. \tag{2.9.1}$$

For example, the response time of the eye is about 100 ms, or it has a bandwidth of approximately 10 Hz. Table 2.9.1 lists the response times and bandwidths of other typical detectors. Since optical frequencies are in the neighborhood of 10^{14}, it is obvious that the detectors cannot follow the fast swing of the electric field. The detection is performed by some form of nonlinearity (most of the time a square nonlinearity) in the detector which measures the intensity. Let us consider that the detector output is proportional to the square of the electric field. That is,

$$\text{output} \propto \frac{1}{T} \int_0^T |E|^2 \, dt \tag{2.9.2}$$

(where T is the response time), or

$$\text{output} \propto \frac{1}{T} \int_0^T E_0^2 \cos^2(\omega t - kz) \, dt,$$

where the incident wave is considered to be propagating in the z direction and $T \gg 2\pi/\omega$, or

$$\text{output} \propto \frac{E_0^2}{2} + \frac{E_0^2}{2T} \cos(2\omega t - 2kz)$$

$$\approx \frac{E_0^2}{2} = \frac{1}{2}(EE^*) = I. \tag{2.9.3}$$

The second harmonic term is negligible as the detector cannot respond at that frequency and is filtered out. Thus, we see that the detectors do measure the intensity, and the output of the detectors is proportional to the square of the electric field magnitude.

Let us also consider a situation where the light is modulated with some information to be transmitted, for example, in an optical communication system. In that case, the electric field can be written as

$$E(t) = E_0 m(t) \cos(\omega t - kz), \tag{2.9.4}$$

Table 2.9.1. Detectors and their properties.*

Type	Spectral response λ in microns	Responsivity amp or volts/watt of incident power	Threshold/noise equivalent power or D* noise equivalent power is in W/Hz$^{1/2}$ D* in cm Hz$^{1/2}$ W	Frequency response (in Hz)
Thermal				
bolometer	0.25–30	5×10^{-5} to 125 V/W	4×10^{-8} to 9×10^{-5} W/Hz$^{1/2}$	175 to $> 10^6$
far infrared	50–5000	10^3 V/W	1 mW	20
calorimeter	0.25–35	0.07–0.1 V/W	10^{-5}–100 W	0.01–0.2
pyroelectric	0.1–1000	15–2400 V/W or 3×10^4 V/J	3×10^{-10} to 5×10^{-8} W/Hz$^{1/2}$ (2.5×10^{-10} W)	10^{-2}–10^9
thermopile	0.2–35	4–55 V/W	3×10^8 cm Hz$^{1/2}$/W (10^{-6} W/cm^2)	5–10
Photomultiplier	0.25–0.9 (depends on photocathode)	0.0014–0.105 A/W		10^8–10^9
Phototube	0.185–1.1 (depends on photocathode)	0.0025–0.080 A/W		10^9 to 3.5×10^{11}
Vacuum photodiode	0.16–1.1 (depends on photocathode)	0.0001–0.001 A/W		3×10^9
Semiconductor				
Germanium				
photoconductive	0.5–1.8	0.15–18 A/W	$> 10^{12}$ cm Hz$^{1/2}$/W	5×10^8 to 5×10^9
avalanche	0.8–1.8	0.2 A/W	10^{-10} W/Hz$^{1/2}$	2×10^9
photon drag	4–22	1.2×10^{-6} V/W		$> 3 \times 10^8$
copper doped	2–30	0.1–3 A/W	2×10^{10} cm Hz$^{1/2}$/W	3.5×10^6 to 10^9
gallium doped	10–130	10^4 V/W	0.1 μW	10^7
gold doped	1.5–11	0.1–0.6 A/W	$(0.15–0.7) \times 10^{10}$ cm Hz$^{1/2}$/W	3.5×10^6 to 3×10^8
mercury doped	6–10.6	0.03–3 A/W	$(1–2) \times 10^{10}$ cm Hz$^{1/2}$/W	3.5×10^6 to 10^9
zinc doped	28–37	0.1–0.5 A/W	$(1–2) \times 10^{10}$ cm Hz$^{1/2}$/W	3.5×10^6 to 3×10^7
Indium antimonide	1–5.5	1.5–2.5 A/W	$(2–30) \times 10^{10}$ cm Hz$^{1/2}$/W	2×10^5 to 10^8
Far infrared	50–5000	10^3 V/W	10 nW	10^7
Indium–arsenide	1–3.8	0.6–1 A/W	$(0.5–60) \times 10^{10}$ cm Hz$^{1/2}$/W	10^6–10^{10}
Indium–gallium–arsenide–phosphide	0.9–1.65	0.7 A/W	$< 10^{-12}$ W/Hz$^{1/2}$	9×10^9
Lead selenide				
ambient	2–5	0.004–0.03 A/W	$\geq 0.1 \times 10^{10}$ cm Hz$^{1/2}$/W	70×10^3

Table 2.9.1 (*continued*)

Type	Spectral response λ in microns	Responsivity amp or volts/watt of incident power	Threshold/noise equivalent power or D* noise equivalent power is in W/Hz$^{1/2}$ D* in cm Hz$^{1/2}$ W	Frequency response (in Hz)
intermediate	3–5	0.01–0.05 A/W	$\geq 1 \times 10^{10}$ cm Hz$^{1/2}$/W	7×10^3
low temperature	3–6	0.04–0.4 A/W	$\geq 1 \times 10^{10}$ cm Hz$^{1/2}$/W	6×10^3
Lead sulfide				
ambient	1.8–2.8	0.5–4 A/W	7×10^{10} cm Hz$^{1/2}$/W	350–8000
intermediate	1.8–2.8	1–10 A/W	40×10^{10} cm Hz$^{1/2}$/W	100–200
low temperature	1.8–3.3	1–10 A/W	10×10^{10} cm Hz$^{1/2}$/W	50–100
Lead–tin–telluride	5–18		$(0.5–20) \times 10^{10}$ cm Hz$^{1/2}$/W	7×10^5 to 1.4×10^6
Mercury–cadmium–telluride	2–20	0.0002–30 A/W	$(2–10) \times 10^{10}$ cm Hz$^{1/2}$/W	5×10^3 to 1.6×10^8
photoconductive				
photovoltaic	8–12	5 A/W	0.1 μW	10^8 to 2×10^9
photoelectro-magnetic	2–12	0.001 V/W	0.1 μW	8×10^8
Silicon				
photoconductive	0.18–1.13	0.1–0.6 A/W	$(1–1000) \times 10^{-14}$ W/Hz$^{1/2}$	10^5–10^{10}
avalanche	0.45–1.1	0.4–80 A/W	1 nW	3.5×10^8 to 2×10^9
photovoltaic	0.185–1.15	0.08–0.65 A/W	$(0.7–50) \times 10–14$ W/Hz$^{1/2}$	10^5 to 3×10^7
antimony doped	21–27	1–5 A/W	$\geq 1.5 \times 10^{10}$ cm Hz$^{1/2}$/W	$> 3.5 \times 10^6$
arsenic doped	16–23.5	1–5 A/W	$\geq 2.5 \times 10^{10}$ cm Hz$^{1/2}$/W	$> 3.5 \times 10^6$
gallium doped	12–15.5	1–3 A/W	$\geq 1.5 \times 10^{10}$ cm Hz$^{1/2}$/W	$> 3.5 \times 10^6$

* This table is from *Lasers and Applications*, p. 38, 1982.

where $m(t)$ is the modulating function which has a bandwidth less than the bandwidth of the detector. For this case, we obtain the detector output as

$$\text{output} \propto |E_0|^2 |m(t)|^2. \tag{2.9.5}$$

However, if the function $m(t)$ itself, and not its squared magnitude, is to be recovered, then we use homodyne detection where part of the light carrier itself is also incident on the detector. Then the incident electric field on the detector is given by

$$E(t) = E_0 \cos(\omega t - kz) + E_1 m(t) \cos(\omega t - kz). \tag{2.9.6}$$

For this case, the output of the detector is given by

$$\text{output} \propto |E_0|^2 + |E_1|^2 + 2E_0 E_1 m(t). \tag{2.9.7}$$

The first two terms are d.c. terms and can easily be filtered out.

Heterodyne detection is used when the two light frequencies are very near one another. If one of the light frequencies is ω_2 and the other is ω_1, and if the difference between them is less than the bandwidth of the detector, i.e.,

$$\omega_2 - \omega_1 < 2\pi \Delta f,$$

then we can determine this difference frequency, $\omega_2 - \omega_1$. The incident light for this case is given by

$$E(t) = E_1 \cos(\omega_1 t - k_1 z) + E_2 \cos(\omega_2 t - k_2 z). \tag{2.9.8}$$

The detector output will be given by

$$\text{output} \propto |E_1|^2 + |E_2|^2 + 2E_1 E_2 \cos[(\omega_2 - \omega_1)t - (k_2 - k_1)z] \tag{2.9.9}$$

(where $\omega_2 \sim \omega_1, k_2 - k_1 = 0$), and so the difference frequency can be observed. A use of this technique is in the determination of frequency differences in different modes of laser oscillations, which will be discussed in Part III of this book.

A typical heterodyne receiver is shown in Fig. 2.9.1(a) and a homodyne receiver is shown in Fig. 2.9.1(b). Note that the same source is used for the homodyne case to maintain coherence. Because of this, homodyne detection is also sometimes referred to as interferometric detection.

For both homodyne and heterodyne detection, it is important that the two sources are perfectly aligned. A typical misalignment is shown in Fig. 2.9.2 where ψ is the angle of misalignment. For a rectangular aperture it can be shown that the detector output will be reduced by a factor, f_d, given by

$$f_d = \frac{\sin \psi_x}{\psi_x} \cdot \frac{\sin \psi_y}{\psi_y}, \tag{2.9.10}$$

where

$$\psi_x = \frac{\pi d \sin \psi_{xR}}{\lambda_1} - \frac{\pi d \sin \psi_{xL}}{\lambda_2}, \tag{2.9.11}$$

$$\psi_y = \frac{\pi d \sin \psi_{yR}}{\lambda_1} - \frac{\pi d \sin \psi_{yL}}{\lambda_2}, \tag{2.9.12}$$

where d is the aperture size, λ_1 is the received light wavelength, and λ_2 is the local oscillator or reference wavelength. Thus, for proper detection, the relative misalignment should be such that

$$\sin \psi_{2L} < \frac{\lambda_2}{\pi d}, \tag{2.9.13}$$

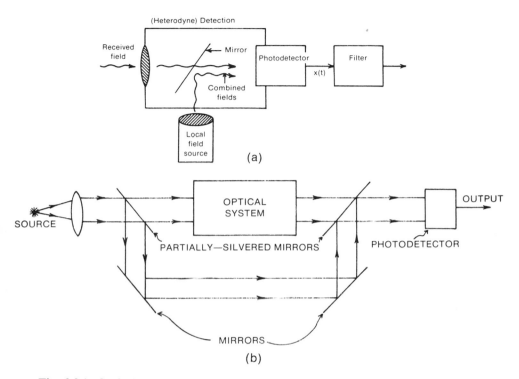

Fig. 2.9.1. Optical coherent detection schemes: (a) heterodyne detection and (b) homodyne detection.

and

$$\sin \psi_{yL} < \frac{\lambda_2}{\pi d}. \tag{2.9.14}$$

If separate lenses are used to focus the received and reference lights on the photodetector, then the diffraction effects might be different and this should also be taken into account.

It is of interest to consider the signal to noise ratios for different schemes of detection for the simple case of unmodulated light. For the intensity detection case, the output signal power is given by

$$P_S = i_S^2 R_L = (R_D P_0)^2 R_L, \tag{2.9.15}$$

where i_S is the detector current, R_L is the detector load resistance, R_D is the detector responsivity, and P_0 is the light power incident.

The shot noise component of the noise power output is given by

$$P_N = 2i_S q(\Delta f) R_L$$
$$= 2(R_D P_0) q(\Delta f) R_L, \tag{2.9.16}$$

where Δf is the detector bandwidth.

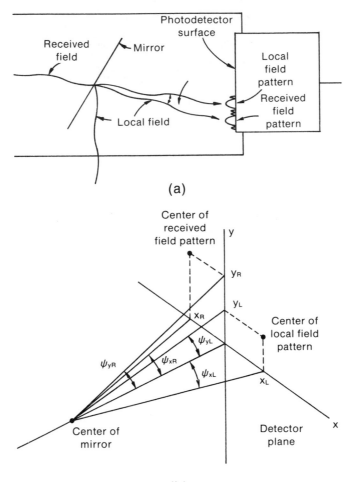

Fig. 2.9.2. Misalignment for coherent detection: (a) field pattern and (b) geometry.

We obtain for signal to noise ratio

$$\left(\frac{S}{N}\right)_{\text{intensity}} = \frac{P_S}{P_N} = \frac{R_D P_0}{2q\Delta f}. \tag{2.9.17}$$

For the heterodyne detection case

$$P_S = 2R_D^2 P_0 P_R R_L, \tag{2.9.18}$$

where P_R is the reference light power. The noise power for this is given by

$$P_N = 2R_D q(\Delta f) R_L (P_0 + P_R). \tag{2.9.19}$$

We obtain

$$\left(\frac{S}{N}\right)_{\text{heterodyne}} = \frac{R_D P_0 P_R}{q(\Delta f)(P_0 + P_R)}. \qquad (2.9.20)$$

For $P_R \gg P_0$, we obtain

$$\left(\frac{S}{N}\right)_{\text{heterodyne}} = \frac{R_D P_0}{q(\Delta f)}. \qquad (2.9.21)$$

For the homodyne case

$$P_S = 4R_D^2 P_R P_0 R_L, \qquad (2.9.22)$$

and

$$\left(\frac{S}{N}\right)_{\text{homodyne}} = \frac{2R_D P_R P_0}{q(\Delta f)(P_0 + P_R)} \qquad (2.9.23)$$

$$\approx \frac{2R_D P_0}{q\Delta f} \quad \text{for} \quad P_R \gg P_0. \qquad (2.9.24)$$

The reason P_S is larger by a factor of 2 in the homodyne case, is that the reference and received light add coherently.

2.9.2. Coherency

Up until now we have implicitly assumed that the light waves are monochromatic having a single frequency only, or that the Fourier transform of the light wave at a particular point in space is a delta function. The wave is given by

$$e^{j(\omega_0 t - kz)} = e^{j2\pi(f_0 t - z/\lambda)},$$

and at a fixed point its time dependence is given by $e^{j2\pi f_0 t}$ and the Fourier transform is $\delta(f - f_0)$.

This is graphically shown in Fig. 2.9.3. However, we notice immediately that for light waves to have just one frequency, f_0, they must exist in perfect phase coherence from $t = -\infty$ to $+\infty$. Because, if the duration is any smaller, say T_0, then the mathematical representation of the wave form is actually given by

$$E = e^{-j2\pi f_0 t} \text{rect}[T_0]. \qquad (2.9.25)$$

The frequency components are then given by

$$T_0[\text{sinc } T_0(f - f_0)]$$

with an approximate width given by

$$\Delta f \simeq \frac{1}{T},$$

which has been discussed in the preceding section (2.9.1). Thus, we see that even if the waveform is a single frequency, there is no way to prove it; because

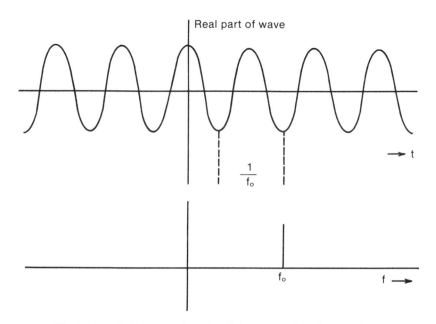

Fig. 2.9.3. Infinitely long duration light and its Fourier transform.

we have to wait an infinite time to check that it is truly a single frequency. This concept is really related to Heisenberg's uncertainty principle, as some of the readers familiar with quantum mechanics will recognize.

But irrespective of this problem of measurement, electromagnetic waves at lower frequencies can be generated which have very small bandwidths. For example, some carrier frequencies of radio waves are kept constantly within a few hertz. It is interesting to note if a frequency of a time waveform is Δf, then within a time period given by

$$T_c \approx \frac{1}{\Delta f} \tag{2.9.26}$$

there is no detectable change in the amplitude or phase of the wave. All the waveforms are in phase coherence. Thus, we call this time, T_c, the coherence time and the wave can be assumed coherent over the time T_c. We know that if an electromagnetic wave travels with velocity v, then we can also define the coherence length, given by

$$l_c = vT_c. \tag{2.9.27}$$

Thus wavefronts within a length l_c along the direction of propagation will be in phase coherence. Now let us consider the case of white light. Its wavelength ranges from 0.4 μm to 0.8 μm. Thus, the frequency span is approximately

$$\Delta f \approx \tfrac{3}{8} \times 10^{15} \text{ s,}$$

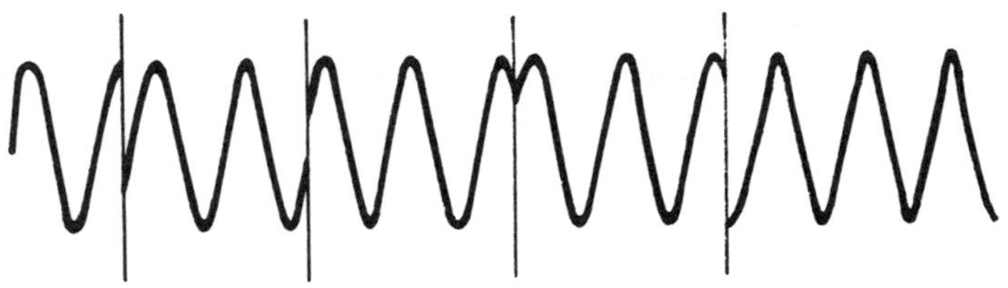

Fig. 2.9.4. Noise-like incoherent light.

or the coherence time is

$$T_c = \tfrac{8}{3} \times 10^{-15}\ \text{s},$$

and the coherence length in free space is

$$l_c = (3 \times 10^8\ \text{m/s})(\tfrac{8}{3} \times 10^{-15}\ \text{s})$$

$$l_c = 8 \times 10^{-15}\ \text{m} = 80\ \mu\text{m}.$$

Thus, we see that only for a length of 80 μm, we can consider that the wavefront of white light is in phase coherence.

We shall see in Part III how light is generated by the transition of electrons from one energy level to another. Because of its nature the light is ordinarily very incoherent as discussed in connection with the white light illustration. Even if we choose a particular color by filtering or using a source like sodium light, it is found that the waveforms are continuous but have phase discontinuities, as shown in Fig. 2.9.4. This makes the ordinary light more or less incoherent and Δf large.

However, as we shall see in Part III using lasers, we can generate light waves which are approximately monochromatic with $\Delta f \sim 1$ MHz.

The subject of coherency in optics is a complex subject. What we have done is to give the reader a glimpse of it so that he can understand the rest of the book. The serious reader should consult the references given at the end of this part.

2.10. Interference

Historically, interference has probably played the most important role in convincing people that light rays are really waves. In this book, we have assumed that light is an electromagnetic wave, but it took scientists many thoughtful experiments to arrive at this conclusion. We have already referred to interference in connection with diffraction gratings, and if we assume the wave nature of light and if we have available a coherent light source, then it is true that the subject of interference becomes a special case of diffraction with discrete sources. However, even 50 years ago, scientists did not have coherent sources, and still they designed experiments to convince everyone that light is really an electromagnetic wave. Some of the experiments discussed in the following sections cover this material.

2.10.1. Young's Experiment

Let us consider an experiment where we have two pinholes and two separate sources of light, as shown in Fig. 2.10.1. Then we know that if both sources do emit absolutely phase coherent wavefronts, which can be represented by

$$e^{j(\omega t - kr)},$$

then, as discussed in Section 2.8.2 and (2.8.19), a detector in the detector plane will detect, as a function of x,

$$I(x) = I_0 4 \cos^2 \pi f_x h = 2\left[1 + \cos 2\pi \frac{xh}{\lambda D}\right], \qquad (2.10.1)$$

where I_0 is the normalization constant. However, for two incoherent sources, the detector cannot respond to the difference frequencies $\omega_2 - \omega_1$, and so the only thing observable will be a constant amplitude. This is easily seen by noting that the light from the two sources arriving, respectively, at the detector plane are really

$$E_1(\mathbf{r}) = \frac{E_{10}}{|\mathbf{r} - \mathbf{r}_1|} e^{j(\omega_1 t - k_1 |\mathbf{r} - \mathbf{r}_1|)}, \qquad (2.10.2)$$

and

$$E_2(\mathbf{r}) = \frac{E_{20}}{|\mathbf{r} - \mathbf{r}_2|} e^{j(\omega_2 t - k_2 |\mathbf{r} - \mathbf{r}_2|)}.$$

However, when these two electric fields are incident on the detector, which cannot respond fast enough with respect to $2\pi/(\omega_2 - \omega_1)$, the output will simply be

$$\text{output} \propto \frac{E_{10}^2}{|\mathbf{r} - \mathbf{r}_1|^2} + \frac{E_{20}^2}{|\mathbf{r} - \mathbf{r}_2|^2}, \qquad (2.10.3)$$

which is a constant and no interference is observed. To avoid this dilemma,

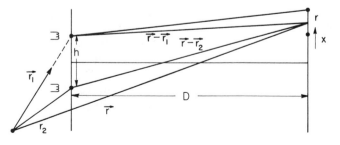

Fig. 2.10.1. Young's experiment with two separate light sources.

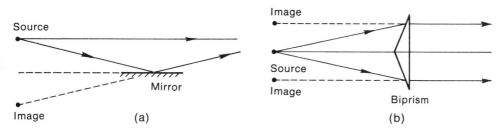

Fig. 2.10.2. Different setup to perform Young's experiment: (a) Lloyd's mirror and (b) Fresnel biprism.

Young used the same light source located on the optical axis for both pinholes, as shown in Fig. 2.10.2. For this case, although the frequency ω is not perfectly coherent and fluctuating continuously, identical frequencies are incident on the detectors through two different paths, and they can be detected since the beat frequency is zero. For this case

$$E_1 = \frac{E_{10}}{|\mathbf{r} - \mathbf{r}_1|} e^{j(\omega t - k|\mathbf{r} - \mathbf{r}_1|)},$$

and

$$E_2 = \frac{E_{20}}{|\mathbf{r} - \mathbf{r}_2|} e^{j(\omega t - k|\mathbf{r} - \mathbf{r}_2|)}. \tag{2.10.4}$$

The electric fields can be approximated as follows, if $z \gg x$, y, and $x_1 y$, $x_2 y_2$:

$$E_{10} = \frac{E_{10}}{z} e^{j\omega t} \cdot e^{-k|\mathbf{r} - \mathbf{r}_1|},$$

$$E_{20} = \frac{E_{20}}{z} e^{j\omega t} \cdot e^{-k|\mathbf{r} - \mathbf{r}_2|}.$$

The output of the detector, although the source is not coherent, now is

$$\text{output} \propto \frac{E_{10}^2}{z^2} + \frac{E_{20}^2}{z^2} + 2\frac{E_{10}E_{20}}{z_2} \cos k[|\mathbf{r}_2 - \mathbf{r}| - |\mathbf{r}_1 - \mathbf{r}|], \tag{2.10.5}$$

using the approximations

$$|\mathbf{r}_2 - \mathbf{r}| = \left[\left(x + \frac{h}{2}\right)^2 + D^2\right]^{1/2} \approx D\left\{1 + \frac{1}{2}\frac{(x + h/2)^2}{D^2}\right\},$$

$$|\mathbf{r}_1 - \mathbf{r}| = \left[\left(x - \frac{h}{2}\right) + D^2\right]^{1/2} \approx D\left\{1 + \frac{1}{2}\frac{(x - h/2)^2}{D^2}\right\},$$

where h is the separation between the pinholes and D is the distance between the pinholes and the detector plane, we obtain

$$\text{output} \propto A_1^2 + A_2^2 + 2A_1 A_2 \cos\left(\frac{kxh}{D}\right). \tag{2.10.6}$$

Thus we see that, for an approximately monochromatic but incoherent source, we will obtain the interference fringes. Of course, if there is more than one wavelength present in the light source, each one will produce its own fringes.

In the above discussion we have assumed that, over the path difference between the two waves reaching the detector (in this case xh/D), the wave must be coherent, or the condition is

$$\frac{xh}{D} < l_c. \tag{2.10.7}$$

Actually, if we start moving the detector beyond the x_{max} given by

$$x_{max} = \frac{l_c D}{h}$$

we will not observe the distinct fringes. Eventually, for $x \gg x_{max}$, the fringes disappear completely. Thus, by measuring the quantity

$$\gamma = \frac{I_{max} - I_{min}}{I_{max} - I_{min}}, \tag{2.10.8}$$

we can determine the coherence length of the source. Because, for perfect coherence or $x < x_{max}$,

$$\gamma \rightarrow 1,$$

whereas, for no coherence or $x \gg x_{max}$,

$$\gamma \rightarrow 0.$$

So the value of x when $\gamma \rightarrow 0$ gives a clue to the magnitude of l_c. It is obvious that the quantity γ also gives an idea about the state of coherency.

A practical way to perform Young's experiment is to use a Lloyd mirror or Fresnel biprism arrangement—these are illustrated in Fig. 2.10.2.

2.10.2. Interference due to the Dielectric Layer

Two effective sources derived from the same physical source can be obtained by double reflection from a dielectric slab, as shown in Fig. 2.10.3(a). The dielectric slab has a refractive index of n and has parallel surfaces separated by a distance "d". As we have seen previously, the important quantity to be determined for an interference experiment is the phase difference between the two equivalent sources or the path difference. The path difference is related to the phase difference by the equation

$$\Delta = \text{phase difference} = \frac{2\pi}{\lambda} \sum_i (\text{path difference})_i n_i, \tag{2.10.9}$$

where the summation includes all the path differences in medium i with refractive index n_i.

For the sake of generality, consider Fig. 2.10.3(b) where the dielectric slab has a refractive index n_2 and the surrounding medium has a refractive index

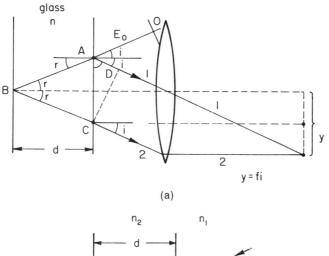

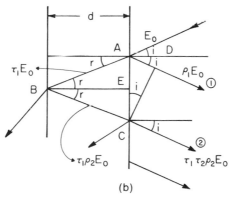

Fig. 2.10.3. Interference due to a dielectric slab: (a) two-reflection interference and (b) geometry for a path-difference calculation.

n_1. The phase difference between the first reflected ray 1 and the second reflected ray 2 is given by

$$\Delta = \frac{2\pi}{\lambda}[(AB + BC)n_2] - \frac{2\pi}{\lambda}\cdot n_1 AD.$$

At first glance we might think that in the above expression only the first term should be there. However, for a ray incident at an angle i, the reflected beam will be in the direction of AD. For this ray and the second ray at C the constant phasefront is not AC but CD where $\sphericalangle ADC = 90°$. Now from the figure

$$BC = AB = \frac{d}{\cos r},$$

$$AC = 2d \tan r,$$

$$AD = AC \sin i = 2d \tan r \sin i, \qquad (2.10.10)$$

and

$$BC + AB = \frac{2d}{\cos r},$$

where r is the angle of refraction. Thus the phase shift becomes

$$\Delta = \frac{2\pi}{\lambda} \cdot 2d \left[\frac{n_2}{\cos r} - n_1 \tan r \sin i \right].$$

Using Snell's law $n_2 \sin r = n_1 \sin i$, we get

$$\Delta = \frac{2\pi}{\lambda} 2d \left[\frac{n_2}{\cos r} - \frac{n_2 \sin^2 r}{\cos r} \right]$$

$$= \frac{2\pi}{\lambda} 2dn_2 \left(\frac{1}{\cos r} \right)(1 - \sin^2 r)$$

$$= \frac{2\pi}{\lambda} \cdot 2dn_2 \cos r$$

or

$$\Delta = \frac{4\pi}{\lambda} dn_2 \cos r. \qquad (2.10.11)$$

It is important to notice that the phase difference is independent of the refractive index, n_1. Now, if these two rays are made to interfere with the help of a lens at the focal plane, they will produce constructive interference when

$$\Delta = 2\pi m \quad \text{where } m \text{ is any integer.}$$

If parallel rays are incident, then at the focal plane we will obtain bright and dark interference rings. The bright rings will occur for

$$\frac{4\pi}{\lambda} nd \cos r_m = 2\pi m,$$

or

$$\frac{4\pi}{\lambda} nd \sqrt{1 - \sin^2 r_m} = 2\pi m, \qquad (2.10.12)$$

where we have used $n_2 = n$.

Using paraxial approximation we obtain that the radius of the mth ring at the focal plane is

$$x_m = fi_m = fnr_m,$$

where i_m and r_m are the incident and refracted angles, respectively, for the formation of the mth bright ring. From (2.10.12) we obtain

$$\frac{2nd}{\lambda} (1 - r_m^2)^{1/2} = m,$$

or

$$1 - \frac{r_m^2}{2} = \frac{m\lambda}{2nd} \qquad \text{(using binomial expansion)},$$

or

$$r_m^2 \approx 2\left(1 - \frac{\lambda m}{2nd}\right),$$

or

$$r_m = \sqrt{2}\left(1 - \frac{\lambda m}{2nd}\right)^{1/2}.$$

We notice that the maximum value of m, m_{max}, is obtained when $r_{max} = 0$ or $m_{max} = 2nd/\lambda$. Thus r_m can be written as

$$r_m = \sqrt{2}\left(1 - \frac{m}{m_{max}}\right)^{1/2}.$$

The mth radius is given by

$$x_m = f_n\sqrt{2}\left(\frac{m_{max} - m}{m_{max}}\right)^{1/2}$$

$$= f_n\sqrt{2}\left(\frac{p}{m_{max}}\right)^{1/2} = f_n\sqrt{2}\left(\frac{p\lambda}{2nd}\right)^{1/2}$$

or

$$x_p = x_m = f\left(\frac{pn\lambda}{d}\right)^{1/2}. \tag{2.10.13}$$

Here p is the number of bright rings counted from the center of the focal plane and we have relabeled x_m as x_p. The area for the mth ring is given by

$$A(m) = \pi x_m^2 = \pi f^2\left(\frac{pn\lambda}{d}\right).$$

Thus

$$A(m + 1) - A(m) = \Delta A = \frac{\pi f^2 n\lambda}{d}, \tag{2.10.14}$$

which is constant and independent of p or m. However, for higher values of the radius, the bright rings are much nearer each other. A drawing of the rings is shown in Fig. 2.10.4(b). These rings are also sometimes called Haidinger interference fringes. A practical way of performing this experiment is shown in Fig. 2.10.4(a).

2.10.3. Michaelson's Interferometer

As shown in Fig. 2.10.5, Michaelson's interferometer also forms Haidinger fringes. However, in place of a dielectric slab, two highly silvered mirrors are

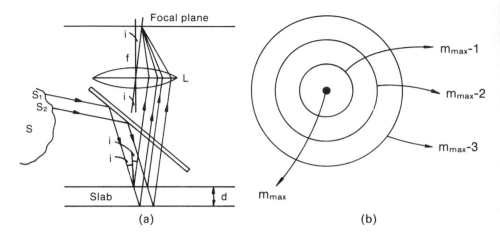

Fig. 2.10.4. (a) Practical way of performing an interference ring experiment. (b) Interference rings.

used. The light from the source splits up equally at the half-silvered glass plate, and after being reflected from the two mirrors is collected by the lens. A compensator plate is used in one branch to compensate for the difference in optical path lengths in the two paths.

The difference between the distances of these mirrors and the half-silvered mirror is equivalent to the "dielectric slab" for this interferometer. Thus, if these distances are exactly equal, then the equivalent thickness is zero. Again,

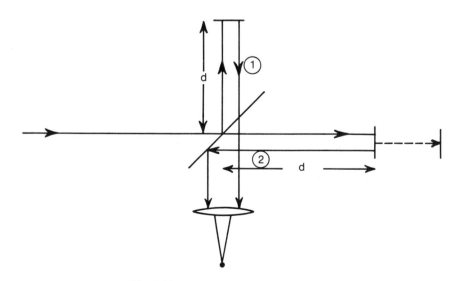

Fig. 2.10.5. Michaelson's interferometer.

the formulas derived in the previous section apply equally here, provided we replace n by 1.

Michaelson's interferometer can be used to measure the coherence length easily because, for large values of $d > l_c$, the rings will disappear. Thus, the arguments given for the coherence and visibility of the rings discussed in Section 2.10.1 also apply here.

2.10.4. Interference by Multiple Reflections and the Fabry–Perot Interferometer

In Fig. 2.10.3 we considered only one reflected beam. However, if two highly silvered mirrors are used, as shown in Fig. 2.10.6, then multiple reflection takes place. If all these beams are collected together by a lens, they give rise to interference by multiple reflections. This multiple reflection interference forms the basis for the Fabry–Perot interferometer. As shown in Fig. 2.10.6, let us

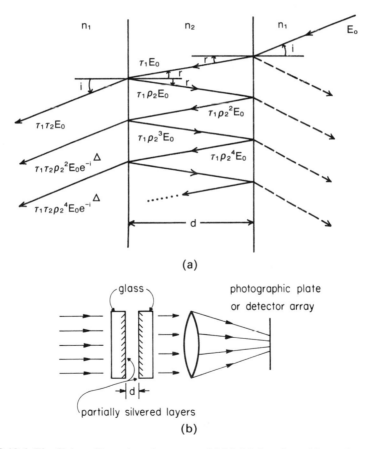

Fig. 2.10.6. The Fabry–Perot interferometer: (a) Multiple reflected beam interference and (b) a typical Fabry–Perot setup.

consider the incident beam with electric field E_0. The angles of incidence and refraction are i are r, respectively. This incident beam is first reflected at interface 1, producing a reflected beam of amplitude $\rho_1 E_0$ where ρ_1 is the reflection coefficient between media 1 and 2. There is a transmitted beam also of amplitude $\tau_1 E_0$ where τ_1 is the transmission coefficient. This transmitted beam in turn produces a reflected and transmitted beam when it arrives at interface 2. The transmitted beam has an E field given by amplitude E_1. The reflected beam, after a reflection at the other surface, produces another transmitted beam whose E value is E_2. This process goes on to infinity. Thus the total electric field transmitted, and if made to interfere, is given by

$$E_{\text{tot}} = E_1 + E_2 + E_3 + E_4 + \cdots$$

$$= \tau_1 \tau_2 E_0 e^{-j\delta_0} + \tau_1 \tau_2 E_0 e^{-j\delta_0} \rho_2^2 e^{-j\Delta} + \cdots$$

$$= \tau_1 \tau_2 E_0 e^{-j\delta_0} [1 + \rho_2^2 e^{-j\Delta} + (\rho_2^2 e^{-j\Delta})^2 + \cdots]. \tag{2.10.15}$$

Here δ_0 is the phase difference between the incident E field and the E_1 beam and $\Delta = (4\pi/\lambda)nd \cos r$, ρ_2 and τ_2 are the complex reflection and transmission coefficients, respectively, between the media 2 and 1, and Δ is given by (2.10.9).

E_{tot} can also be written as

$$E_{\text{tot}} = \frac{\tau_1 \tau_2 E_0 e^{-j\delta_0}}{1 - \rho_2^2 e^{-j\Delta}}.$$

The total transmission coefficient, T, is defined as

$$T = \left| \frac{E_{\text{tot}}}{E_0} \right|^2 = \frac{|\tau_1 \tau_2|^2}{|1 - \rho_2^2 e^{-j\Delta}|^2}. \tag{2.10.16}$$

If we write $\rho_2^2 = (\rho_0 e^{j\theta})^2 = \rho_0^2 e^{j2\theta}$, where ρ_0 is the magnitude and θ is the phase angle of the complex reflection coefficient ρ_2, then

$$T = \frac{|\tau_1 \tau_2|^2}{|1 - \rho_0^2 e^{-j(\Delta - 2\theta)}|^2}. \tag{2.10.17}$$

The denominator of (2.10.17) can be written as

$$|1 - \rho_0^2 e^{-j(\Delta - 2\theta)}|^2 = (1 - \rho_0^2 e^{-j(\Delta - 2\theta)})(1 - \rho_0^2 e^{+j(\Delta - 2\theta)})$$

$$= 1 + \rho_0^4 - \rho_0^2(e^{j(\Delta - 2\theta)} + e^{-j(\Delta - 2\theta)})$$

$$= 1 + R_1^2 - 2R_1 \cos \delta,$$

where $R_1 = \rho_0^2$ and $\delta = \Delta - 2\theta$. Thus T can be written as

$$T = \frac{|\tau_1 \tau_2|^2}{1 + R_1^2 - 2R_1 \cos \delta}. \tag{2.10.18}$$

The maximum value of T occurs when $\cos \delta = +1$. This maximum value, T_{max}, is given by

$$T_{\text{max}} = \frac{|\tau_1 \tau_2|^2}{(1 - R_1)^2}.$$

In terms of T_{max}, (2.10.18) can be written as

$$T = T_{max} \frac{(1 - R_1)^2}{1 + R_1^2 - 2R_1 \cos \delta}, \tag{2.10.19}$$

$$T = T_{max} \frac{(1 - R_1)^2}{(1 - R_1)^2 + 2R_1(1 - \cos \delta)}$$

$$= \frac{T_{max}(1 - R_1)^2}{(1 - R_1)^2 + 4R_1 \sin^2 (\delta/2)}$$

$$= \frac{T_{max}}{1 + F \sin^2 (\delta/2)},$$

where $F = 4R_1/(1 - R_1)^2$ and is called the contrast.

The minimum value of T can be written as

$$T_{min} = \frac{T_{max}}{1 + F}, \tag{2.10.20}$$

or

$$F = \frac{T_{max}}{T_{min}} - 1. \tag{2.10.21}$$

We notice that for $T_{max} = T_{min}$, $F = 0$. On the other hand, for $T_{max} \gg T_{min}$, the value of F or contrast is very high. This is illustrated in Fig. 2.10.7 where for two different values of F, T is plotted as a function of δ. It is obvious from this figure why F is called contrast.

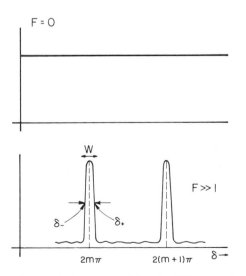

Fig. 2.10.7. Plot of transmission versus δ for the different values of contrast.

Now let us consider the case where F is rather large. In this case, we would like to show that the plot of T versus δ has large values only near the maxima. The maxima occurs when

$$\sin \frac{\delta}{2} = 0,$$

or

$$\delta = 2m\pi \quad \text{where } m \text{ is an integer.}$$

Near the mth maximum we can write

$$\sin^2 \frac{\delta}{2} = \sin^2 \left(\frac{\delta - 2\pi m}{2} \right)$$

$$\approx \left(\frac{\delta - 2\pi m}{2} \right)^2 .$$

Thus, near the mth maximum we can write

$$T \approx \frac{T_{\text{max}}}{1 + F[(\delta - 2\pi m)/2]^2}$$

$$= \frac{T_{\text{max}}/F}{1/F + [(\delta - 2\pi m)/2]^2}. \tag{2.10.22}$$

From the above expression we notice that unless δ is near $2\pi m$ the value of T is negligibly small because $F \gg 1$. A typical plot of T versus δ is shown in Fig. 2.10.7. It is also observed that the shape of the curve near the maximum is Lorentzian.

The half-transmission points, denoted by δ_+ and δ_-, are given by

$$\left(\frac{\delta - 2\pi m}{2} \right)^2 = \frac{1}{F},$$

or

$$\frac{\delta - 2\pi m}{2} = \pm \frac{1}{\sqrt{F}},$$

or

$$\delta_+ = \pi m + \frac{2}{\sqrt{F}},$$

and

$$\delta_- = \pi m - \frac{2}{\sqrt{F}}.$$

Thus the half-width w is given by

$$w = \delta_+ - \delta_- = \frac{4}{\sqrt{F}}. \tag{2.10.23}$$

Remember that, in the plot of T in Fig. 2.10.7, the abscissa is δ.

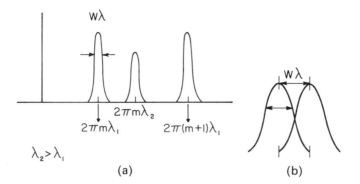

$\lambda_2 > \lambda_1$

(a)   (b)

Fig. 2.10.8. (a) T versus δ for two wavelengths and (b) the Rayleigh criterion.

The highly silvered mirrors discussed in Fig. 2.10.6(a) are used as a Fabry–Perot interferometer, as shown in Fig. 2.10.6(b). If different wavelengths of light are incident, then for different angles the maxima occurs for the same value of m. Thus, we should plot T as a function of $4\pi nd \cos r$. This is shown in Fig. 2.10.8 for two incident wavelengths. The maxima now occurs at

$$4\pi nd \cos r = -2\theta\lambda + 2\pi m\lambda$$

(note that 2θ, being a constant, can be neglected). Thus,

$$4\pi nd \cos r \approx 2\pi m \quad \text{for maxima.}$$

To be an effective spectrometer, the interferometer should have large resolution. To find the resolution, using the discussion of the Rayleigh criterion in Section 2.8.2, we obtain, for the unambiguous determination of $\delta\lambda(\lambda_2 = \lambda + \delta\lambda)$ from Fig. 2.10.8(b),

$$2\pi m\lambda_2 - 2\pi m\lambda = w\lambda.$$

The resolution, $\delta\lambda$, is given by $\delta\lambda = (\lambda_2 - \lambda)$ or

$$\delta\lambda = \frac{w\lambda}{2\pi m}. \tag{2.10.24}$$

The resolving power R is then given by

$$R = \frac{\lambda}{\delta\lambda} = \frac{2\pi m}{2} = \frac{\pi m\sqrt{F}}{2}. \tag{2.10.25}$$

If λ_2 is very large then the mth maxima of λ_2 can coincide with the $(m + 1)$th maxima of λ_1—we should avoid it for an unambiguous determination of λ_2. Thus the free spectral range $\Delta\lambda$ is defined when this coincidence takes place with $\lambda_2 = \lambda_1 + \Delta\lambda$. Thus,

$$2\pi m(\lambda_1 + \Delta\lambda) = 2\pi(m + 1)\lambda_1,$$

or

$$\Delta\lambda = \frac{\lambda_1}{m} \approx \frac{\lambda}{m}. \tag{2.10.26}$$

We define finesse, $\mathscr{F}$, as

$$\mathscr{F} = \frac{\Delta\lambda}{\delta\lambda} = \frac{\pi}{2}\sqrt{F} = \frac{2\pi}{w}. \tag{2.10.27}$$

Physically, finesse means how many wavelengths can be unambigously determined, and is a kind of quality factor of the Fabry–Perot interferometer.

Until now we have not discussed what happens at the focal plane of the lens. Corresponding to the maxima of T, a bright ring is produced. Thus, we see a set of interference fringes at the focal plane. The radii of these bright fringes in the focal plane can be derived in a fashion similar to that done for the Haidinger fringes in Section 2.10.2. Using (2.10.13) we thus obtain

$$x_p = f\left(\frac{p\lambda n}{d}\right)^{1/2}, \tag{2.10.28}$$

where f = focal length of the lens,
         n = refraction index of the media between the plates,
         d = separation between the mirrors, and
         p = number of rings starting from the center.

Figure 2.10.9 illustrates these rings. It also compares the Haidinger fringes and the Fabry–Perot fringes showing the sharpness of fringes for the latter case.

It can easily be shown that the half-width of these rings are given by

$$\Delta x \approx f\left(\frac{\lambda n}{d}\right)^{1/2} \frac{\sqrt{p+1} - \sqrt{p}}{\mathscr{F}}$$

$$= f\left(\frac{\lambda n}{d}\right)^{1/2} \frac{w}{2\pi}(\sqrt{p+1} - \sqrt{p}). \tag{2.10.29}$$

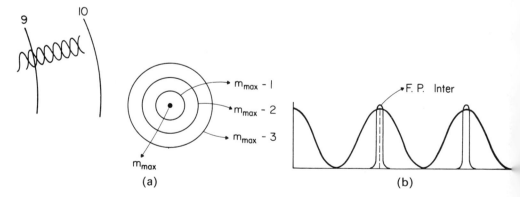

Fig. 2.10.9. (a) Fabry–Perot interference fringes. (b) Difference between two beam (Michaelson) and multiple beam (Fabry–Perot) interference. Transmission function is plotted as a function of position.

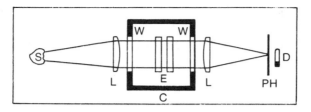

Fig. 2.10.10. A scanning Fabry–Perot interferometer.

A very practical way of using the Fabry–Perot interferometer is in the form of its scanning version—this is shown in Fig. 2.10.10. Here, instead of moving the detector in the focal plane, either the d or n of the Fabry–Perot interferometer is slowly varied. The separation d is generally varied by mounting the mirrors on a piezoelectric plate. When voltage is applied to this piezoelectric plate, the separation between the plates changes. The refractive index can be varied by pumping out the air between the plates and slowly letting the air in through a leak valve. The refractive index is a function of the density of gases between the plates which alters the effective path difference.

The scanning Fabry–Perot interferometer is a very valuable tool. For spectroscopic purposes, it competes with the diffraction grating discussed earlier.

A Numerical Example

A Fabry–Perot interferometer is to be designed which can resolve two wavelengths 1 Å apart. If the free spectral region has to be 1000 Å, then calculate the following quantities of the interferometer:

(a) the finesse;
(b) the contrast;
(c) the reflection coefficient of the mirrors used;
(d) if a lens of focal length 30 cm is used to observe the rings, what is the half-intensity width of the tenth bright ring if the mirrors are separated by 1 mm?

For this problem, the resolution is $\delta\lambda = 1$ Å and the free spectral range is $\Delta\lambda = 1000$ Å. Thus finesse, $\mathscr{F}$, is given by $\mathscr{F} = \Delta\lambda/\delta\lambda = 1000$. As $\mathscr{F} = (\pi/2)\sqrt{F}$, $F = 4.05 \times 10^5$. Also $\mathscr{F} \approx \pi/(1 - R_1)$, thus $R_1 \approx 0.997 = (\rho_0)^2$ or $\rho_0 \approx 0.9985$. Using (2.10.28), we obtain $x_{10} = 21.2132$ mm and $x_{11} = 22.2486$ mm. Thus,

$$\Delta x \approx 1.0354 \times 10^{-3} \text{ mm}$$

$$= 1.0354 \ \mu\text{m}.$$

Note that $x_9 = 20.1246$. Thus, if we consider $\Delta x \approx (x_{10} - x_9)/100$, then $\Delta x \approx 1.0886 \ \mu\text{m}$.

2.11. Holography

Holography was discovered by Dennis Gabor in 1947, in connection with three-dimensional viewing of x-ray images. However, only recently, due to the use of laser light and improvements performed by many researchers (especially Leith and Upatnieks (1967)), has holography become very useful and popular. To most people, holography is like three-dimensional photography. However, other aspects of holography, such as the storage of optical information, are probably more important from the scientific point of view. Holography has been treated in many books and is still the subject of active research. The purpose of this section is to introduce holography in terms of the Fourier optics developed in this book and to discuss some applications. For a thorough knowledge, the reader should refer to other books listed in the References.

2.11.1. Photography

Before we start discussing holography, it is worthwhile to review photography. In photography, in general, a light-sensitive, silver-compound-based film is used. As discussed, in connection with detectors in Section 2.9.1, this film has the property of recording the square of the incident light amplitude. Typical information that we are interested in recording can be written as

$$\vec{E}(x, y) = \vec{a}(x, y)e^{-j\varphi(x,y)} = \vec{a}. \qquad (2.11.1)$$

We will use arrows over quantities that are complex. For example, if we look through a window we see a scenery of trees, birds, and mountains, etc. The reason we can see this scenery is because an electric field wavefront, defined in (2.11.1), exists in the plane of the window and carries complete information about the scenery. I want to stress the word *complete*. At a particular instant, all the information about the outside scenery that can be obtained is present in that two-dimensional electric field distribution. It has both an amplitude part and a phase part. The phase part carries some of the three-dimensional aspects of the scenery. For example, if there is an object, A, behind an object, B, and we look directly at B, we cannot see A (Fig. 2.11.1(a)). However, if we move a little, as shown in Fig. 2.11.1(b), so that A is not obstructed by B, we can see A. This information, by looking from different lines of sight we see different aspects of scenery, is part of the so-called three-dimensional photography.

In ordinary photography, we place the film near the window (actually, using a camera lens we project the electric field at the window onto the film) and record the square of the electric field. This is done by first exposing the film to the incident radiation for a specified length of time. This radiation generates a photochemical reaction which is proportional to the square of the electric field. After exposure the film is chemically processed resulting in the

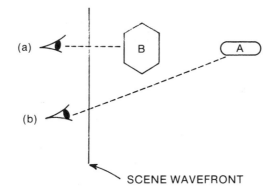

Fig. 2.11.1. Viewing of object (A) in the presence of obstruction (B).

transmission function of the film given by

$$
\begin{aligned}
t(x, y) &= t_0 + \beta |\vec{E}|^2 \\
&= t_0 + \beta [\vec{a}\vec{a}^*] \\
&= t_0 + \beta |\vec{a}|^2 \\
&= t_0 + \beta |\vec{a}(x, y)|^2,
\end{aligned}
\tag{2.11.2}
$$

where t_0 and β are constants determined by the film and processing used. Then the transmission function of the exposed and developed film contains the information $|(\vec{a}(x, y)|^2$. Of course, this film can be printed or viewed by shining light through it to obtain a picture of the scenery.

Thus, we see that although our goal was to obtain the information $\vec{a}(x, y)e^{-j\varphi(x, y)}$, or to recreate the wavefront itself, what we have obtained is $|(\vec{a}(x, y)|^2$ by using photography. We have lost the phase information. Thus, if we took a picture in which a cat was behind a tree, photography will never reveal it. To obtain this information, we must somehow be able to reproduce the E field of the window itself, with all its amplitude and phase variations. That is, the whole wavefront has to be reconstructed. This wavefront reconstruction can be performed using holograms.

How can we record the phase of a light beam? A clue to this question can be found in the discussion on interference. As shown in Fig. 2.11.2(a), consider two point sources denoted by A and B. If we record photographically at the detector plane (in the Fraunhofer zone), we record for each point A or B a uniform light distribution. However, if both are present simultaneously, we obtain interference fringes. In this case, the detected light intensity is given by

$$
E(x) \propto E_A^2 + E_B^2 + 2E_A E_B \cos \alpha_0,
\tag{2.11.3}
$$

where α_0 is the phase difference between the two electric fields at the detector point. Thus, we see that we can obtain the phase information with respect to another source. For example, the period of the recording in the detector plane

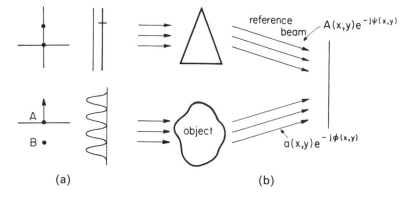

Fig. 2.11.2. (a) The use of interference to obtain phase information. (b) The making of a hologram.

gives information about the separation between the two sources, and the amplitude on the optical axis gives information about the phase difference between the two.

By recording the square of the E field, in the presence of a reference source, the phase information can be recorded. Of course, this discussion assumes that both the sources are coherent. Thus, we see that in holography we need coherent sources and a reference beam, as will be discussed in the next section.

2.11.2. The Making of a Hologram

As shown in Fig. 2.11.2(b), we are interested in the imaging of the object which is illuminated with a coherent source. The light rays carrying information about the object has an electric field at the recording plate denoted by

$$\vec{E}(x, y) = \vec{a}(x, y)e^{-j\varphi(x,y)} = \vec{a}.$$

We shall denote this total electric field by $\vec{a}$ for the sake of brevity. A reference beam is also incident on the recording medium at the same time. This electric field is denoted by

$$\vec{E}_{\text{ref}} = \vec{A}(x, y)e^{-j\psi(x,y)} = \vec{A}. \tag{2.11.4}$$

As shown in the figure, the light shining on the object and the reference beam come from the same laser source, to keep them coherent, and so they interfere over larger areas. This has been performed in Fig. 2.11.2 by using a beam splitter and a prism.

One of the simplest cases arises when the reference beam is parallel and uniform and is incident at an angle θ on the recording medium. In this case*

$$\vec{A} = \vec{A}_0 e^{j2\pi f_0 x}, \tag{2.11.5}$$

---

* Note the sign convention: $f_{i0} = \sin\theta/\lambda$ is again measured as discussed in *Fourier Optics*, p. 67.

where $f_0 = \sin\theta/\lambda$. It is obvious that this reference beam has a more or less single spatial frequency which is determined by the incident angle θ. This reference beam frequency is called the carrier frequency in analogy with radio-engineering terminology. The significance of this will be discussed in detail later.

The total incident electric field on the recording plane is given by

$$\vec{E}_{\text{tot}} = \vec{a} + \vec{A},$$

and the transmission function of the recording film after proper processing will be given by

$$
\begin{aligned}
t(x, y) &= t_0 + \beta[\vec{E}_{\text{tot}}\vec{E}_{\text{tot}}^*] \\
&= t_0 + \beta[\vec{a}\vec{a}^* + \vec{A}\vec{A}^* + \vec{a}\vec{A}^* + \vec{a}^*\vec{A}] \\
&= t_0 + \beta[|\vec{a}|^2 + |\vec{A}|^2 + 2|\vec{A}(x, y)|\,|\vec{a}(x, y)|\cos(\varphi(x, y) - \psi(x, y))].
\end{aligned}
$$

$$(2.11.6)$$

Thus we see that although we have recorded the square of the electric field magnitude, we have been able to keep the phase information of the object, $\varphi(x, y)$, due to the presence of the reference beam which interferes with the beam scattered by the object.

Remember that in photography, the recording process would have been the same with the exception that the reference beam is absent. In photography, we could have looked at this transparency and would have observed some resemblance to the object—not so for the case of holograms. What we see in holograms is a gibberish of different interference fringes which has absolutely no resemblance to the object at all. Thus in a hologram, the viewing or reconstruction of the holographic field is a separate and essential process.

2.11.3. Reconstruction of a Hologram

To view a hologram, another coherent source illuminates the hologram. For the sake of simplicity, let us consider that the wavelengths of light for the reference beam and the viewing beam are identical. The effect of these being different will be discussed later.

If the viewing beam electric field at the hologram plane is denoted by B, then we obtain the electric field of the light emergent from the hologram to be given by

$$\vec{E}_{\text{out}} = t_0\vec{B} + \beta[|\vec{a}|^2 + |\vec{A}|^2]\vec{B} + \beta\vec{a}\vec{A}^*\vec{B} + \beta\vec{a}^*\vec{A}\vec{B}. \qquad (2.11.7)$$

The output light thus consists of five individual terms. Again, for simplicity, let us consider that the viewing and the reference beams are identical in spatial frequency, that is,

$$\vec{B} = \vec{i}_B\vec{B}_0 e^{j2\pi f_{x0}x}. \qquad (2.11.8)$$

For this viewing field we can write the output E field consisting of five terms as

$$\vec{E}_{\text{tot}} = \vec{E}_1 + \vec{E}_2 + \vec{E}_3 + \vec{E}_4 + \vec{E}_5, \qquad (2.11.9)$$

where

$$\vec{E}_1 = \vec{i}_B t_0 \vec{B}_0 e^{j2\pi f_{x0}x},$$
$$\vec{E}_2 = \vec{i}_B \beta |\vec{a}|^2 \vec{B}_0 e^{j2\pi f_{x0}x},$$
$$\vec{E}_3 = \vec{i}_B \beta |\vec{A}_0|^2 \vec{B}_0 e^{j2\pi f_{x0}x},$$
$$\vec{E}_4 = \vec{i}_B \beta |\vec{A}_0| |\vec{B}_0| \cdot \vec{a},$$
$$\vec{E}_5 = \vec{i}_B \beta |\vec{A}_0| |\vec{B}_0| \vec{a}^* \cdot e^{-j2\pi 2 f_{x0}x}.$$

We notice immediately that the five beams come out at three different angles without any overlap, provided the spatial bandwidth of the field $\vec{a}$ is not too large—these are shown in Fig. 2.11.3. We see that E_1, E_2, and E_3 come out at an angle θ with respect to the hologram, E_4 can be viewed directly, whereas E_5 comes at an angle which is approximately 2θ. Beam 4, except for a constant term $\beta A_0 B_0$, is an exact reproduction of the electric field associated with the object at the plane of the hologram when it was recorded. It contains all the amplitude and phase information—all the possible information contained in

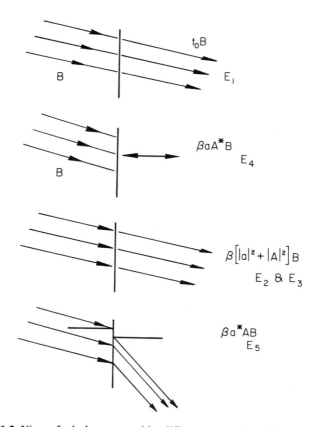

Fig. 2.11.3. View of a hologram and its different transmitted light components.

it when the hologram was made. Thus we can see it, as if the object was entirely recreated. Beam 5 has all the information about the object too. However, it is a phase conjugate picture.

In the above discussion, a very important assumption was made. Beams 4 and 5 do not overlap nor do they overlap with beams 1, 2, and 3. We can see that if they do overlap, then while looking directly, for example, we will see not only $\vec{a}$ but other E fields as well. This separation of beams, or the use of the carrier frequency to make the beams come out in different directions, was the contribution of Leith and Upatnieks in their classic paper. As we shall see later, the Gabor hologram did not have this separation and for that reason it was of much poorer quality.

Let us assume that the spatial bandwidth associated with the object is Δf_x. For simplification, let us consider one dimension only. This means that the frequency components of $\{E_{\text{object}}\}$ are limited between the frequency spread,

$$-\frac{\Delta f_x}{2} < f < \frac{\Delta f_x}{2}.$$

Thus we can associate with any object electric field a spatial bandwidth which denotes how fast the spatial variation amplitudes and phases are. Figure

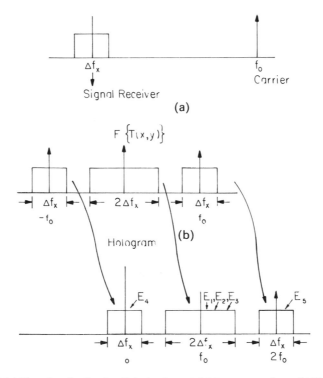

Fig. 2.11.4. (a) Signal and reference light in the spatial frequency plane. (b) Transmission function of the hologram in the spatial frequency plane.

2.11.4(a) shows the signal and the reference (carrier) in the spatial frequency plane. Figure 2.11.4(b) shows the spatial frequency components of the hologram transmission function. In the spatial frequency plane, we can plot E_1 to E_5, all of the output beams as shown in Fig. 2.11.4(c). Note that the difference between (b) and (c) is that the horizontal axis is shifted by f_0, the carrier frequency. This is due to the fact that we have reconstructed using a beam of spatial frequency f_0 (same as the reference beam). We see for the beams not to overlap, the following condition must be satisfied:

$$f_{x0} > \tfrac{3}{2}(\Delta f_x).\tag{2.11.10}$$

2.11.4. The Gabor Hologram

For the case of the Gabor hologram, the reference beam is incident parallel to the optical axis. That is, for this case, $f_0 = 0$. Thus, in this case, there is no angular separation. However, if we make $|\vec{A}_0| \gg |\vec{a}|$, then beam 2 is negligible. Beams 1 and 3 are nothing but uniform beams. So, for this case, we can still view the hologram as shown in Fig. 2.11.5. It is to be mentioned that beam 5 forms a virtual image, whereas beam 4 forms a real image.

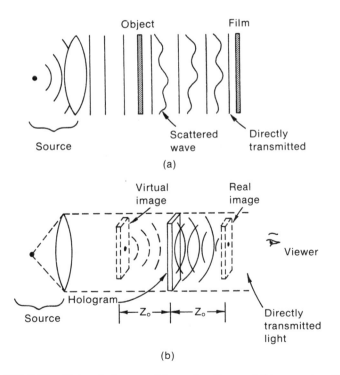

Fig. 2.11.5. The Gabor hologram: (a) construction and (b) reconstruction.

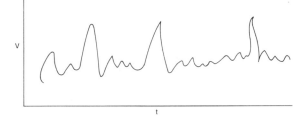

Fig. 2.11.6. Typical radio waveform.

2.11.5. Analogy with Radio and Information Storage

In radio engineering our interest is to transmit and receive signals. These signals are functions of time, for example, it could be the music of an orchestra. An illustrative signal is shown in Fig. 2.11.6. This signal has a bandwidth of Δf, for example, most radio signals have $\Delta f \approx 30$ kHz. If we want to send this radio signal directly via an electric wire or through a medium by radiation, we need a separate wire or medium for each signal. For example, we cannot send the signals of two orchestras via the wire directly, because at the output we receive the sum of the two time signals and there is no way of separating them. However, if we modulate this signal with different carrier frequencies f_1, $f_2 \gg \Delta f$, then even if we send them through the same wire or the same medium we can separate them by frequency filters after we mix them with carrier frequencies again. Thus, this frequency multiplexing and demultiplexing is essential to the operation of radio and TV.

Now let us look at holography. For this case, we have spatial frequencies, although two dimensional. We also have a bandwidth of the signal, the E field due to the object, a. If we do not add any carrier frequency, we have a situation somewhat similar to photography. However, if we multiplex with different carrier frequencies, then we can store and view many objects or pictures by the same hologram, just as if we were sending different audio signals through the same wire.

To do this, let us consider the bandwidth of each picture to be Δf_x. Then we can record one page or one picture with a carrier frequency, f_{x0}, and the next picture or page with a carrier frequency $f_{x0} + 2\Delta f_x$, and so on, as shown in the figure. Notice all the pictures are recorded on the same film. However, when viewing it with different carrier lights, or with lights incident at different angles, we see the different pictures without any overlap or distortion. Thus we see that enormous amounts of information can be stored on one piece of film. For example, it is possible to record, say, a whole book in just one square of film, provided the resolution of the film is high enough. This is shown in Fig. 2.11.7.

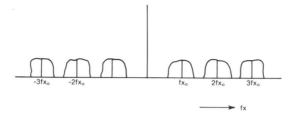

Fig. 2.11.7. Multiplexing in holograms.

2.11.6. Some Comments About Holograms

(i) If in (2.11.7), $B \neq B_0 e^{j2\pi f_0 x}$, but has a different spatial frequency given by $B = B_0 e^{j2\pi f_B x}$, then the output E field emerging from the hologram will be arriving at angles θ_-, θ_B, and θ_+, respectively, given by

$$\frac{\sin \theta_B}{\lambda_B} = f_B, \qquad \frac{\sin \theta_-}{\lambda_B} = f_B - f_0 \qquad \text{and} \qquad \frac{\sin \theta_+}{\lambda_B} = f_B + f_0,$$

where λ_B is the wavelength of the viewing light.

(ii) We have not discussed how bright the image is when viewed through the holographic process. It turns out that the diffraction efficiency of thin film holograms is not very large; thus, the image may not be bright. However, we can increase the diffraction efficiency through the use of a thick-film hologram. In thick-film holograms, the efficiency may be large enough so that no laser light is needed; for viewing, a simple white light might be sufficient. Because the different colors will be diffracted at different angles, and if the separation between them is enough, we can view what is known as a white-light hologram.

(iii) Another form of the white-light hologram has also been discovered. In this type, no thick film is used. However, the diffraction efficiency is increased at the cost of the perspective in one direction. To make this hologram, we first of all must make a hologram of the object. This hologram is then illuminated with a viewing light, and another reference beam is used to make a second hologram using the real image of the first hologram. This second hologram, when viewed with white light, produces bright single-color images at different angles.

(iv) Up until now we might have implied that there is no difference between holographic recording film and that used in photography. From the photochemical point of view, there is no difference. However, the photographic film needs to record only a bandwidth of Δf_x, whereas the holographic film must be able to record the carrier frequency. Thus, the holographic film needs to be, in general, of much higher resolution. Because, usually, the higher resolution films need to be exposed for a longer time for the same intensity of incident light, the light sources must be coherent over this recording time period. This might be a restriction difficult to satisfy. However, we can use a high-power laser source so that the exposure time is smaller.

2.11.7. Hologram Using Point-Source References

Up until now we have considered only the holograms made using a light beam which was uniform and parallel. However, it is possible to use point sources both for viewing and recording. This is illustrated in Fig. 2.11.8, where for simplicity we have used the object to be a point source also. The incident field at the recording plane, using the Fresnel approximation (see Section 2.6.5), can be written as

$$E_{\text{tot}} = A + a,$$

where

$$A = A_o e^{-j\frac{\pi}{\lambda_1 z_r}}[(x - x_r)^2 + (y - y_r)^2], \qquad (2.11.11)$$

and

$$a = a_o e^{-j\frac{\pi}{\lambda_1 z_o}}[(x - x_o)^2 + (y - y_o)^2].$$

The reference source and the object are situated at x_r, y_r, z_r and x_o, y_o, z_o, respectively. The recording medium is situated at $z = 0$ and the recording plane is designated by (x, y). It is assumed that the wavelength of the reference and object beam is λ_1.

Let us assume that the viewing source has wavelength λ_2 and is situated at x_p, y_p, and z_p. We want to find the conditions when a holographic image is obtained. The viewing field is given by

$$B = B_0 e^{-j\frac{\pi}{\lambda_2 z_p}}[(x - x_p)^2 + (y - y_p)^2]. \qquad (2.11.12)$$

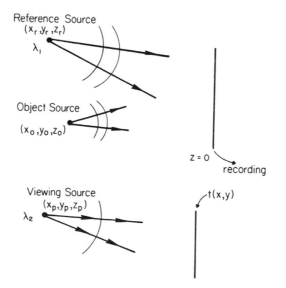

Fig. 2.11.8. A hologram using a point source.

The important term to consider in the output field is $aA*B$, this is given by

$$aA*B = A_0 B_0 a_0 e^{-j\pi}[\ \ ], \tag{2.11.13}$$

where the terms in brackets are

$$[\ \ ] = \frac{1}{\lambda_1 z_o}[(x - x_o)^2 + (y - y_o)^2]$$

$$+ \frac{1}{\lambda_2 z_p}[(x - x_p)^2 + (y - y_p)^2]$$

$$- \frac{1}{\lambda_1 z_r}[(x - x_r)^2 + (y - y_r)^2]. \tag{2.11.14}$$

If the holographic image is formed at x_i, y_i, and z_i, then the above expression from (2.11.13) must equal

$$E \propto e^{-j}\frac{\pi}{\lambda_2 z_i}[(x - x_i)^2 + (y - y_i)^2]. \tag{2.11.15}$$

Thus we obtain

$$\frac{1}{\lambda_2 z_i} = \frac{1}{\lambda_1 z_o} + \frac{1}{\lambda_2 z_p} \pm \frac{1}{\lambda_1 z_r},$$

or

$$z_i = \left[\frac{1}{z_p} + \frac{\lambda_2}{\lambda_1 z_o} \pm \frac{\lambda_2}{\lambda_1 z_o}\right]^{-1}. \tag{2.11.16}$$

The lower sign is for the virtual image and the upper sign is for the real image which is inverted (i.e., mirror image). We also observe that

$$x_i = x_o\frac{\lambda_2 z_i}{\lambda_1 z_o} + x_p\frac{z_i}{z_p} - x_r\frac{\lambda_2}{\lambda_1}\frac{z_i}{z_r}, \tag{2.11.17}$$

and

$$y_i = y_o\frac{\lambda_2 z_i}{\lambda_1 z_o} + y_p\frac{z_i}{z_p} - y_r\frac{\lambda_2}{\lambda_1}\frac{z_i}{z_r}. \tag{2.11.18}$$

The magnification is given by

$$M = \left|\frac{\Delta x_i}{\Delta x_o}\right| = \left|\frac{\Delta y_i}{\Delta y_o}\right|$$

$$= \left|1 - \frac{z_o}{z_r} \mp \frac{\lambda_1}{\lambda_2}\frac{z_o}{z_p}\right|^{-1}, \tag{2.11.19}$$

where the same sign convention applies for the (positive) real and (negative) virtual images.

We can easily obtain the results obtained in earlier sections by considering $z_r \to \infty$ and $z_p \to \infty$. Then

$$z_i = \frac{\lambda_1}{\lambda_2} z_o \qquad (2.11.20)$$

and

$$M = 1.$$

Thus, there is no change in the size of the image with respect to the object for this case. However, the position of the object is moved if $\lambda_2 \neq \lambda_1$.

2.12. Physical Optics

There are many topics of scientific and technological interest relating to optical wave propagation which we will not be able to cover in this book. However, they are of significant interest for the later sections of the book. In this section, we simply give the results of these topics with no derivation and a somewhat concise explanation.

2.12.1. Total Internal Reflection and Optical Tunneling

Let a wave be incident from a medium of higher refractive index to a medium of lower refractive index, i.e., $n_1 > n_2$ and $\theta_t > \theta_i$ (Fig. 2.12.1). As the angle of refraction is larger than the angle of incidence, if we go on increasing the angle of incidence we then reach a point when $\theta_T = 90°$. This happens for $\theta_i = \theta_c$, given by

$$\sin \theta_c = \frac{n_2}{n_1} < 1,$$

or

$$\theta_c = \sin^{-1} \left(\frac{n_2}{n_1} \right), \qquad (2.12.1)$$

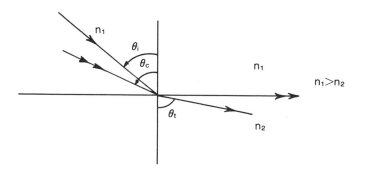

Fig. 2.12.1. Geometry for total internal reflection.

As the transmitted angle cannot be greater than 90°, all the light will be reflected for $\theta_i \geq \theta_c$. The angle θ_c is also known as the critical angle of total internal reflection. This description of total internal reflection is only true when the wavefront is infinite and the depth of the medium 2 is also infinite. To understand these comments we note that the transmitted wave is proportional to

$$e^{j(\omega t - \mathbf{K}_t \cdot \mathbf{r})} = e^{j(\omega t - K_2 \sin \theta_t x + K_2 \cos \theta_t z)},$$

where

$$\cos \theta_t = \sqrt{1 - \sin^2 \theta_t}$$

$$= \sqrt{1 - \left(\frac{n_1}{n_2}\right)^2 \sin^2 \theta_i} = \sqrt{1 - \frac{\sin^2 \theta_i}{\sin^2 \theta_c}}. \qquad (2.12.2)$$

Thus, for $\theta_i > \theta_c$, $\cos \theta_t = j\alpha$ where

$$\alpha = \sqrt{\frac{\sin^2 \theta_i}{\sin^2 \theta_c} - 1} = \text{real.} \qquad (2.12.3)$$

The transmitted wave for this case becomes

$$e^{-\alpha K_2 z} e^{j(\omega t - K_2 \sin \theta_t x)}.$$

The wave amplitude decays as a function of z. Thus if medium 2 is not very large, as shown in Fig. 2.12.2, we satisfy the condition for total internal reflection at the upper boundary, but light will nevertheless be transmitted in medium 3. This phenomenon is called optical tunneling and has some practical applications in fabricating narrowband optical filters.

On the other hand, even when medium 2 is infinite, if a finite wavefront is incident, as shown in Fig. 2.12.3, almost all the light energy is reflected. However, there is then a lateral beam displacement Δz given by

$$\Delta z \approx \frac{1}{\alpha K_2} \qquad (2.12.4)$$

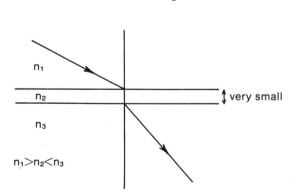

n_1

n_2 $\updownarrow$ very small

n_3

$n_1 > n_2 < n_3$

Fig. 2.12.2. Optical tunneling.

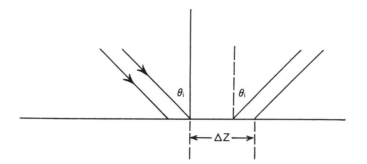

Fig. 2.12.3. Lateral displacement of total internally reflected beam.

for the case of parallel polarization. This lateral displacement was experimentally verified by Goos and Hänchen and is generally known as the Goos–Hänchen effect.

2.12.2. Reflection and Transmission Coefficients

Snell's law only predicts the direction of propagation for the reflected and the transmitted beams. It does not specify the magnitude of the reflected and transmitted electric field vectors. Consider Fig. 2.12.4 where the incident, reflected, and transmitted beams are given by

$$\mathbf{E}_i = \mathbf{i}_i E_{i0} e^{j(\omega t - \mathbf{K}_i \cdot \mathbf{r})}, \tag{2.12.5}$$

$$\mathbf{E}_r = \mathbf{i}_r E_{r0} e^{j(\omega t - \mathbf{K}_r \cdot \mathbf{r})}, \tag{2.12.6}$$

$$\mathbf{E}_t = \mathbf{i}_t E_{t0} e^{j(\omega t - \mathbf{K}_t \cdot \mathbf{r})}. \tag{2.12.7}$$

Here $\mathbf{i}_i$, $\mathbf{i}_r$, and $\mathbf{i}_t$ are the unit vectors and $\mathbf{K}_i$, $\mathbf{K}_r$, and $\mathbf{K}_t$ are the propagation vectors. We define the reflection coefficient, Γ, and transmission coefficient, T,

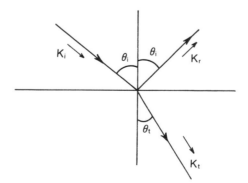

Fig. 2.12.4. Wave vectors for reflection and transmission.

as follows

$$\Gamma = \frac{E_{r0}}{E_{i0}}, \tag{2.12.8}$$

$$T = \frac{E_{t0}}{E_{i0}}. \tag{2.12.9}$$

Note that both Γ and T are, in general, complex quantities. To solve for Γ and T we note that

$$i_i E_{i0} = i_y E_{i\perp} + i_\| E_{i\|}, \tag{2.12.10}$$

where we have assumed that the plane of incidence is the xz-plane. Thus the unit vector $i_\|$ is in the xz-plane but perpendicular to $\mathbf{K}_i$. To solve this general case, where the incident angles θ_i and $\mathbf{E}_i$ have components parallel and perpendicular to the plane of incidence, we can write the boundary conditions at the interface and solve for Γ and T. However, it is convenient and customary to subdivide the general case problem into two separate cases: perpendicular polarization ($E_{i\perp} \neq 0$, $E_{i\|} = 0$) and parallel polarization ($E_{i\perp} = 0$, $E_{i\|} \neq 0$).

For the perpendicular polarization we obtain

$$\Gamma_\perp = \frac{Z_1 \sec \theta_i - Z_2 \sec \theta_t}{Z_1 \sec \theta_i + Z_2 \sec \theta_t} = \frac{n_1 \cos \theta_i - n_2 \cos \theta_t}{n_1 \cos \theta_i + n_2 \cos \theta_t}, \tag{2.12.11}$$

where

$$Z_1 = \sqrt{\frac{\mu_1}{\varepsilon_1}} = \frac{Z_0}{n_1},$$

$$Z_2 = \sqrt{\frac{\mu_2}{\varepsilon_2}} = \frac{Z_0}{n_2}, \tag{2.12.12}$$

$$n_1 \sin \theta_i = n_2 \sin \theta_t,$$

$$T_\perp = 1 + \Gamma_\perp = \frac{2n_1 \cos \theta_i}{n_1 \cos \theta_i + n_2 \cos \theta_t}. \tag{2.12.13}$$

For the parallel polarization case we obtain

$$\Gamma_\| = \frac{Z_1 \cos \theta_i - Z_2 \cos \theta_t}{Z_1 \cos \theta_i + Z_2 \cos \theta_t} = \frac{n_1 \cos \theta_t - n_2 \cos \theta_i}{n_1 \cos \theta_t + n_2 \cos \theta_i}, \tag{2.12.14}$$

$$T_\| = 1 + \Gamma_\| = \frac{2n_1 \cos \theta_t}{n_1 \cos \theta_t + n_2 \cos \theta_i}. \tag{2.12.15}$$

For the general case, where both the parallel and perpendicular components are present, we calculate separately, the parallel and perpendicular reflected and transmitted components and add them vectorially to obtain the final result.

Let us consider the parallel case and try to find an incident angle for this case when there is no reflection. Denoting the incident angle at which this happens by θ_B, we have

$$n_1 \cos \theta_t = n_2 \cos \theta_B. \tag{2.12.16}$$

Using

$$n_2 \sin \theta_t = n_1 \sin \theta_B, \tag{2.12.17}$$

we easily obtain

$$\sin^2 \theta_B = \frac{n_2^2}{n_2^2 + n_1^2} = \frac{\varepsilon_2}{\varepsilon_1 + \varepsilon_2}, \tag{2.12.18}$$

or

$$\tan \theta_B = \sqrt{\frac{\varepsilon_2}{\varepsilon_1}} = \frac{n_2}{n_1}.$$

This incident angle, θ_B, for which the reflection coefficient is zero and the transmission coefficient is unity, is called the Brewster angle. The Brewster angle plays a very important role for lasers for which the cavity mirrors are outside the amplifying media. To minimize losses, the Brewster angle is used at both ends of the amplifying media. This is shown in Fig. 2.12.5 for a typical laser.

Now we might ask the obvious question. Is there also an angle like the Brewster angle for the perpendicular polarization case? For this problem, we must have

$$\Gamma_\perp = 0,$$

or

$$\frac{n_1 \cos \theta_1 - n_2 \cos \theta_2}{n_1 \cos \theta_1 + n_2 \cos \theta_2} = 0,$$

or

$$n_1 \cos \theta_1 = n_2 \cos \theta_2. \tag{2.12.19}$$

However, from Snell's law

$$n_1 \sin \theta_1 = n_2 \sin \theta_2.$$

Thus

$$n_1^2 \cos^2 \theta_1 = n_2^2 \cos^2 \theta_2$$
$$= n_2^2(1 - \sin^2 \theta_2),$$

or

$$n_1^2 - n_1^2 \sin^2 \theta_1 = n_2^2 - n_1^2 \sin^2 \theta_1,$$

Fig. 2.12.5. Laser with a Brewster angle plasma tube and an external mirror.

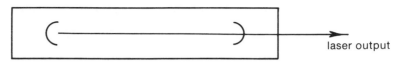

Fig. 2.12.6. Laser with internal mirrors.

or

$$n_1 = n_2. \tag{2.12.20}$$

So there is no equivalent Brewster angle for the perpendicular polarization case.

Note that the radiation coming out of the laser (shown in Fig. 2.12.5) which uses the Brewster angle is always polarized with parallel polarization, because the perpendicular polarization has higher loss and is less likely to oscillate. However, if the mirrors are inside the lasing media, as shown in Fig. 2.12.6, the output of the laser is unpolarized. It is to be mentioned that eqs. (2.12.11), (2.12.13)–(2.12.15) are also known as the Fresnel equations for reflection and transmission coefficients.

2.12.3. Polarization

The polarization of the light wave is defined by the orientation of the electric field. If the **E** field is always in one plane, it is called the plane wave. So far, we have only considered plane polarized waves. As Maxwell's equations are linear, and any linear combinations of elementary solutions are possible, we can easily construct the general elliptically polarized wave, as shown in Fig. 2.12.7. For a z propagating elliptically polarized wave, the electric field is given by

$$E_x = E_{x0} \cos(\omega t - kz), \tag{2.12.21}$$

$$E_y = E_{y0} \sin(\omega t - kz). \tag{2.12.22}$$

Thus, the x and y components have a phase difference of $90°$. We note that,

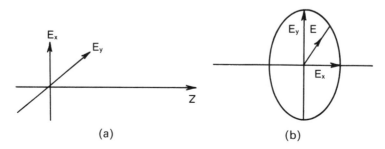

(a)   (b)

Fig. 2.12.7. Polarization of an optical wave.

if we observe the electric field vector in the plane transverse to the direction of propagation, the tip of the E field vector follows an elliptical contour given by

$$\left(\frac{E_x}{E_{x0}}\right)^2 + \left(\frac{E_y}{E_{y0}}\right)^2 = 1. \tag{2.12.23}$$

For circular polarization

$$E_{x0} = E_{y0}. \tag{2.12.24}$$

In Section 2.12.5 we will discuss different polarizations and how they are obtained when we consider light propagation through anisotropic media.

2.12.4. Phase Velocity, Group Velocity, and Ray Velocity

For a plane electromagnetic wave propagating through an isotropic medium, the phase velocity, v_p, or the velocity with which the constant phase front advances, is given by

$$\mathbf{v}_p = \frac{\omega}{|k|}\,\mathbf{i}_k. \tag{2.12.25}$$

If the medium is dispersive, i.e., $\varepsilon = \varepsilon(\omega)$, the phase velocity will, in general, be a function of frequency.

Group velocity is given by

$$\mathbf{v}_g = \nabla_k \omega(\mathbf{k}). \tag{2.12.26}$$

This group velocity describes the propagation for the envelope of a wave consisting of a group of plane waves having frequencies in the range ω and $\omega + d\omega$. Note that group velocity is an important quantity because it represents how energy is transferred, i.e., how the information is propagated. Note that, theoretically, a wave having a single frequency component ω must exist for all times, i.e., for t from $-\infty$ to $+\infty$. Thus we will never know whether, in fact, it has existed or not. Because of the detection process, we change the wave by a small amount, which in turn makes the wave to be represented by a group of waves with finite width $d\omega$.

It is interesting to note that if the relationship between ω and k is nonlinear, then

$$v_g \neq v_p.$$

Actually, it is quite possible to envisage a case where the group velocity is negative whereas the phase velocity is positive. However, for the isotropic homogeneous case, the ω versus k curve is linear and

$$v_g = v_p = \frac{1}{\sqrt{\mu\varepsilon}} = \frac{v_0}{n}. \tag{2.12.27}$$

The ray velocity, or energy velocity, V_e is defined as

$$v_e = v_p \cos\psi, \qquad \mathbf{V}_e = \frac{\mathbf{P}}{u_{av}}, \tag{2.12.28}$$

where P is the Poynting vector $= \frac{1}{2}(\mathbf{E} \times \mathbf{H}^*)$ and u_{av} is the peak stored electromagnetic energy.

It can be shown that

$$v_e = v_p \cos \psi, \tag{2.12.29}$$

where ψ is the angle between $\mathbf{k}$ and $\mathbf{P}$. For isotropic material, $\psi = 0$. For a lossless medium, v_g and v_e are identical.

2.12.5. Propagation in Anisotropic Media

An anisotropic medium is characterized by a dielectric tensor defined as follows:

$$\begin{pmatrix} D_x \\ D_y \\ D_z \end{pmatrix} = \begin{pmatrix} \varepsilon_{xx} & \varepsilon_{xy} & \varepsilon_{xz} \\ \varepsilon_{yx} & \varepsilon_{yy} & \varepsilon_{yz} \\ \varepsilon_{zx} & \varepsilon_{zy} & \varepsilon_{zz} \end{pmatrix} \begin{pmatrix} E_x \\ E_y \\ E_z \end{pmatrix}. \tag{2.12.30}$$

The dielectric tensor is symmetric. Thus

$$\varepsilon_{ij} = \varepsilon_{ji}$$

and there are only six independent elements. The axes we have chosen for x, y, and z are not unique. We can choose a new set of axes, represented by x', y', and z'. In this coordinate system, the symmetrical real dielectric matrix can always be made to be diagonal. Thus, in this new system, denoted here by x, y, and z again for simplicity from now on, eqn. (2.12.30) simplifies to

$$\begin{pmatrix} D_x \\ D_y \\ D_z \end{pmatrix} = \begin{pmatrix} \varepsilon_x & 0 & 0 \\ 0 & \varepsilon_y & 0 \\ 0 & 0 & E^z \end{pmatrix} \begin{pmatrix} E_x \\ E_y \\ E_z \end{pmatrix}. \tag{2.12.31}$$

Note that the axes in the crystal, along which the dielectric matrix becomes diagonal, is called the principal axes.

In any direction of propagation, $\mathbf{i}_k$, in general, there are two refractive indices, n_1 and n_2, corresponding to two different phase and group velocities. The displacement vectors, $\mathbf{D}_1$ and $\mathbf{D}_2$, are orthogonal to each other. To obtain $n_1, n_2, \mathbf{D}_1, \mathbf{D}_2, \mathbf{E}_1, \mathbf{E}_2, \mathbf{H}_1, \mathbf{H}_2$ and all other properties of these two propagating waves, we use the method of index ellipsoid. The equation of the index ellipsoid is given by

$$\frac{x^2}{\varepsilon_x'} + \frac{y^2}{\varepsilon_y'} + \frac{z^2}{\varepsilon_z'} = 1, \tag{2.12.32}$$

where

$$\varepsilon_x' = \frac{\varepsilon_x}{\varepsilon_0}; \qquad \varepsilon_y' = \frac{\varepsilon_y}{\varepsilon_0} \quad \text{and} \quad \varepsilon_x' = \frac{\varepsilon_z}{\varepsilon_0}. \tag{2.12.33}$$

Equation (2.12.32) is plotted in Fig. 2.12.8. To obtain the values of n_1 and n_2 for a particular direction of propagation $\mathbf{i}_k$, find the plane passing through the origin of the ellipsoid and which is perpendicular to $\mathbf{i}_k$. The intersection of this

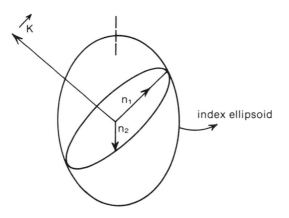

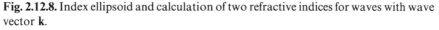

Fig. 2.12.8. Index ellipsoid and calculation of two refractive indices for waves with wave vector **k**.

plane and the index ellipsoid give us an ellipse. The two major axes of this ellipse correspond to $2n_1$ and $2n_2$, respectively. The corresponding D_1 and D_2 are parallel to these major axes of the ellipse. To calculate $\mathbf{E}_1$ and $\mathbf{E}_2$, we use the relationship

$$\mathbf{D} = n^2\varepsilon_0[\mathbf{E} - \mathbf{i}_k(\mathbf{i}_k \cdot \mathbf{E})]. \tag{2.12.34}$$

The magnetic field vector, **H**, can then be obtained from

$$\mathbf{H} = \frac{n}{\mu c}\mathbf{i}_k \times \mathbf{E}. \tag{2.12.35}$$

The phase velocity and group velocity are not collinear. Actually, they are given by

$$v_g = v_p \cos \psi, \tag{2.12.36}$$

where ψ is the angle between $\mathbf{i}_k$ and **P**. Group velocity is in a direction normal to the E field as shown in Fig. 2.12.9.

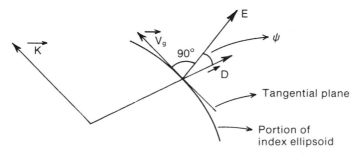

Fig. 2.12.9. The E field, D field, and $\mathbf{V}_g$ for a wave with wave vector **k**.

2.12.6. Double Refraction and Polarizers

Let us consider the problem of incident light on an anisotropic crystal, as shown in the Fig. 2.12.10. As k_t, the transmitted wave vector, can have two values given by the wave-vector surface we have, in general, two transmitted beams given by

$$\sin \theta = n_1(\theta_{t1}) \sin \theta_{t1}, \tag{2.12.37}$$

$$\sin \theta = n_2(\theta_{t2}) \sin \theta_{t2}. \tag{2.12.38}$$

Note that n_1 and n_2 are functions of the direction of propagation itself. Thus, in general, the usual Snell's law does not hold good. For a biaxial crystal, we have to solve numerically eqs. (2.12.37) and (2.12.38) to obtain θ_{t1} and θ_{t2}. This is done using the wave-vector surface in the plane of incidence, as shown in Fig. 2.12.11.

For a uniaxial crystal the problem becomes somewhat simpler. For this case

$$\varepsilon'_x = \varepsilon'_y = n_o^2, \tag{2.12.39}$$

and

$$\varepsilon'_z = n_e^2, \tag{2.12.40}$$

where n_o and n_e are defined to be the ordinary and extraordinary refractive index, respectively. If $n_o < n_e$, it is called positive uniaxial; otherwise, if $n_o > n_e$, it is called negative uniaxial. The wave-vector surface corresponding to n_o is a sphere, whereas the one corresponding n_e is a spheriod. Some wave-vector surface cross sections are shown in Fig. 2.12.12. Because the ordinary wave-vector surface is a sphere, it has a constant value of refractive index n_o, and the transmitted ray for this case is called the ordinary ray and it obeys the

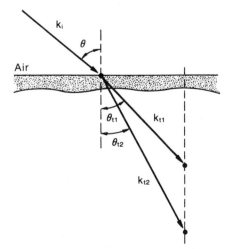

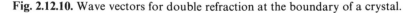

Fig. 2.12.10. Wave vectors for double refraction at the boundary of a crystal.

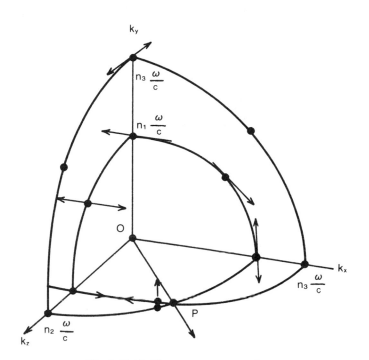

Fig. 2.12.11. The wave-vector surface.

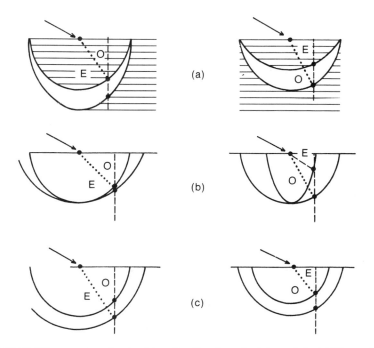

Fig. 2.12.12. Wave vectors for double refraction in uniaxial crystals. (a) The optic axis parallel to the boundary and parallel to the plane of incidence. (b) The optic axis perpendicular to the boundary and parallel to the plane of incidence. (c) The optic axis parallel to the boundary and perpendicular to the plane of incidence.

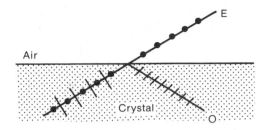

Fig. 2.12.13. Refraction of an unpolarized beam propagating through a negative uniaxial crystal and incident on an air interface.

normal Snell's law. However, the extraordinary ray can be obtained numerically. Some cases of double refraction for uniaxial crystals are shown in Fig. 2.12.12. It is to be noted that the two refracted beams are polarized orthogonal to each other, as discussed previously. This property and the fact that $n_1(\theta_1) \neq n_2(\theta_2)$ is used to make practical polarizers.

Let us consider the case of a negative uniaxial crystal through which an unpolarized beam is propagating and is incident on an air interface, as shown in Fig. 2.12.13. We note that the total internal reflection for the ordinary ray takes place beyond the critical angle, θ_{co}, given by

$$n_o \sin \theta_{co} = 1. \tag{2.12.41}$$

However, the critical angle for the extraordinary ray, θ_{ce}, is larger than θ_{co} as

$$n_e \sin \theta_{ce} = 1 \quad \text{and} \quad n_o > n_e. \tag{2.12.42}$$

Thus, if the incident light has its angle of incidence, θ, between θ_{co} and θ_{ce}, the ordinary ray will be totally reflected whereas part of the extraordinary ray will be transmitted. The transmitted wave is thus plane polarized even though the incident light is unpolarized. Using this property we can fabricate polarizers such as the Glan prism and the Nicol prism.

The Glan prism consists of two identical calcite prisms mounted as shown in Fig. 2.12.14. The space between the two prisms contains either air or a transparent material, such that the ordinary ray suffers total internal reflection. The output consists of only the extraordinary ray which is linearly

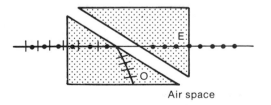

Air space

Fig. 2.12.14. Configuration for the Glan prism.

Canada balsam cement

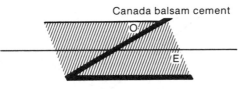

Fig. 2.12.15. (a) Separation of the extraordinary and ordinary rays at the boundary of a crystal in the case of internal refraction. (b) Construction of the Glan polarizing prism. (c) The Nicol prism.

polarized. Note that the prisms have their optic axes parallel to the corner edges. The Nicol prism, shown in Fig. 2.12.15, is in the form of a rhomb and the principle of operation is very similar to the Glan prism.

The Glan and Nicol prisms use total internal reflection of the ordinary ray as the main function of the polarizer, whereas the Wallaston, Rochon, and Sevarmont prisms use the separation between the extraordinary and ordinary rays. Because angles of refraction are different for the two rays, as shown in Fig. 2.12.16(a), (b) and (c), the extraordinary and ordinary rays will exit the prisms separately and thus can be used as polarizers.

Finally, in Fig. 2.12.17, the most general problem of double refraction is shown, where light is incident from media 1 to media 2 where both media are anisotropic. In general, there will be two refracted and two reflected beams, the directions of which can be determined using the wave-vector surfaces for the two media. However, to calculate the amount of light reflected or refracted in the individual beams, we need to apply the boundary conditions discussed earlier.

For light which does not propagate along the optic axis, there are two fixed polarizations of light which do propagate. As discussed before, these polarizations are determined by the axes of the ellipse, formed by the normal to the direction of propagation and the index ellipsoid. However, if the propagation direction is along the optical axis, then the ellipse is really a circle, as shown in Fig. 2.12.18, and thus all the polarizations are possible. Thus, if a narrow unpolarized light beam is incident normally on a plane parallel crystalline plate with the normal along the optic axis, then the light ray inside

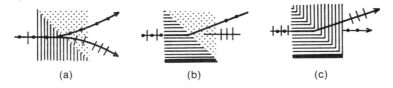

(a)   (b)   (c)

Fig. 2.12.16. Three types of prisms for separating unpolarized light into two divergent orthogonally polarized beams: (a) the Wollaston prism, (b) the Rochon prism, and (c) the Sevarmont prism. All prisms shown are made with uniaxial positive material (quartz).

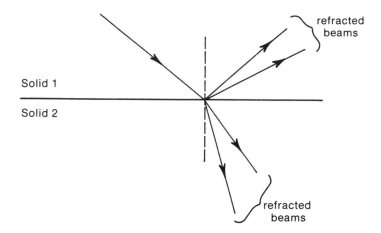

Fig. 2.12.17. Double refraction and reflection for two anisotropic solids.

the plate forms a hollow cone and emerges as a hollow cylinder. The light at each point on the cylinder is linearly polarized. Note that the cone and cylinder are formed as the direction of the group velocity or ray is different from the direction of the phase velocity. This formation of cone and cylinder is known as internal conical refraction.

2.12.7. The Electro-Optic Effect

The application of an electric field changes the dielectric tensor of a material, however small. The electro-optic effect is, in general, defined through the change in the refractive index rather than through the change in the dielectric constant, because of the usefulness of the index-ellipsoid method in solving problems. Thus the change in the index ellipsoid, due to an applied electric field, is generally written as

$$\Delta\left(\frac{1}{n^2}\right)_{ij} = r_{ijq}E_q + R_{ijpq}E_pE_q, \qquad i, j, p, q \rightarrow x, y, z, \qquad (2.12.43)$$

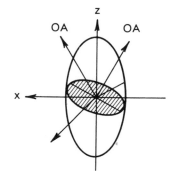

Fig. 2.12.18. Index ellipsoid.

where r_{ijq} is the linear (Pockels) electro-optic coefficient and R_{ijpq} is the quadratic (Kerr) electro-optic coefficient. Note that the summation convention for the repeated indices is used in (2.12.43). The summation convention means that any time two of the same variables occur in the subscript twice, it must be summed over x, y, and z. For a centrosymmetric crystal, the Pockels coefficients go to zero. Thus, for an isotropic material, there is only the Kerr effect.

The index ellipsoid, given by (2.12.32), can be written as

$$\frac{x^2}{n_x^2} + \frac{y^2}{n_y^2} + \frac{z^2}{n_z^2} = 1. \tag{2.12.44}$$

In the presence of the electro-optic effect, the above equation is modified to

$$x^2\left[\frac{1}{n_x^2} + \Delta\left(\frac{1}{n^2}\right)_{xx}\right] + y^2\left[\frac{1}{n_y^2} + \Delta\left(\frac{1}{n^2}\right)_{yy}\right] + z^2\left[\frac{1}{n_z^2} + \Delta\left(\frac{1}{n^2}\right)_{zz}\right]$$
$$+ xy\left[\Delta\left(\frac{1}{n^2}\right)_{xy} + \Delta\left(\frac{1}{n^2}\right)_{yx}\right] + xz\left[\Delta\left(\frac{1}{n^2}\right)_{xz} + \Delta\left(\frac{1}{n^2}\right)_{zx}\right]$$
$$+ yz\left[\Delta\left(\frac{1}{n^2}\right)_{yz} + \Delta\left(\frac{1}{n^2}\right)_{zy}\right] = 1. \tag{2.12.45}$$

It is customary to rewrite (2.12.45) as

$$x_1^2\left[\frac{1}{n_1^2} + \Delta\left(\frac{1}{n^2}\right)_1\right] + x_2^2\left[\frac{1}{n_2^2} + \Delta\left(\frac{1}{n^2}\right)_2\right] + x_3^2\left[\frac{1}{n_3^2} + \Delta\left(\frac{1}{n^2}\right)_3\right]$$
$$+ 2x_2x_3\Delta\left(\frac{1}{n^2}\right)_4 + 2x_1x_3\Delta\left(\frac{1}{n^2}\right)_5 + 2x_1x_2\Delta\left(\frac{1}{n^2}\right)_6 = 1, \tag{2.12.46}$$

where $x \to x_1$, $y \to x_2$, $z \to x_3$, $xx \to 1$, $yy \to 2$, $zz \to 3$, $yz \to 4$, $zx \to 5$, and $xy \to 6$. In this notation, (2.12.43) can be rewritten as

$$\Delta\left(\frac{1}{n^2}\right)_i = r_{ij}E_j + R_{ipq}E_pE_q. \tag{2.12.47}$$

The Pockels coefficients are uniquely determined by the point group symmetry of the crystal.

It is of interest to consider some typical examples. For KH_2PO_4, considering only the linear term, (2.12.47) becomes

$$\begin{vmatrix} \Delta(1/n^2)_1 \\ \Delta(1/n^2)_2 \\ \Delta(1/n^2)_3 \\ \Delta(1/n^2)_4 \\ \Delta(1/n^2)_5 \\ \Delta(1/n^2)_6 \end{vmatrix} = \begin{vmatrix} 0 & 0 & 0 \\ 0 & 0 & 0 \\ 0 & 0 & 0 \\ r_{41} & 0 & 0 \\ 0 & r_{41} & 0 \\ 0 & 0 & r_{63} \end{vmatrix} \begin{vmatrix} E_1 \\ E_2 \\ E_3 \end{vmatrix}. \tag{2.12.48}$$

We have observed that the index-ellipsoid equation (2.12.44), defined with the principle axes as the coordinate axes, becomes (2.12.46) under the influence of the electro-optic effect. Thus the new principle axes must be obtained to solve the problem.

Let us consider an example where an electric field is applied in the z direction $(E_3 = E_z)$ for KH_2PO_4. For this case, the index ellipsoid equation becomes

$$\frac{x^2}{n_0^2} + \frac{y^2}{n_0^2} + \frac{z^2}{n_e^2} + 2r_{63}E_z xy = 1. \tag{2.12.49}$$

Note that KH_2PO_4 is positive uniaxial.
The new directions for the major axes are given by

$$x' = \frac{x}{\sqrt{2}} + \frac{y}{\sqrt{2}} = x \cos 45° + y \sin 45°,$$

$$y' = \frac{x}{\sqrt{2}} - \frac{y}{\sqrt{2}} = x \sin 45° - y \cos 45°, \tag{2.12.50}$$

$$z' = z.$$

Note that the new principle axes, x' and y', are at an angle of 45° with respect to the crystal axes. In the new system of axes, (X'), the index-ellipsoid equation is given by

$$x'^2 \left(\frac{1}{n_0^2} + r_{63}E_z \right) + y'^2 \left(\frac{1}{n_0^2} - r_{63}E_z \right) + \frac{z'^2}{n_e^2} = 1, \tag{2.12.51}$$

or

$$\frac{x'^2}{n_{x'}^2} + \frac{y'^2}{n_{y'}^2} + \frac{z'^2}{n_e^2} = 1. \tag{2.12.52}$$

Let us consider the problem where light with polarization in the x direction, and propagating in the z direction, is incident on a plate of KH_2PO_4 having thickness l as shown in Fig. 2.12.19. We are interested in the output light. The input light can be decomposed into two components given by

$$\mathbf{E} = E_{x'}\mathbf{i}_{x'} + E_{y'}\mathbf{i}_{y'}, \tag{2.12.53}$$

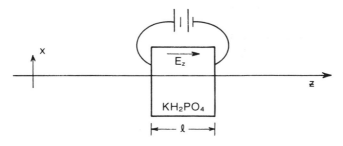

Fig. 2.12.19. Configuration for electro-optic effect.

where

$$E_{x'} = \frac{A}{\sqrt{2}} e^{j(\omega t - (\omega/c)n_{x'}z)}, \tag{2.12.54}$$

and

$$E_{y'} = \frac{A}{\sqrt{2}} e^{j(\omega t - (\omega/c)n_{y'}z)}, \tag{2.12.55}$$

where A is the input amplitude.

Note that

$$\frac{1}{n_{x'}^2} = \frac{1}{n_0^2} + r_{63}E_z, \tag{2.12.56}$$

or

$$n_{x'} \approx n_0 - \frac{n_0^3}{2} r_{63}E_z, \tag{2.12.57}$$

provided

$$r_{63}E_z \ll n_0^{-2}.$$

Similarly,

$$n_{y'} \approx n_0 + \frac{n_0^3}{2} r_{63}E_z. \tag{2.12.58}$$

The phase difference between $E'_x(z = l)$ and $E'_y(z = l)$ is given by

$$\Gamma = \frac{\omega}{c}[n_{y'} - n_{x'}]E_z l$$

$$= \frac{\omega}{c} n_0^3 r_{63} V, \tag{2.12.59}$$

where $V = E_z l$ is the applied voltage across the crystal. If we define the voltage required to change the phase by π as V_π, often referred to as the half-wave voltage, then

$$V_\pi = \frac{\pi c}{\omega} \frac{1}{n_0^3 r_{63}}. \tag{2.12.60}$$

Thus (2.12.59) can be rewritten as

$$\Gamma = \frac{\pi V}{V_\pi}. \tag{2.12.61}$$

2.12.8. The Acousto-Optic Effect

Acoustic, elastic, or ultrasonic waves propagating through a solid or a fluid cause periodic perturbations of the refractive index. Light propagating through this periodic grating is diffracted. To analyze properly the acousto-optic interaction in solids, we must start with the index ellipsoid and include its perturbation due to the sonic field, to obtain finally the change in the refractive

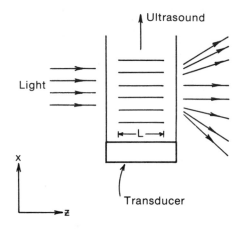

Fig. 2.12.20. Acousto-optic interaction in an isotropic solid.

index. This situation is very similar to the one mentioned under the electro-optic effect and will be discussed later.

Before we consider the more complex case, let us first consider the simple case shown in Fig. 2.12.20, where the ultrasound is propagating in the z direction and the material is isotropic. Thus, for this case, we can write the ultrasonically induced change in the refractive index, $\Delta n(x, t)$, as

$$\Delta n(x, t) = \Delta n_0 \sin(\omega_s t - K_s z + \delta), \tag{2.12.62}$$

where δ is a constant phase difference and Δn_0, the change in refractive index, is given by

$$\Delta n_0 \approx \frac{n_0^3}{2} ps, \tag{2.12.63}$$

and K_s is the wave vector of ultrasound, ω_s is the radian frequency of ultrasound, p is the relevant photoelastic constant, and s is the compressional strain.

If we consider the width of the ultrasonic beam, L, rather small, then we can consider that the effect of the ultrasonic beam is to form a very thin equivalent phase grating whose phase dependence is given by

$$\Delta\varphi(x, t) = \frac{2\pi}{\lambda_0} (\Delta n_0 L) \sin(\omega_s t - K_s x + \delta). \tag{2.12.64}$$

If the incident light is represented by an electric field, E_i, given by

$$\mathbf{E}_i = \mathbf{E}_0 e^{j(\omega_l t - K_l z)}, \tag{2.12.65}$$

then the output light is given by

$$\mathbf{E}_{\text{out}} = \mathbf{E}_i e^{j\Delta\varphi(x, t)}$$
$$= \mathbf{E}_0 e^{j[\omega_l t - K_l z + \Delta\varphi(x, t)]}. \tag{2.12.66}$$

Equation (2.12.66) can be expanded in Bessel functions given by

$$e^{j\Delta\varphi(x,t)} = \sum_{q=-\infty}^{+\infty} J_q\left(\frac{2\pi}{\lambda_0}\Delta n_0 L\right) e^{j[q(\omega_s t - K_s x + q]}, \tag{2.12.67}$$

thus E_{out} can be written as

$$\begin{aligned}
\mathbf{E}_{out} &= \mathbf{E}_0 \sum_{q=-\infty}^{+\infty} J_q\left(\frac{2\pi}{\lambda_0}\Delta n_0 L\right) e^{j(\omega_l + q\omega_s t)} : e^{jq\delta} e^{j(-k_l z - qK_s x)} \\
&= \mathbf{E}_1 + \mathbf{E}_2 + \mathbf{E}_3 + \cdots \\
&= \sum_{-\infty}^{+\infty} \mathbf{E}_q, \tag{2.12.68}
\end{aligned}$$

where

$$\mathbf{E}_q = \mathbf{E}_0 J_q\left(\frac{2\pi}{\lambda}\Delta n_0 L\right) e^{j(\omega_l + q\omega_s)t} \cdot e^{jq\delta} e^{-j(K_l z + qK_s x)}. \tag{2.12.69}$$

For $\mathbf{E}_q$ we note that its amplitude is proportional to the incident electric field, as well as to the qth-order Bessel function. The frequency of the qth-order diffracted light is given by

$$\omega_q = \omega_l + q\omega_s \tag{2.12.70}$$

and the wave vector is given by

$$\mathbf{K}_q = \mathbf{i}_z K_l + \mathbf{i}_x qK_s = \mathbf{i}_K K_l. \tag{2.12.71}$$

From the wave vector diagram, shown in Fig. 2.12.21, we obtain

$$K_l \sin \theta_q = qK_s,$$

or

$$\sin \theta_q = q\left(\frac{\lambda_l}{\lambda_s}\right). \tag{2.12.72}$$

Note that in (2.12.72) we have assumed $\omega_s \ll \omega_l$, so that $|K_q| \sim |K_l|$. The polarization of the light is also maintained at diffraction. It is of interest to consider the first-order diffraction given by

$$\mathbf{E}_{\pm 1} = \mathbf{E}_0 J_1\left(\frac{2\pi}{\lambda}\Delta n_0 L\right) e^{j(\omega_l \pm \omega_s t)} e^{-jK_l[\mathbf{i}_z \cos \theta_{\pm 1} z + \mathbf{i}_x \sin \theta_{\pm 1} x]}. \tag{2.12.73}$$

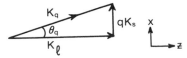

Fig. 2.12.21. Wave vector matching for acousto-optic interaction.

The ratio of intensities between the zeroth order and the first order are given by

$$\frac{I_1}{I_0} = \frac{J_1((2\pi/\lambda_0)\Delta n_0 L)}{J_0((2\pi/\lambda_0)\Delta n_0 L)}. \tag{2.12.74}$$

Thus for

$$\frac{2\pi}{\lambda_0}\Delta n_0 L \ll 1,$$

$$\frac{I_1}{I_0} \simeq \left(\frac{2\pi}{\lambda_0}\Delta n_0 L\right)^2$$

$$= \left(\frac{2\pi}{\lambda_0}\frac{n_0^3}{2}psL\right)^2$$

$$= \pi^2\left(\frac{L}{\lambda_0}\right)^2 (n_0^6 p^2 s^2). \tag{2.12.75}$$

Noting that the acoustic power, p_A, is given by

$$p_A = \tfrac{1}{2}\rho v_s^3 s^2, \tag{2.12.76}$$

where ρ is the density, we obtain

$$\frac{I_1}{I_0} = \pi^2\left(\frac{L}{\lambda_0}\right)^2 n_0^6 p^2 \frac{2p_A}{\rho v_s^3}$$

$$= (2\pi^2)\left(\frac{L}{\lambda_0}\right)^2 \left(\frac{n_0^6 p^2}{\rho v_s^3}\right) p_A. \tag{2.12.77}$$

The total acoustic power, p_{tot}, is given by

$$p_{tot} = p_A(LH), \tag{2.12.78}$$

where H is the other dimension of the transducer. Thus, (2.12.78) can be rewritten as

$$\frac{I_1}{I_0} = (2\pi^2)\left(\frac{L}{H}\right)\left(\frac{n^6 p^2}{\rho v_s^3}\right)\left(\frac{p_{tot}}{\lambda^2}\right)$$

$$= (2\pi^2)\left(\frac{L}{H}\right)\left(\frac{p_{tot}}{\lambda^2}\right) M_2, \tag{2.12.79}$$

where M_2 is defined as a figure of merit

$$M_2 = \frac{n^6 p^2}{\rho v_s^3}. \tag{2.12.80}$$

In the thin grating approximation discussed above, light is diffracted in different orders, as expected from a phase grating. However, as the grating is moving with the velocity of sound, there is also a Doppler shift in the frequency of light for each order. Thin grating diffraction is also known as Raman–Nath diffraction. If the interaction length, L, is large, then we can consider it to

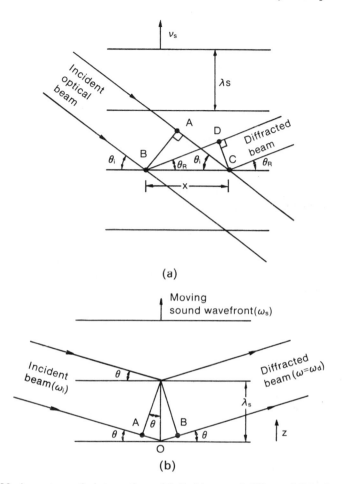

Fig. 2.12.22. Acousto-optic interaction with incident and diffracted light beams. The horizontal lines separated by the acoustic wavelength λ_s represent the moving sound beam.

consist of many thin gratings. For this thick grating case, there is multiple diffraction at every plane and only that one which is phase matched can be diffracted. This is also understood from Fig. 2.12.22(a) where the phase grating is shown. Note that this grating is moving with a much slower velocity compared to light. If we consider the incident wavefront AB and the diffracted waveform CD, then the optical path difference $AC - BD$ is given by

$$AC - BD = x(\cos \theta_i - \cos \theta_r). \qquad (2.12.81)$$

For constructive interference, the path difference must be an integer multiple of λ or

$$x(\cos \theta_i - \cos \theta_r) = m\lambda = \frac{m\lambda_0}{n}, \qquad (2.12.82)$$

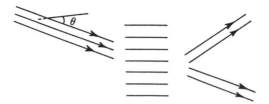

Fig. 2.12.23. Bragg diffraction geometry.

where m is an integer. The above equation can be satisfied for all values of x if the following condition is satisfied:

$$\theta_i = \theta_r. \tag{2.12.83}$$

We also note that the diffraction from planes parallel to the acoustic phase-front and separated by λ_s must add up. Thus, the path difference $(AO + OB)$ in Fig. 2.12.22(b) must equal λ/n. Thus,

$$2\lambda_s \sin \theta_B = \frac{\lambda_0}{n} = \lambda_l, \tag{2.12.84}$$

where θ_B is the Bragg angle. Thus, in the Bragg diffraction case, as shown in Fig. 2.12.23, light is incident at the Bragg angle, and only the first order of light is diffracted out at the Bragg angle also. The dividing region between the Raman–Nath and the Bragg regimes is determined by a quantity Q defined as

$$Q = 2\pi \left(\frac{L\lambda}{\lambda_s^2} \right) n_0. \tag{2.12.85}$$

For $Q \ll 1$ and $\theta_i \approx 0$, we obtain the Raman–Nath regime; for $Q \gg 1$ and $\theta_i = \theta_B$, we obtain the Bragg regime. In the Bragg regime, the diffracted light intensity is given by

$$I_1(L) = I_0 \sin^2 \left[\frac{\pi(\Delta n)_0 L}{\lambda} \right], \tag{2.12.86}$$

where I_0 is the incident power. Thus, with Bragg diffraction, it is possible to diffract 100% of the incident light in the first order.

Bragg diffraction can also be understood from the phonon–photon interaction picture. The electromagnetic wave with angular frequency ω and wave vector K can be considered as consisting of photons, with energy and momentum given by

$$\text{energy} = \hbar\omega = hf, \tag{2.12.87}$$

$$\text{momentum} = \hbar K = \frac{h}{\lambda} \mathbf{i}_K, \tag{2.12.88}$$

where h is the Planck constant.

Similarly, for the acoustic wave, we have phonons with energy and momentum given by

$$\text{energy} = \hbar\omega_s = hf_s, \tag{2.12.89}$$

$$\text{momentum} = \hbar\mathbf{K}_s = \frac{h}{\lambda_s}\mathbf{i}_{K_s}. \tag{2.12.90}$$

If we considers acousto-optic interaction as photon–phonon interaction, then due to the conservation of energy, we have

$$\hbar\omega_i = \hbar\omega_d - \hbar\omega_s, \tag{2.12.91}$$

or

$$f_i = f_d - f_s. \tag{2.12.92}$$

Note that, in general, $f_s \ll f_d, f_i$. Similarly, to satisfy the momentum conservation law, we have

$$\hbar\mathbf{K}_i = \hbar\mathbf{K}_d - \hbar\mathbf{K}_s, \tag{2.12.93}$$

or

$$\mathbf{K}_i = \mathbf{K}_d - \mathbf{K}_s. \tag{2.12.94}$$

Note that

$$|K_i| = \frac{2\pi}{\lambda_0}n_i; \qquad |K_d| = \frac{2\pi}{\lambda_0}n_d\left(\frac{f_i + f_s}{f_i}\right) \simeq \frac{2\pi}{\lambda_0}n_d; \tag{2.12.95}$$

$$|K_s| = 2\pi\frac{f_s}{v_s}, \tag{2.12.96}$$

where n_i and n_d are the refractive indices corresponding to the incident and diffracted wave, and v_s is the velocity of sound. Thus (2.12.94) becomes

$$n_i\mathbf{i}_{K_i} = n_d\mathbf{i}_{K_d} - \frac{f_s\lambda_0}{v_s}\mathbf{i}_{K_s}. \tag{2.12.97}$$

If we choose the angle between the normal to $\mathbf{i}_{K_s}$ and $\mathbf{i}_{K_i}$ as θ_i, and similarly for $\mathbf{i}_{K_d}$ as θ_d as shown in Fig. 2.12.24, we have, equating the parallel and normal components,

$$n_i \sin\theta_i = n_d \sin\theta_d - \frac{f_s\lambda_0}{v_s}, \tag{2.12.98}$$

$$n_i \cos\theta_i = n_d \cos\theta_d. \tag{2.12.99}$$

Note that $\mathbf{K}_i$, $\mathbf{K}_d$, and $\mathbf{K}_s$ have to be coplanar.

Solving for $\sin\theta_i$ and $\sin\theta_d$ from (2.12.98) and (2.12.99), we obtain

$$\sin\theta_i = \frac{1}{2n_i}\frac{\lambda_0 f_s}{v_s}\left[1 + \left(\frac{v_s}{\lambda_0 f}\right)^2(n_i^2 - n_d^2)\right], \tag{2.12.100}$$

$$\sin\theta_d = \frac{1}{2n_d}\frac{\lambda_0 f_s}{v_s}\left[1 - \left(\frac{v_s}{\lambda_0 f}\right)^2(n_i^2 - n_d^2)\right]. \tag{2.12.101}$$

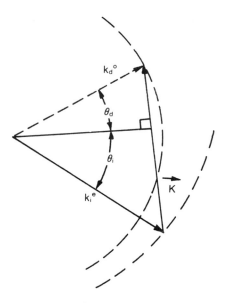

Fig. 2.12.24. Wave vector construction, describing Bragg diffraction, in a positive uniaxial crystal when the incident optical wave is extraordinarily polarized.

For the isotropic case, when $n_i = n_d$, we have

$$\sin \theta_B = \sin \theta_i = \sin \theta_d = \frac{\lambda_1}{2\lambda_s}, \tag{2.12.102}$$

where $\lambda_1 = \lambda_0/n$ is the wavelength of the light in the material itself. The vector diagram for this case is shown in Fig. 2.12.25.

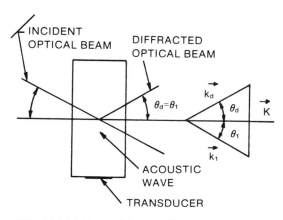

Fig. 2.12.25. Normal Bragg diffraction geometry.

So far we have not discussed how to calculate exactly the value of (Δn). To calculate this we note that, similar to the electro-optic effect, any elastic deformation causes change in the dielectric tensor. To a first approximation, the change in the index ellipsoid is denoted as

$$\Delta\left(\frac{1}{n^2}\right)_{ij} = p_{ijkl}S_{kl}, \qquad (2.12.103)$$

where p_{ijkl} are called the Pockel elasto-optic coefficients and S_{kl} are the strain components. For example, in an isotropic solid for compressional wave propagation along the x direction we have

$$\Delta\left(\frac{1}{u^2}\right) = p_{11}S_1x^2 + p_{12}S_1y^2 + p_{12}S_1z^2, \qquad (2.12.104)$$

where S_1 is the compressional wave and p_{11} and p_{12} are the Pockel coefficients.* For this case, if the incident light is polarized in the x direction, then

$$\Delta n \approx -\frac{n_0^3}{2}p_{11}S_1. \qquad (2.12.105)$$

For y- or z-polarized light

$$\Delta n \approx -\frac{n_0^3}{2}p_{12}S_1. \qquad (2.12.106)$$

In general, however, in an anisotropic solid, the analysis is quite complex.

2.12.9. Optical Activity and Magneto-Optics

Many crystals (e.g., quartz) have the property that they can change the plane of polarization for an incident linearly polarized light. This is known as optical activity. Optical activity can be physically explained by considering that, in an optically active crystal, right circularly polarized light and left circularly polarized light have different refractive indices n_R and n_l, respectively. For a thickness of the crystal, l, the polarization is rotated by an angle, θ, given by

$$\theta = (n_R - n_l)\frac{\pi l}{\lambda} = \delta l, \qquad (2.12.107)$$

where δ is called the specific rotary power. The optical activity of a crystal corresponds to an effective dielectric tensor given by

$$\varepsilon = \varepsilon_0 \begin{bmatrix} \varepsilon_{11} & j\varepsilon_{12} & 0 \\ -j\varepsilon_{12} & \varepsilon_{11} & 0 \\ 0 & 0 & \varepsilon_{33} \end{bmatrix}. \qquad (2.12.108)$$

---

* Note that compressed notation, discussed in connection with (2.12.47), has been used.

For this case

$$n_R = \sqrt{\varepsilon_{11} + \varepsilon_{12}},$$ (2.12.109)

$$n_l = \sqrt{\varepsilon_{11} - \varepsilon_{12}},$$ (2.12.110)

and

$$\delta = \frac{\varepsilon_{12}\pi}{n_0 \lambda},$$ (2.12.111)

where n_0 is the ordinary index of refraction.

Application of a magnetic field to material causes a change in the dielectric tensor and this in turn causes a change in the electromagnetic wave propagation. Basically, magneto-optics can be divided into three parts:

Faraday effect,
Voigt effect,
Kerr effect.

Application of a magnetic field to a crystal sometimes makes it optically active. This is generally known as the Faraday effect. For the applied magnetic induction, B, and the crystal thickness, l, the amount of rotation of the plane of polarization, θ, is

$$\theta = VBl,$$ (2.12.112)

where V is a constant known as the Verdet constant and depends on the material used. For this case, the specific rotary power is given by

$$\delta = VB.$$ (2.12.113)

The Faraday effect can easily be explained by the model of the movement of the electronic charge with the resultant change in the displacement vector.

If we include the effect of absorption in the medium, then a linearly polarized light, after propagating a distance l, will be elliptically polarized. This is generally known as the Voigt effect.

The Kerr effect refers to magneto-optic effects in reflection. This can be subdivided into three categories depending on the magnetic field directions, as shown in Fig. 2.12.26. These are:

Polar Kerr effect,
Longitudinal Kerr effect,
Equatorial Kerr effect.

The polar Kerr effect corresponds to the situation where the magnetic field H is in the crystal plane and in the plane of incidence, whereas for the equatorial Kerr effect, H is perpendicular to the plane of incidence.

References

[1] J.W. Goodman, *Introduction to Fourier Optics*, McGraw-Hill, 1968.
[2] R.J. Collier, C.B. Burckhardt, and L. H. Lin, *Optical Holography*, Academic Press, 1971.

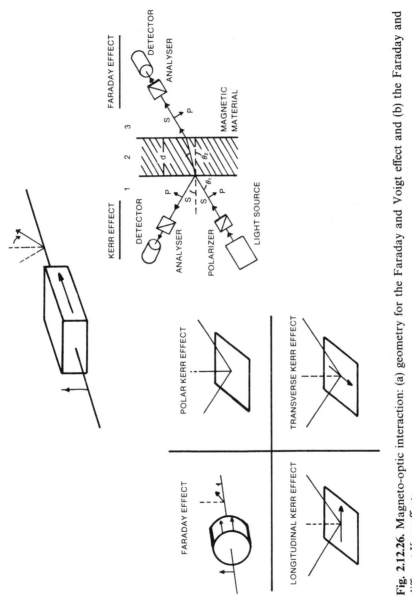

Fig. 2.12.26. Magneto-optic interaction: (a) geometry for the Faraday and Voigt effect and (b) the Faraday and different Kerr effects.

[3] A. Papoulis, *The Fourier Integral and its Applications*, McGraw-Hill, 1962.

[4] M. Born and E. Wolf, *Principles of Optics*, Pergamon Press, 1964.

[5] D. Gabor, A new microscope principle, *Nature*, **161**, 777, 1948.

[6] E.N. Leith and J. Upatneiks, Reconstructed wavefronts and communication theory, *J. Opt. Soc. Am.*, **52**, 1123, 1962.

[7] P. Das, *Optical Signal Processing*, Springer-Verlag, 1990.

[8] R. Gagliardi and S. Karp, *Optical Communications*, Wiley, 1976.

[9] M.V. Klein and R. Furtak, *Optics*, 2nd ed., Wiley, 1990.

[10] G.R. Fowles, *Introduction to Modern Optics*, Holt, Rinehart and Winston, 1968.

[11] A. Yariv, *Optical Electronics*, 2nd ed., Holt, Rinehart, and Winston, 1976.

[12] A.H. Cherin, *An Introduction to Optical Fibers*, McGraw-Hill, 1983.

[13] J.T. Verdeyen, *Laser Electronics*, Prentice-Hall, 1981.

[14] J.D. Jackson, *Classical Electrodynamics*, Wiley, 1967.

[15] M.K. Barnoski, *Introduction to Integrated Optics*, Plenum, 1973.

[16] M.J. Fraiser, A survey of magneto-optic effects, *IEEE Trans. Magnetics*, **MAG-4**, 152–161, 1968.

[17] W.R. Klein and B.D. Cook, *IEEE Trans. Sonics Ultrason.*, **SU-14**, 123–134.

[18] M. Gottlieb, C. Ireland, and J. M. Ley, *Electro-Optic and Acousto-Optic Scanning and Deflection*, Marcel Dekker, 1983.

PART III

Lasers

3.1. Introduction

Of all the light sources, the laser was discovered most recently, in 1960; however, it is probably the most important one. The word laser is an acronym for the following words:

Light Amplification by Stimulated Emission and Radiation.

Thus, as defined, the laser is an amplifier, but it is really an oscillator, as we shall soon see; however, it is obvious why we do not replace the word amplification by oscillation. As any good taxpayer knows, the Defense Department is not going to fund any research program whose title is LOSER.

Actually, the MASER. (M stands for microwave) was discovered by Townes and this was followed by the discovery of the laser. It was Maiman who, in 1960, first demonstrated experimentally a working laser using a ruby rod. Since then nearly every element in the periodic table has been found to lase. If the emitted radiation is infared, then a laser is sometimes called an IRASER.

There are several ways in which we can classify the different types of laser. First of all, it can be according to what material or element is responsible for the light amplification; thus, for example, He–Ne laser, the ruby laser, and the YAG laser. Some of these important lasers are listed in Table 3.1.1. The highest power that can be achieved is also an important quantity. Of course, for this quantity, the numbers change continuously as new research is performed to improve system performance. We also consider whether the laser operates in pulse mode or in a continuous (CW) fashion. Efficiency of the laser is also an important parameter. Of course, some of these lasers are especially suited for some specific application; these are also mentioned in Table 3.1.1. For example, both the CO_2 and the Nd–YAG lasers, being the highest power-output lasers, appear to be promising candidates for a possible laser fusion project.

It is of interest to point out that although peak power can be very high, because of the short duration of the pulse and the low-repetition rate, the total average power per second can be rather modest. For example, for a peak

Table 3.1.1.

Gas lasers			
Gas used	Wavelength range	Excitation	Comments
He–Ne	3.39 μm 0.6328 μm (0.543 μm) 1.15 μm	Electrical	Pulsed
CO_2	10.6 μm (9.17–10.91 μm) 9.6 μm	Electrical highest efficiency and power	CW/pulsed
Ar^+	0.4880 μm 458 514	Electrical	CW
Nitrogen	0.337 μm	High power	Pulsed
Far infrared different gases	40 μm–1.2 mm	CO_2 pumped	CW and pulsed
Kr^+	0.675 μm 0.647 μm 0.58 μm	Electrical	
Xe	2.02 μm	Electrical	
He–Se	24 visible lines		
He–Cd	0.442 μm 0.325 μm	Electrical	CW

Other lasers			
Type	Wavelength range	Excitation	Comments
Ruby	0.7 μm	Optical flash lamp	Pulsed
Nd^{3+} : YAG	1.06 μm 0.53 (using doubler)	Optical flash lamp	Pulsed and CW
Junction lasers			
GaAs	0.8 μm	Electrical	CW/pulsed
InP	0.9 μm		
InAs	3.1 μm		
GaInAsP	1.5 μm		
Organic dye dye solvent	0.217–0.96 μm	Optical laser pumped	CW/pulsed shortest pulse width $\sim 10^{-15}$
Excimer laser			
KrF	0.248 μm	Electrical/optical	Pulsed
KrCl	0.222 μm		
ArF	0.193 μm		
XeF	0.351 μm		
XeCl	0.308 μm		
Ar_2^*			
K_2^*			
Metal vapor laser gold and copper	0.628 μm UV–IR	Electrical	CW

Table 3.1.1 (*continued*)

	Other lasers		
Type	Wavelength range	Excitation	Comments
Free-electron laser	Infrared–ultraviolet	High energy electron beam in a magnetic field	Wavelength continuously variable
Glass doped with Neodymium	1.064 μm	CW	Pulsed, very high peak power
Alexandrite Ti: Sapphire	730–780	Optical	Continuously tunable solid state laser
I	1.3 μm	Chemical	CW
HF	2.6–3.5 μm	Chemical	CW
HCl	3.5–4.1 μm	Chemical	CW
DF	3.5–4.1 μm	Chemical	CW
HBr	4.0–4.7 μm	Chemical	CW
CO	4.9–5.8 μm	Chemical	CW
CO_2	10–11	Chemical	CW
X-Ray	150 Å–200 Å	Nuclear	

power of 10^{11} W, a pulse duration of 1 ns, and a repetition rate of one per second the average power is only 100 W. Some of the numbers which have been achieved for the highest power are astounding. For example, a pulse power of the order of $\sim 10^{12}$ W, using an Nd–glass laser, has been reported, and higher levels are projected. For the CO_2 laser, a CW power of hundreds of kilowatts has been reported, and even higher power is rumored (they are still classified for possible use as laser weapons).

Lasers or light oscillators need an active medium which can amplify light; this amplifier, using a suitable cavity for feedback, becomes an oscillator. This will be discussed in detail in the next and later sections. Here we want to point out that, in a strict sense, an amplifier is nothing but an energy converter from one form of energy to another. For example, consider an ordinary transistor or an integrated circuit amplifier as shown in Fig. 3.1.1. Say it can amplify input power from 10 mW to 1000 mW at radio frequency: Where is this extra power coming from? As any student of electronics knows, this power comes from the power supply which is d.c. power. Thus, the amplifier in a sense is converting d.c. power to r.f. power. The same happens in a laser amplifier, that is, every laser must have what we call a pump; this pump power through the

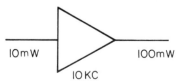

Fig. 3.1.1. Electronic amplifier.

laser is converted to light energy. For example, a typical He–Ne laser uses electric discharge as its pump, which comes from the d.c. or r.f. power supply. It turns out that there are different pumps or energy sources from which the laser can be built. These are:

(1) electrical (d.c. or r.f.),
(2) chemical,
(3) optical,
(4) thermal,
(5) nuclear,
(6) accelerator or electron-beam pumped.

Most of the common lasers are pumped either electrically or optically. A large number have been reported to be pumped chemically, and some are pumped thermally; nuclear pumping has also been achieved. The newest pumping mechanism is by acceleration of the electrons through the accelerators—sometimes called electron-beam pumped.

Before we go to the next section, to discuss some of the properties of amplifiers and oscillators, we should mention something about spontaneous are stimulated emission. All sources of light, except the laser, emit spontaneous emission only. This is like noise in electrical engineering terminology—highly fluctuating in amplitude, and if not in amplitude, certainly in phase; whereas stimulated emission is phase coherent—like that coming out of an electronic generator or oscillator.

3.2. Amplifier and Oscillator

As any student of electronics knows, an amplifier can always be made to oscillate if a suitable feedback is provided; this is shown in Fig. 3.2.1. The amplifier has a gain of A, thus

$$\frac{E_{\text{out}}}{E'_{\text{in}}} = A, \tag{3.2.1}$$

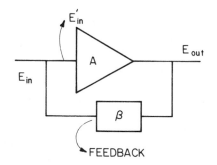

Fig. 3.2.1. Feedback oscillator.

where E'_{in} is the input to the amplifier and E_{out} is the output. However, E'_{in} is given by

$$E'_{in} = E_{in} + \beta E_{out}, \qquad (3.2.2)$$

where E_{in} is the actual input and β is the feedback factor, denoting the fraction of E_{out} fed back to the input. Thus

$$E_{out} = AE'_{in} = A(E_{in} - \beta E_{out}),$$

or

$$\frac{E_{out}}{E_{in}} = \frac{A}{1 - A\beta}. \qquad (3.2.3)$$

We define the condition for oscillation as

$$A\beta = 1, \qquad (3.2.4)$$

because in that case the effective gain goes to infinity, or we need not apply any input. A small noise signal will start and grow until a steady state situation arises. Although we have found the condition for oscillation we need to know how the frequency is determined and how much output power is produced. The oscillation frequency is determined from (3.2.4). All the frequencies which satisfy this equation can oscillate. However, both A and β are functions of frequency, thus by proper choice of their frequency dependence we can select the frequency of oscillation.

For output power determination, we need to consider the nonlinearity of the amplifier. As shown in Fig. 3.2.2, any physical amplifier must eventually saturate. Thus, as the amplitude increases, the gain decreases. Initially, if the value of $A\beta$ is greater than 1, the amplitude of oscillation goes on increasing; however, at a particular value of output power, (3.2.4) will again be satisfied. That will be the output power of the oscillator because at that value of the output a steady state is reached.

The reason for discussing this electronic oscillator is that a very similar situation also happens for the case of a light oscillator or laser. However, since in general, the high-frequency wave cannot be confined in electric wires, different kinds of feedback mechanisms are needed. As the frequency is raised to the microwave region, we use cavities for microwave generators, as shown

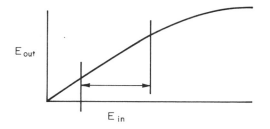

Fig. 3.2.2. Saturation of an amplifier.

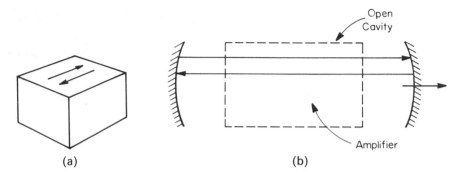

Fig. 3.2.3. (a) Microwave cavity and (b) open cavity.

in Fig. 3.2.3(a). These cavities are completely enclosed metal boxes in which a standing wave can be produced. In general, the dimensions of the cavity are of the order of a wavelength and thus, for $f = 30$ GHz, the dimensions are of the order of 1 cm. For masers, we also use this kind of cavity.

As we try to increase the frequency, f, of the electromagnetic wave oscillators from microwave to light, we realize that, using the closed cavity argument, we must have a cavity whose dimensions should be of the order of 1 μm, which is nearly impossible to fabricate or use. At the beginning of laser research, this was an important hurdle; however, scientists realized that there is no necessity for a closed cavity. We can use an open cavity which, for the case of lasers, is nothing other than two mirrors; and, actually, all lasers do use this form of cavity, as shown in Fig. 3.2.3(b). The lasing medium acting as the light amplifier is placed inside the cavity. As in the case of an ordinary electronic amplifier, if the condition $A\beta = 1$ is satisfied, then the laser can start from noise and oscillate at the proper frequency. The frequency and shape of the output beam is mostly dependent on this cavity. We are going to study the properties of the open laser cavity structure in the next few sections and then discuss the light amplifier. As we shall see, most of the properties of the laser output are related to the cavity structure. Thus if we are interested in the use of lasers for engineering applications, then these sections are most important. However, for a complete understanding of the laser itself, the physical process of light amplification has to be mastered.

Before we get involved in the cavity properties, we shall consider, in the next section, the simplest laser, the Fabry–Perot laser.

3.3. The Fabry–Perot Laser

The Fabry–Perot laser is a Fabry–Perot interferometer which includes an active medium between the two mirrors; of course, the active medium should be able to amplify light. Let this amplifier, which is a distributed amplifier rather than the discrete amplifier discussed in the last section, have an amplification constant, γ. This means that if the electric field at $x = 0$ is $E(x = 0)$, then the electric field at $x = x$ is given by

$$E(x) = E(0)e^{\gamma x}. \tag{3.3.1}$$

In the absence of any active medium, there will be losses due to scattering, absorbtion, etc., by the medium itself and for other reasons. Generally, all these losses can be lumped together and denoted by a loss factor, α. Thus, due to these loss mechanisms only, we shall have

$$E(x) = E(0)e^{-\alpha x}. \tag{3.3.2}$$

Thus, due to the loss and gain mechanisms in the medium, the wave will propagate with a propagation constant given by

$$k' = k_0 + \Delta k + j\gamma - j\alpha, \tag{3.3.3}$$

where k_0 is the propagation constant in the absence of the loss or gain mechanism. The reason γ and α appear with a "j" preceding is that ordinarily the wave propagation is given by

$$E(x, t) \propto e^{j(\omega t - k_0 x)}.$$

However, in the presence of the loss or gain mechanism it should be given by

$$E(x, t) = e^{-\alpha x} e^{\gamma x} e^{j(\omega t - k_0 x)},$$

or

$$E(x, t) = e^{j[\omega t (-k_0 + j\gamma - j\alpha)x]}. \tag{3.3.4}$$

The reason the Δk term appears in (3.3.3) is due to what is known as causality; that is, any physical process, if started at $t = 0$, cannot have an effect at any time $t < 0$. This physical restriction can be shown to impose a condition on the real and imaginary parts of the propagation constant. The condition is that the real and imaginary parts must be related by Hilbert transforms; or the term Δk must appear if α or γ is nonzero. For further details the reader is referred to reference [2]. In any case, the reader must have recognized, by this time, that the real part of k' determines the wavelength, and the imaginary part denotes the amplification or attenuation of the wave. Thus, in general, for complex k,

$$k = k_r + jk_{im}, \tag{3.3.5}$$

where $k_r = 2\pi/\lambda$ and k_{im} is the amplification factor if positive, or the attenuation factor for negative values.

Let us consider the input–output relations for the Fabry–Perot laser shown in Fig. 3.3.1. Let the incident light on the back side of mirror 1 be E_i.* The transmitted beam will be $t_1 E_i$ where the t's are the transmission coefficients and the r's are the reflection coefficients, as discussed in Section 2.10.4. This beam, when it arrives at mirror 2, is given by $t_1 E_1 e^{-jk'L}$, as shown schematically in Fig. 3.3.1(b). Here L is the length of the optical cavity. The

---

* Note that we consider, for this case, the incident angle to be zero (i.e., $r = 0$ for the Fabry–Perot interferometer discussed in Section 2.10.4).

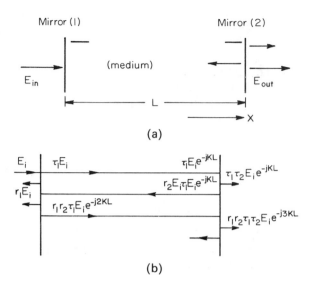

Fig. 3.3.1. A Fabry–Perot laser: (a) laser cavity dimensions and (b) reflected and transmitted beam components.

first beam transmitted from the output mirror, mirror 2, is given by

$$E_1 = t_2 t_1 E_i e^{-jk'L}. \qquad (3.3.6)$$

Part of this beam will be reflected from mirror 2, then reflected from mirror 1, and then transmitted through mirror 2 again. This second transmitted beam will be given by

$$E_2 = (t_2 t_1 E_1 e^{-jk'L})(r_1 r_2 e^{-jk'2L}). \qquad (3.3.7)$$

Similarly, there will be a third beam, and a fourth beam, and so on to infinity. The output transmitted beam can be written as

$$E_{\text{out}} = E_1 + E_2 + E_3 + \cdots$$
$$= t_1 t_2 e^{-jk'L} E_i [1 + \beta + \beta^2 + \cdots],$$

where

$$\beta = r_1 r_2 e^{-jk'2L} < 1. \qquad (3.3.8)$$

Thus

$$\frac{E_{\text{out}}}{E_i} = \frac{t_1 t_2 e^{-jk'L}}{1 - r_1 r_2 e^{-jk'2L}}. \qquad (3.3.9)$$

This equation is similar to (3.2.3), derived for the discrete amplifier case. Here the amplifier is distributed and the feedback is built into the device.

Thus, the condition for oscillation will be given by

$$r_1 r_2 e^{-jk'2L} = 1. \qquad (3.3.10)$$

This is a complex equation. Both the real and imaginary parts

$$r_1 r_2 e^{2L(\gamma-\alpha)} e^{-j2L(k_0+\Delta k)} = 1 \cdot e^{-j2\pi p}$$

(where p is any integer), must be individually satisfied. Equating the real part we obtain*

$$r_1 r_2 e^{2(\gamma-\alpha)L} = 1. \tag{3.3.11}$$

As we shall see later, in a laser the amplification factor, γ, will be dependent on how hard the laser medium is pumped. Eventually, as the pump power is slowly increased, a value of γ will be reached, called γ_{th}, the threshold value when the laser starts oscillating. This γ_{th} will be given by

$$r_1 r_2 e^{2(\gamma_{th}-\alpha)L} = 1,$$

or

$$\gamma_{th} = \alpha - \frac{1}{2L} \ln r_1 r_2. \tag{3.3.12}$$

Thus the value of γ must be at least γ_{th}, the threshold value for oscillations to start. If the value is larger, the waves grow and the amplifier reaches saturation due to some kind of nonlinearity. This lowers the value of γ and eventually an equilibrium value is reached at γ_{th}.

Equating the imaginary parts of (3.3.10), we obtain

$$e^{-j(k_0+\Delta k)2L} = e^{-j2p\pi}, \tag{3.3.13}$$

where p is an integer. The wavelength corresponding to p, called λ_p, is given by

$$\frac{2\pi}{\lambda_p} = (k_0 + \Delta k)_p = \frac{2p\pi}{2L}, \tag{3.3.14}$$

or

$$\lambda_p = \frac{2L}{p}.$$

The corresponding frequencies will be given by

$$f_p = \frac{v}{\lambda_p} = \left(\frac{v}{2L}\right) p, \qquad f_0 = \frac{v}{2L}, \tag{3.3.15}$$

where v is the velocity of light in the lasing medium. Thus we see that the laser cannot oscillate at all frequencies in order to satisfy the condition for oscillation. It can only oscillate at some discrete frequencies—at multiples of $(v/2L)$. For a cavity of 1 m long, we obtain

$$f_0 = \frac{v}{2L} \approx 150 \text{ MHz},$$

---

* Here the reflection coefficients have been assumed to be real and are denoted by r and not p.

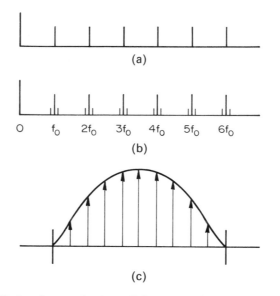

$$O \quad f_0 \quad 2f_0 \quad 3f_0 \quad 4f_0 \quad 5f_0 \quad 6f_0$$
(b)

(c)

Fig. 3.3.2. Oscillation frequencies for a Fabry–Perot laser: (a) possible longitudinal modes; (b) possible modes; and (c) actual modes including a frequency-dependent gain variable.

if the refractive index of the media, n, ≈ 1; these frequencies are shown schematically in Fig. 3.3.2(a). As we have taken only a one-dimensional case, we see only these so-called longitudinal modes. However, when we consider the three-dimensional case, we see that we really have three mode numbers, m, n, and p, and the frequencies will be denoted by f_{mnp}, and Fig. 3.3.2(a) will be modified as shown in Fig. 3.3.2(b). These are other modes and are generally known as transverse modes. We see from (3.3.15) that the laser can oscillate at any of these infinite frequencies. However, at what exact frequency or frequencies it does oscillate will be determined by the frequency characteristics of the light amplifier. For example, a typical case is shown in Fig. 3.3.2(c) where the envelope is the frequency variation of the gain constant "γ". For all the modes in which $\gamma \gg \gamma_{\text{th}}$, they can oscillate. It is of interest to note that the longitudinal mode numbers are very high. For example, for

$$L = 1 \text{ m,}$$

$$v = 3 \times 10^8 \text{ m/s,}$$

$$f_0 = \frac{3 \times 10^8 \text{ m/s}}{2(1) \text{ m}} = 150 \text{ MHz,}$$

$$\lambda = 0.5 \times 10^{-6} \quad \text{and} \quad f = 6 \times 10^{14},$$

$$f_p = f,$$

$$p = f_p/f_0,$$

or

$$p = 4 \times 10^6.$$

This is quite different from the case in microwave cavities where, to restrict oscillation to one mode only, we generally choose $p = 1$, 2, etc., or very low values of p. A price we pay for using the open cavity is that we, in general, get multimode oscillation, unless some other techniques are used to obtain a single mode. The techniques for the suppression of undesired modes will be discussed in later sections.

As discussed in Section 2.10.4, the transmission function, T, of a high-contrast Fabry–Perot interferometer, on which light is incident mostly parallel to the axis, is given by

$$T = \frac{T_{\max}}{1 + F(\delta - 2\pi p)/2)^2}, \tag{3.3.15a}$$

where $\delta = ((4\pi/\lambda)nd)$ and p is an integer. The expression for the contrast, F, is modified due to the presence of scattering losses and is given by

$$F = \frac{4R_1}{(1 - R_1)^2}, $$

where $R_1 = |\gamma_1\gamma_2 e^{-2\alpha L}|$. Actually, the diffraction losses are to be included in R_1 and will be discussed later.

Equation (3.3.15a) can be rewritten as

$$T(f) = \frac{T_{\max}}{1 + \pi^2 F(f/f_0 - p)^2}. \tag{3.3.15b}$$

If this is plotted in Fig. 3.3.2(d) as a function of frequency, it peaks around pf_0 where p is the longitudinal mode number. The width of the lines, Δf, is given by

$$\Delta f = \frac{f_0}{\mathscr{F}}, \tag{3.3.15c}$$

where $\mathscr{F} = (\pi/2)\sqrt{F}$ = finnesse. Thus we see that the laser cavity not only determines the resonant frequencies in which a laser can oscillate but also the bandwidth of the oscillations.

3.4. Laser Cavity

In the previous section we considered a simple laser cavity consisting of two parallel plane mirrors. However, in general, the mirrors can be curved having radii of curvature R_1 and R_2, as shown in Fig. 3.4.1(a). Actually, in between the mirrors we can have lenses, prisms, and other mirrors, as shown in Fig. 3.4.1(b). These prisms and mirrors, etc., are in the propagation path; however, in most cases, the feedback is given by the two end mirrors.

The properties of the cavity consisting of two mirrors should be studied

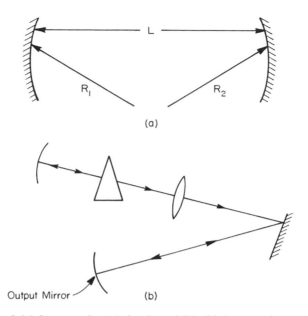

Fig. 3.4.1. Laser cavity: (a) simple and (b) with lenses, prisms, etc.

using the diffraction integral. This will be done in a later section. However, simple geometrical optics arguments, using the matrix method developed earlier in Part I, give important insight. Thus this will be discussed in the next section.

3.4.1. Cavity Stability Using Geometrical Optics

Consider a wavefront reflecting back and forth between two mirrors. This wavefront can be represented by a ray, as discussed in Section 1.1. If the ray remains within a finite transverse dimension of the cavity, as it bounces back and forth, we say the cavity is stable. However, if the beam "walks away" after many bounces or is not confined within a finite transverse direction, as shown in Fig. 3.4.2, we call it an unstable cavity. In general, stable cavities are preferred for laser construction because we, of course, have a finite amount of active medium. However, unstable cavities are of great interest in connection with very high gain lasing mediums, like that of the CO_2 laser, and will be discussed later.

To study the stability of the cavity using geometrical optics, we need to know the position of the ray from the optical axis as it bounces between the mirrors. In place of going back and forth between the mirrors, it is more convenient to "unfold" these beams, as shown in Fig. 3.4.3. We are interested only in the value of x as the rays bounce, in a sense it does not matter between the actual case and the equivalent case shown in Fig. 3.4.3. The propagation from one mirror to the other can be represented by the translational matrix

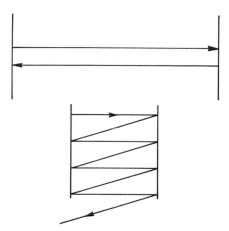

Fig. 3.4.2. Stable and unstable cavity.

$T(L)$. Similarly, reflection by the mirrors, for the equivalent case, can be represented by the two lenses having focal lengths given by

$$f_1 = \frac{R_1}{2},$$

and   (3.4.1)

$$f_2 = \frac{R_2}{2}.$$

(Note that for reflection we use an effective dielectric constant, as $n = -1$ in the lens designer's formula.) The infinite set of lenses is a periodic set with the unit cell, given by Fig. 3.4.4, repeating. It consists of a lens of focal length $2f_2$, a lens of focal length f_1 at a distance L from that lens, and another lens of focal length $2f_2$ located at a distance L from the lens having focal length f_1.

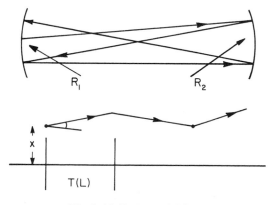

Fig. 3.4.3. Beam unfolding.

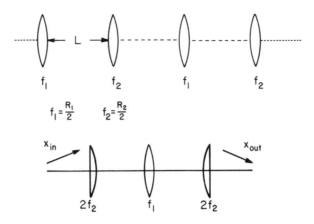

Fig. 3.4.4. Lens equivalent and unit cell.

This unit cell is symmetrical and that is the reason it is chosen in this fashion. Of course, other unit cells are equivalent and can be chosen.

The equivalent system matrix for the unit cell is given by

$$M = M(2f_2)T(L)M(f_1)T(L)M(2f_2)$$

$$= \begin{pmatrix} 1 & 0 \\ -1/2f_2 & 1 \end{pmatrix}\begin{pmatrix} 1 & L \\ 0 & 1 \end{pmatrix}\begin{pmatrix} 1 & 0 \\ -1/f_1 & 1 \end{pmatrix}\begin{pmatrix} 1 & L \\ 0 & 1 \end{pmatrix}\begin{pmatrix} 1 & 0 \\ -1/2f_2 & 1 \end{pmatrix}. \quad (3.4.2)$$

Remember that in the above equivalent system we use the column matrix for X as

$$\begin{pmatrix} x \\ \theta \end{pmatrix}.$$

However, in this case, it is convenient to represent a somewhat different one whose elements have the same dimensions. This one is

$$\begin{pmatrix} x \\ L\theta \end{pmatrix}.$$

Defining the equivalent matrix, M_{unit}, as

$$\begin{pmatrix} x_{out} \\ L\theta_{out} \end{pmatrix} = \begin{pmatrix} M_{unit} \end{pmatrix}\begin{pmatrix} x_{in} \\ L\theta_{in} \end{pmatrix}, \quad (3.4.3)$$

we can easily obtain, after some algebraic manipulation,

$$M_{unit} = \begin{pmatrix} 2g_1g_2 - 1 & 2g_2 \\ 2g_1(g_1g_2 - 1) & 2g_1g_2 - 1 \end{pmatrix}, \quad (3.4.4)$$

where

$$g_1 = 1 - \frac{L}{R_1} \quad \text{and} \quad g_2 = 1 - \frac{L}{R_2}. \quad (3.4.5)$$

Notice that all the elements of the matrix M_{unit} are dimensionless. Rewriting (3.4.3) in symbolic form as

$$X_{out} = M_{out} X_{in}, \qquad (3.4.6)$$

we see that the ray X_{in}, after passing through the nth unit cell, becomes

$$X_n = (M_{unit})^n X_{in}. \qquad (3.4.7)$$

We are interested in the value of x_n as n gets very large; however, ordinary matrix multiplication for such a large number of times is rather difficult to carry out. In place of that, it is advantageous to define an eigenray X_r corresponding to a eigenvalue λ_r given by the following equation:*

$$(M_{unit})X_r = \begin{pmatrix} \lambda_r & 0 \\ 0 & \lambda_r \end{pmatrix} X_r. \qquad (3.4.8)$$

Because for the eigenray we see that (3.4.7) simplifies to

$$X_{rn} = \begin{pmatrix} \lambda_r & 0 \\ 0 & \lambda_r \end{pmatrix}^n, \qquad X_r = \begin{pmatrix} \lambda_r^n & 0 \\ 0 & \lambda_r^n \end{pmatrix} X_r.$$

Of course, we have to decompose the input ray, X_{in}, into the eigenray components to complete the solution, as will be discussed later. However, it is of interest to study the equation further as follows:

$$\begin{pmatrix} 2g_1 g_2 - 1 - \lambda & 2g_2 \\ 2g_1(g_1 g_2 - 1) & 2g_1 g_2 - 1 - \lambda \end{pmatrix} \begin{pmatrix} x_r \\ L\theta_r \end{pmatrix} = 0, \qquad (3.4.9)$$

or

$$(2g_1 g_2 - 1 - \lambda)x_r + 2g_2(L\theta_r) = 0,$$

and

$$(3.4.10)$$

$$2g_1(g_1 g_2 - 1)x_r + (2g_1 g_2 - 1 - \lambda)(L\theta_r) = 0.$$

The above equations are only satisfied when the determinant of the matrix on the left-hand side of (3.4.9) is zero. From this condition we obtain, after some algebraic manipulation, the second-order equation in λ_r given by

$$\lambda_r^2 - 2\lambda_r(2g_1 g_2 - 1) + 1 = 0. \qquad (3.4.11)$$

Thus

$$\lambda_r = (2g_1 g_2 - 1) \pm \sqrt{4g_1 g_2(g_1 g_2 - 1)}. \qquad (3.4.12)$$

However, for $0 \le g_1 g_2 \le 1$, the quantity under the square root sign becomes imaginary and can be written as

$$\lambda_{r\pm} = (2g_1 g_2 - 1) \pm j\sqrt{|4g_1 g_2(1 - g_1 g_2)|}. \qquad (3.4.13)$$

Defining $\cos \alpha_r = (2g_1 g_2 - 1)$ we obtain

$$\sin \alpha_r = \sqrt{4g_1 g_2(1 - g_1 g_2)}. \qquad (3.4.14)$$

---

* Note that λ in this section denotes the eigenvalue and not the wavelength of light.

Substituting these in (3.4.13) we get

$$\lambda_{r\pm} = e^{\pm j\alpha_r}, \tag{3.4.15}$$

where

$$\alpha_r = \cos^{-1}(2g_1 g_2 - 1). \tag{3.4.16}$$

Using the two eigenvalues obtained above we derive the corresponding eigen-rays from (3.4.8). It can be shown that these eigenrays represent a complete set. Thus, the input ray can be written as

$$X_{\text{in}} = C_1 \begin{pmatrix} x_{r1} \\ L\alpha_{r1} \end{pmatrix} + C_2 \begin{pmatrix} x_{r2} \\ L\alpha_{r2} \end{pmatrix}, \tag{3.4.17}$$

where C_1 and C_2 are constants and

$$\begin{pmatrix} x_{r1} \\ L\alpha_{r1} \end{pmatrix} \quad \text{and} \quad \begin{pmatrix} x_{r2} \\ L\alpha_{r2} \end{pmatrix}$$

represent the two eigenrays. Thus,

$$X_n = C_1 \begin{pmatrix} \lambda_{r1} & 0 \\ 0 & \lambda_r \end{pmatrix}^n X_{r1} + C_2 \begin{pmatrix} \lambda_{r2} & 0 \\ 0 & \lambda_{r2} \end{pmatrix}^n X_{r2}, \tag{3.4.18}$$

or

$$x_n = C_1 \lambda_{r1}^n x_{r1} + C_2 \lambda_{r2}^n x_{r2}$$

$$= C_1 e^{jn\theta} x_{r1} + C_2 e^{-jn\theta} x_{r2} \quad \text{for} \quad 0 \le g_1 g_2 \le 1. \tag{3.4.19}$$

We thus see that, for the value of $g_1 g_2$ between 0 and 1, the ray will be confined to a finite transverse dimension. Thus the cavity is stable under the condition

$$0 \le g_1 g_2 \le 1. \tag{3.4.20}$$

If this condition is not satisfied, the eigenvalues are real and they continue to grow as n increases. This is shown in Fig. 3.4.5.

Using the condition given by (3.4.20) we can draw a stability diagram as a function of g_1 and g_2. This is shown in Fig. 3.4.6, where the shaded region denotes stability of the cavity. The boundaries of the stable region are given by

$$g_1 = 0,$$

$$g_2 = 0,$$

$$g_1 g_2 = 1.$$

The last equation defines two hyperbolas. Some particular values of R_1 and R_2 are denoted in the diagram and tabulated in Table 3.4.1.

It is of interest to consider the Fabry–Perot cavity for which $R_1 = R_2 = \infty$ or $g_1 = g_2 = 1$; we can see that the Fabry–Perot cavity is on the boundary of the stability diagram. Physically, it means that for only one set of rays, with $\theta = 0$, the cavity is stable; for all other rays, it is unstable. Of course, this is

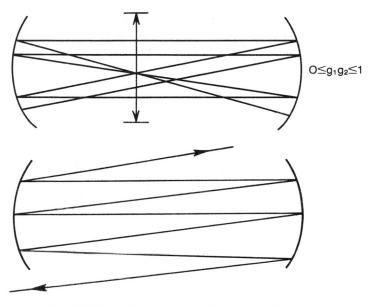

Fig. 3.4.5. Ray diagram of a stable and unstable cavity.

obvious from the ray diagram of the cavity shown in Fig. 3.4.2. Thus for a low-gain active medium, the Fabry–Perot laser is nearly impossible to align, because only under perfect conditions does it lase. However, for other cavities in the stable region, alignment of the laser cavity is much easier. Thus, nearly all the lasers use cavities which are in the stable region, with some exceptions which will be discussed in the next section.

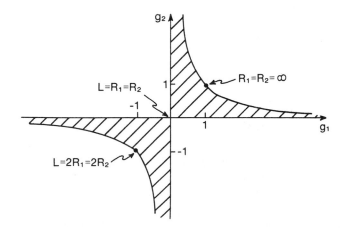

Fig. 3.4.6. Stability diagram.

Table 3.4.1. Different resonator structures.
(From A.E. Siegman, *An Introduction to Lasers and Masers*, McGraw-Hill, New York, 1971.)

1. Plane parallel,
   $R_1 = R_2 = \infty$,
   $g_1 = g_2 = 1$,
   $\cos \theta = 1$,
   $\theta = 0$.

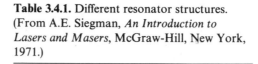

2. Slightly concave,
   $R_1 = R_2 = $ large,
   $g_1 = g_2 \lesssim 1$,
   $\cos \theta = 1 - \delta$,
   $\theta \approx \sqrt{2\delta}$.

3. "Focal" resonator.
   (Focus of each mirror on other mirror.)
   $R_1 = R_2 = 2L$,
   $g_1 = g_2 = \frac{1}{2}$,
   $\cos \theta = -\frac{1}{2}$,
   $\theta = 3\pi/4$.

4. Confocal resonator.
   (Focal points of mirrors coincide.)
   $R_1 = R_2 = L$,
   $g_1 = g_2 = 0$,
   $\cos \theta = -1$,
   $\theta = \pi$.

5. Near concentric,
   $L/2 < (R_1, R_2) < L$,
   $g_1 = g_2 \approx -1$,
   $\cos \theta = -1 + \delta$,
   $\theta \approx \pi + \sqrt{2\delta}$.

6. Concentric (spherical),
   $R_1 = R_2 = L/2$,
   $g_1 = g_2 = -1$,
   $\cos \theta = 1$,
   $\theta = 2\pi$.

3.5. Gaussian Beam Optics

To solve the cavity problem using the diffraction integral is a very difficult task. However, some solutions of this problem are easy to comprehend. We will discuss the so-called simple Gaussian beam first. This will be shown later to be a solution of the electric field both inside and outside the laser cavity. Afterwards, we will show that other complex beams are also solutions of the cavity problem.

Let us consider a beam, propagating in the $+z$ direction and having amplitude distribution in the x and y directions, as Gaussian with a waist

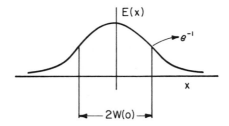

Fig. 3.5.1. Gaussian beam.

$\omega(z = 0)$ denoted by $\omega(0)$. This is given by the following equation, and shown in Fig. 3.5.1:

$$E(x, y) = \sqrt{\frac{2}{\pi}} \frac{1}{\omega(0)} e^{-(x^2+y^2)/\omega^2(0)}. \tag{3.5.1}$$

Remember that up until now we have only considered a beam with rectangular cross section (Section 2.6.1) and circular cross section (Section 2.8.1). It was found that those cases produced diffraction rings in the far-field approximation.

The factor $\sqrt{(2/\pi)}(1/\omega(0))$ in front of the integral is introduced for the purpose of normalization. That is,

$$\int\int_{-\infty}^{+\infty} |E|^2 \, dy \, dx = 1 \tag{3.5.2}$$

as can be proved easily.* Then (3.5.1) can be rewritten as

$$E(x, y, 0) = \sqrt{\frac{2}{\pi}} \frac{1}{\omega(0)} e^{-j\frac{k}{2}\frac{x^2+y^2}{q(0)}}, \tag{3.5.3}$$

where we have defined a new complex variable given by

$$q(0) = q(z = 0) = j\frac{\omega^2(0)\pi}{\lambda}. \tag{3.5.4}$$

Let us consider the Fresnel diffraction of the beam at $z = z$ from the Gaussian shape given by (3.5.1) at $z = 0$. It is given by

$$E(x, y, z) = \frac{1}{j\lambda z} e^{j(\omega t - jkz)} \int\int_{-\infty}^{+\infty} E(x', y', 0) e^{-j(k/2z)[(x-x')^2+(y-y')^2]} \, dx' \, dy'. \tag{3.5.5}$$

This integral has been evaluated in Reference 1. It can be written as

$$E(x, y, z) = \frac{1}{j\lambda z} \sqrt{\frac{2}{\pi}} \frac{1}{\omega(z)} \cdot \exp\left[-j\frac{k}{2}\frac{x^2+y^2}{q(z)}\right] e^{j\{(\omega t - kz) - \psi(z)\}}, \tag{3.5.6}$$

---

* For a known power of P watts, the E field must be multiplied by $\sqrt{2ZP}$ where Z is the characteristic impedance and is given by (2.1.12).

where

$$\frac{1}{q(z)} = \frac{1}{R(z)} - j\frac{\lambda}{\pi\omega^2(z)} = \frac{1}{z + j\pi\omega^2(0)/\lambda} = \frac{1}{q(0) + z},$$

$$\omega(z) = \omega(0)\left\{1 + \left(\frac{z}{z_R}\right)^2\right\}^{1/2}, \tag{3.5.7}$$

$$R(z) = z + \frac{z_R^2}{z}, \qquad z_R = \frac{\pi\omega(0)^2}{\lambda}; \qquad \psi(z) = \tan^{-1}\left(\frac{z}{z_R}\right).$$

In terms of $R(z)$ and $\omega(z)$, the expression for $E(x, y, z)$ can be written as

$$E(x, y, z) = \frac{1}{j\lambda z}\sqrt{\frac{2}{\pi}}\frac{1}{\omega(z)}\exp\left[-\frac{x^2 + y^2}{\omega(z)}\right]\exp\left[-j\frac{k}{2}\frac{x^2 + y^2}{R(z)}\right]e^{j\{\omega t - kz - \psi(z)\}}. \tag{3.5.8}$$

Let us compare the above equation with the equation for a spherical wavefront originating from $x = y = z = 0$. In the Fresnel approximation, it is given by the equation

$$E(x, y, z) = \frac{1}{j\lambda z}\exp\left[-j\frac{k}{2}\frac{x^2 + y^2}{R}\right]e^{j(\omega t - kz)}, \tag{3.5.9}$$

where, of course, $z \approx R$, the distance from the origin to the wavefront. Thus we see that the Gaussian beam has a phase term which is very similar to the spherical wavefront. The radius of curvature of the wavefront is however, given by

$$R(z) = z + \frac{z_R^2}{z}, \tag{3.5.10}$$

We also observe that the Gaussian beam shape is preserved with the new waist size, which is a function of z and is given by

$$\omega(z) = \omega(0)\left\{1 + \left(\frac{z}{z_R}\right)^2\right\}^{1/2}. \tag{3.5.11}$$

In the above two equations, we have defined a quantity, z_R, called the Rayleigh distance, which is given by

$$z_R = \frac{\pi\omega^2(0)}{\lambda}.$$

Equation (3.5.8) is graphically illustrated in Fig. 3.5.2. We see that as the distance increases, the curvature of the wavefront and the waist size increase. If we plot $\omega(z)$ as a function of z (shown in Fig. 3.5.3), we observe that, up to a distance of $z \approx z_R$,

$$\omega(z_R) \approx \sqrt{2}\omega(0). \tag{3.5.12}$$

Thus, up to a distance $z = z_R$ the Gaussian waist size, also called the spot size, is approximately constant. However, beyond z_R, for $z \gg z_R$,

$$\omega(z) \approx \frac{\omega(0)z}{z_R} = \frac{\lambda z}{\omega(0)}\frac{1}{\pi},$$

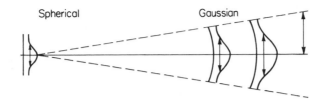

Fig. 3.5.2. Propagation of a Gaussian beam.

and
$$R(z) \approx z.$$

This is expected from the Fraunhofer diffraction of an opening with an approximate radius $\omega(0)$.

From (3.5.7) we note that $q(z)$ is a complex constant which is similar to R in (3.5.10). However, its real part is related to the curvature of the wavefront, and the imaginary part to the Gaussian waist size. As z changes, by noting the change in $q(z)$, we can determine the properties of these spherical Gaussian waves. If we are given a particular wavefront with $q(z)$, we can always use (3.5.6) to retrace it backward or forward to get its waist; this is illustrated in Fig. 3.5.4.

Let us now consider a spherical Gaussian wavefront propagating back and forth between the cavity illustrated in Fig. 3.5.5. If the wavefront exactly matches its radius of curvature with that of the mirror at $z = -z_1$ and again at $z = z_2$, where

$$z_1 + z_2 = L, \tag{3.5.13}$$

we see that the wavefront will be bouncing back and forth without any distortion. Thus we will have a stable situation. This condition will only be satisfied when

$$R_2 = z_2 + \frac{z_R^2}{z_2}, \tag{3.5.14}$$

or

$$R_1 = -z_1 + \frac{z_R^2}{z_1}.$$

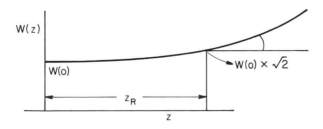

Fig. 3.5.3. Waist size versus z.

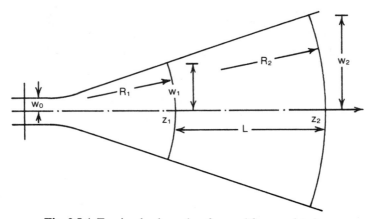

Fig. 3.5.4. Tracing backward or forward from waist size.

The two equations (3.5.13) and (3.5.14) can be solved to obtain the three unknowns, z_R, z_1, and z_2. Remember that we did not know where the origin of the z-axis is or where the Gaussian beam has the minimum waist size. By algebraic manipulation, we obtain

$$z_R^2 = L^2 \frac{g_1 g_2 (1 - g_1 g_2)}{(g_1 + g_2 - 2g_1 g_2)^2} = \left(\frac{\pi \omega^2(0)}{\lambda} \right)^2, \tag{3.5.15}$$

$$z_1 = -\frac{g_2(1 - g_1)}{g_1 - g_2 + 2g_1 g_2} \cdot L, \tag{3.5.16}$$

$$z_2 = \frac{g_1(1 - g_2)}{g_1 + g_2 - 2g_1 g_2} L = z_1 + L.$$

It is of interest to obtain the waist size on the cavity mirrors. These are

$$\omega_1(z = z_1) = \left\{ \frac{L\lambda}{\pi} \sqrt{\frac{g_2}{g_1(1 - g_1 g_2)}} \right\}^{1/2},$$

$$\tag{3.5.17}$$

$$\omega_2(z = z_2) = \left\{ \frac{L\lambda}{\pi} \sqrt{\frac{g_1}{g_2(1 - g_1 g_2)}} \right\}^{1/2}.$$

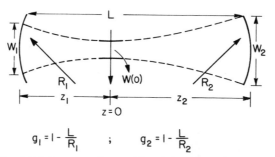

$$g_1 = 1 - \frac{L}{R_1} \quad ; \quad g_2 = 1 - \frac{L}{R_2}$$

Fig. 3.5.5. Laser cavity problem using Gaussian optics.

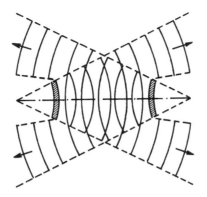

Fig. 3.5.6. Unstable cavity.

These will be the spot sizes of the laser at the two mirrors, if the laser lases with this particular wavefront. We shall see later that this is the only mode, the (0, 0) mode, in which a laser often can lase. The meaning of the symbol (0, 0) will be given later.

From (3.5.17) we see immediately that for the laser to be stable, the spot size must be finite; also, the right-hand side of (3.5.15) can never be imaginary. Thus, we obtain the same stability condition as we obtained before

$$0 \le g_1 g_2 \le 1. \tag{3.5.18}$$

We also note here, however, that the spot size is determined by g_1 and g_2, and most of the time it is rather small. For example, for $R_1 = R_2 = 2$ m and $L = 1$ m, the spot size for $\lambda = 0.5$ μm is $\omega = 428$ μm.

This is sometimes not acceptable for enegineering applications, and for that reason people sometimes use unstable cavities. We might ask how we are going to get lasing started in a finite lasing medium using an unstable cavity. Well, if the laser amplifier has a very high gain, then it is quite possible that in a few passes the gain of the beam is enough to offset the losses. In that case, we can use an unstable cavity. A typical case is shown in Fig. 3.5.6. It is found that for this case the beam size is not determined by the radii of the mirrors. More about unstable resonators can be found in Section 3.6.2.

3.5.1. Gaussian Optics Including Lenses

A Gaussian wave is completely characterized by the quantity q, which is a complex quantity. The real part is related to the radius of curvature of the wavefront and the imaginary part to the spot size. For an optical system whose matrix is given by M, as shown in Fig. 3.5.7(a), we note that

$$x_2 = M_{11}x_1 + M_{12}\theta_1,$$

$$\theta_2 = M_{21}x_1 + M_{22}\theta_1.$$

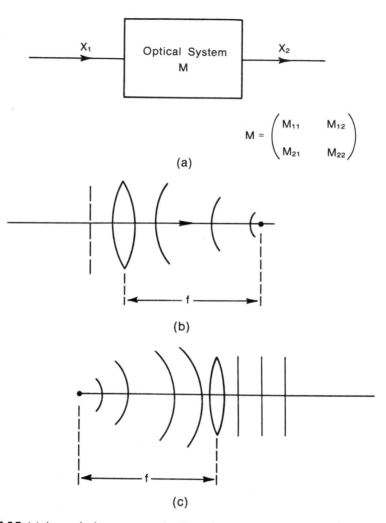

Fig. 3.5.7. (a) An optical system matrix; (b) a plane wave incident on a lens; and (c) a spherical wavefront with radius of curvature f incident on the lens with focal length f.

Thus, the radius of curvature, R_2 is defined as

$$R_2 = \frac{x_2}{\theta_2} = \frac{M_{11}x_1 + M_{12}\theta_1}{M_{21}x_1 + M_{22}\theta_1},$$

or

$$R_2 = \frac{M_{11}R_1 + M_{12}}{M_{21}R_1 + M_{22}}. \tag{3.5.19}$$

Thus, (3.5.19) gives the relationship between an input wave of radius of curvature R_1 and the output wave with radius of curvature R_2. For example, for a simple lens with focal length f

$$M = M(f) = \begin{pmatrix} 1 & 0 \\ -1/f & 1 \end{pmatrix}.$$

Thus, for this case,

$$R_2 = \frac{R_1}{R_1/f + 1} = \frac{fR_1}{f - R_1}.$$

As shown in Fig. 3.5.7(b), if the input $R_1 = \infty$, i.e., the plane wave, then

$$R_2 = -f;$$

or the output wavefront will have a curvature such that it focuses at the focal point. Or as shown in Fig. 3.5.7(c), if $R_1 = f$, then

$$R_2 = \infty;$$

or if the source is at the focal point, the incident wavefront on the lens has $R_1 = f$, and the output wavefront is plane or has a radius of curvature which is infinity.

This gives us a clue as to the desired relationship between q_2 and q_1, as follows:

$$q_2 = \frac{M_{11}q_1 + M_{12}}{M_{21}q_1 + M_{22}}. \tag{3.5.20}$$

Although no formal proof of the above equation exists as yet, it is always found to be true. If we use the simple lens matrix again, and consider the propagation of a Gaussian wavefront, then we have

$$q_2 = \frac{fq_1}{f - q_1}.$$

If $q_1 = q(0) = jz_r$, then it can easily be shown that the spot size is not minimum at $z = f$, but at z_m given by

$$z_m = \frac{f}{1 + (f/z_R)^2}.$$

3.6. Solution of the Cavity Problem

In the previous section, we showed that the Gaussian beam is a solution of the laser cavity problem, if certain conditions defined by (3.5.16) are met. In this section, we shall formulate the problem in general and discuss other solutions. For a wavefront, having an electric field $E(x, y, z)$ as a solution of the electromagnetic wave equation inside the cavity, its wavefronts must match the shape of the mirror as the ray bounces back and forth. Actually, what we require is that the shape of the wavefront remains unchanged as it propagates through a distance z. Because, if the shape remains unchanged, then we can always place two mirrors of the same shape as that of the wavefront to reflect the rays exactly without any disturbance and then obtain a stable situation. In other words, the electric field must satisfy the following integral equation:

$$E(x, y, z) = \frac{1}{j\lambda z} e^{-jkz} \int\int E(x', y', 0) e^{-j(k/2L)[(x-x')^2+(y-y')^2]} \, dx' \, dy'. \quad (3.6.1)$$

Here the functional dependence of E on x, y, and z is unknown, and any function satisfying the above equation will be a solution. The above equation represents a basic integral transformation. It can be shown that its solutions are given by

$$E_{mn}(x, y, z) = \sqrt{\frac{2}{2^{m+n} m! \, n! \, \pi}} \cdot \frac{1}{\omega(z)}$$

$$\times H_m\left(\frac{\sqrt{2}x}{\omega(z)}\right) H_n\left(\frac{\sqrt{2}y}{\omega(z)}\right) e^{-j(k/2) \cdot ((x^2+y^2)/q(z))} e^{-jkz + j(n+m+1)\psi(z)},$$

$$m, n = 0, 1, 2 \dots . \quad (3.6.2)$$

The $H_m(x)$ is known as mth-order Hermite polynomial. The reader familiar with the simple harmonic problem in quantum mechanics has seen these same Hermite polynomials. Some of these Hermite polynomials are listed in Table 3.6.1 and are plotted in Fig. 3.6.1. In general, the nth-order polynomial has $(n + 1)$ number of maxima and minima. Figure 3.6.1 also shows the square of the electric field as a function of x. It is observed that the $(0, 0)$ mode is the one we have discussed in the previous section. However, other modes have complex shapes; for example, the $(3, 2)$ mode has four lumps in the x direction and three lumps in the y direction. Also, the outermost lumps are a little bigger than the other lumps; corresponding to the plots in Fig. 3.6.1, the spot shapes are shown in Fig. 3.6.2. The spot shapes are an equipower contour of the laser beam, in the plane transverse to the direction of propagation. These are obtained, for example, in the x direction only by taking a cross section of the plots of $|H_n(x)e^{-x^2}|$ in Fig. 3.6.1. It is amazing that all these shapes can actually be observed from a laser if it is adjusted properly.

Table 3.6.1. Hermite polynomials.

$H_0(x) = 1$
$H_1(x) = 2x$
$H_2(x) = 4x^2 - 2$
$H_3(x) = 8x^3 - 12x$
$H_4(x) = 16x^4 - 48x^2 + 12$
$H_5(x) = 32x^5 - 160x^3 + 120x$
$H_6(x) = 64x^6 - 480x^4 + 720x^2 - 120$

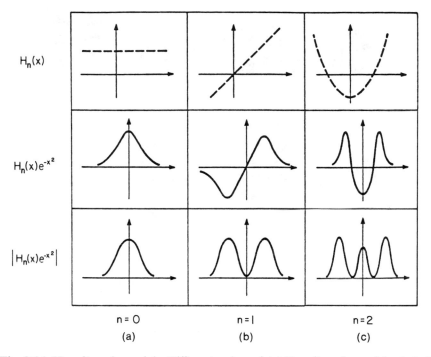

Fig. 3.6.1. Hermite polynomials. Different orders of: (a) Hermite polynomials plotted as a function of x; (b) Gaussian–Hermite polynomials plotted as a function of x; and (c) the magnitude of Gaussian–Hermite polynomials plotted as a function of x.

The Gaussian nature of the beam for all the laser modes is evident from (3.6.2). This also explains why we can get lasing action with a finite open cavity in the transverse direction. Because, by eliminating the outer edges of the mirrors, as the electric field is already very small, the perturbation produced by it will be negligible. The equation is valid for an infinite size radius and, in practice, the limits of the integral should be replaced by the sizes of the mirrors. However, for the particular case of two plane parallel mirrors and for the sizes of the mirrors given by the Fresnel numbers, N, defined as

$$N = \frac{a^2}{L\lambda}, \tag{3.6.3}$$

numerical solution has been performed. Here $2a$ is the mirror diameter, and N represents the number of Fresnel zones on the mirror as viewed from the center of the next mirror. It is found that, because of the finite size of the mirrors, a diffraction loss must be included. This diffraction loss as a function of the Fresnel number is shown in Fig. 3.6.3. This diffraction loss must be included in the calculation of the linewidth of laser oscillation, as discussed in Section 3.3.

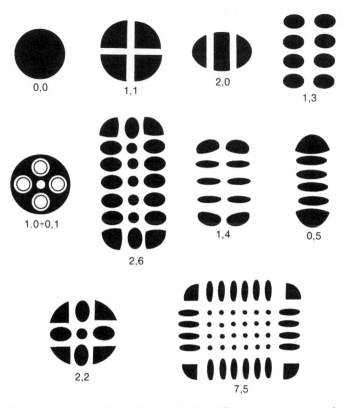

Fig. 3.6.2. Equienergy plots of laser beams having different transverse mode numbers (m, n).

3.6.1. Frequency of Oscillation

The phase term in (3.6.2) is somewhat different from that in (3.5.7) and is dependent on the mode number. We can use the same arguments now to determine the cavity frequencies, as we did in Section 3.3 in (3.3.14). That is, the round trip phase difference between the wavefronts has to be multiples of 2π. In this case, it becomes

$$kz_2 - (n + m + 1)\psi(z_2) - kz_1 + (n + m + 1)\psi(z_1) = p\pi,$$

or

$$kL = p\pi + (n + m + 1)[\psi(z_2) - \psi(z_1)],$$

or

$$f_{mnp} = \frac{v}{2L}\left[p + (n + m + 1)\frac{\psi(z_2) - \psi(z_1)}{\pi}\right], \tag{3.6.4}$$

or

$$\lambda_{mmp} = 2L\left[p + (n + m + 1)\frac{\psi(z_2) - \psi(z_1)}{\pi}\right]^{-1}.$$

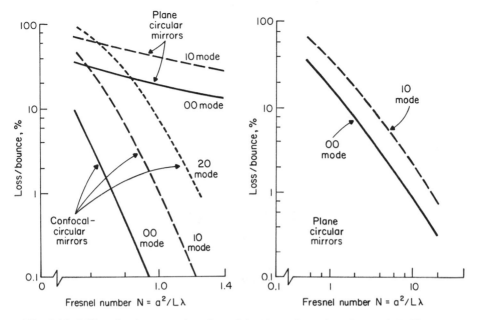

Fig. 3.6.3. Diffraction loss as a function of the Fresnel number. (From A.E. Siegman, *An Introduction to Lasers and Masers*, McGraw-Hill, New York, 1971.)

After some algebraic manipulation it can be shown that

$$[\psi(z_2) - \psi(z_1)] = \cos^{-1} \pm [\sqrt{g_1 g_2}]. \tag{3.6.5}$$

Thus we see that for each longitudinal mode denoted by the number p, we can have a series of transverse modes denoted by the subscripts m and n. Each of these transverse modes has a particular beam shape and a particular frequency—some of these are shown in Fig. 3.6.4.

It is to be mentioned that (3.6.1) is a linear equation; thus a linear combination of individual modes is also a solution to the problem. For example, we can have a shape like that in Fig. 3.6.5, given by the combination of (1, 0) and (0, 1) modes having two different frequencies.

Some other forms of solution have also been found. For example, the solution written in polar coordinates in Gaussian–Laguerre form is given by

$$u_{pl}(r, \theta, z) = \frac{j}{\lambda z} \frac{2}{\sqrt{1 + \delta_{0l}}} \frac{p!}{\pi(l + p)!} \frac{1}{\omega(z)} \left(\frac{\sqrt{2}r}{\omega(z)}\right)^l L_p^l \left(\frac{2r^2}{\omega^2(z)}\right)$$

$$\times \binom{\cos l\theta}{\sin l\theta} e^{-j(k/2)(r^2/q(z))} e^{-jkz + j(2p+l+1)\psi(z)}, \tag{3.6.6}$$

where p and l are different sets of integer indices, L_p^l is the associated Laguerre polynomial, $\delta_{0l} = 1$ for $l = 0$ but $\delta_{0l} = 0$ for $l \neq 0$. Some of these shapes of the beam in the transverse direction are shown in Fig. 3.6.6.

This concludes the discussion on laser cavities. It is interesting to point out that, without ever knowing how a laser amplifier works, we have been able to

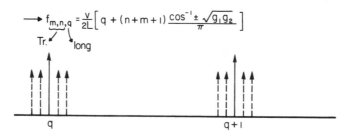

$$f_{m,n,q} = \frac{v}{2L}\left[q + (n+m+1)\frac{\cos^{-1} \pm \sqrt{g_1 g_2}}{\pi} \right]$$

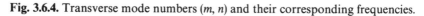

Fig. 3.6.4. Transverse mode numbers (m, n) and their corresponding frequencies.

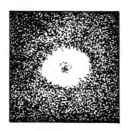

Complex modes

Fig. 3.6.5. Complex mode shape of a laser beam.

TRANSVERSE MODE CONTROL

0.0

0.3

0.4

Fig. 3.6.6. Circular mode shapes of laser beams. (From A.E. Siegman, *An Introduction to Lasers and Masers*, McGraw-Hill, New York, 1971.)

predict the mode shapes and the possible frequencies of the laser beam. In the next section, we discuss the unstable resonators.

3.6.2. Unstable Resonators

Stable resonators are useful for making lasers using lasing material with a very small gain constant. However, as discussed before, the beam size is limited and thus the active region has a very small volume. This translates into rather low-power lasers. Also, to obtain single-mode lasers with stable resonators, we must use small apertures to increase the losses for the higher order modes. However, the positioning and adjustment of the aperture is rather difficult.

As mentioned before, if the gain constant is high we can use an unstable resonator. Consider the unstable resonator shown in Fig. 3.6.7, consisting of two mirrors with radii of curvature given by R_1 and R_2. To analyze it using geometrical optics, consider the two points P_1 and P_2 which are the virtual sources for the waves impinging on the mirrors M_1 and M_2, respectively, this is shown in Fig. 3.6.8. If the separation between the two mirrors is denoted by d, consider the distance between P_1 and the mirror M_1, $\alpha_1 d$; similarly, for the point P_2, it is $\alpha_2 d$. Note that α_1 and α_2 are dimensionless. Consider the wave emanating from P_1 and incident on M_2. Upon reflection, it appears as if it is radiating from P_2. Thus P_1 and P_2 must satisfy the object and image relationship of mirror M_2 given by

$$\frac{1}{(\alpha_1 + 1)d} - \frac{1}{\alpha_2 d} = \frac{2}{R_2}. \tag{3.6.7}$$

Similarly, for mirror M_1, we have

$$\frac{1}{(\alpha_2 + 1)d} - \frac{1}{\alpha_1 d} = \frac{2}{R_1}. \tag{3.6.8}$$

Solving (3.6.7) and (3.6.8) we obtain

$$\alpha_1 = \frac{\sqrt{1 - 1/g_1 g_2} - 1 + 1/g_1}{2 - 1/g_1 - 1/g_2}, \tag{3.6.9}$$

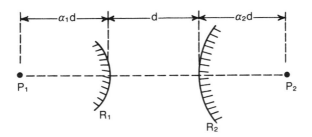

Fig. 3.6.7. Unstable resonator with two virtual point sources, P_1 and P_2.

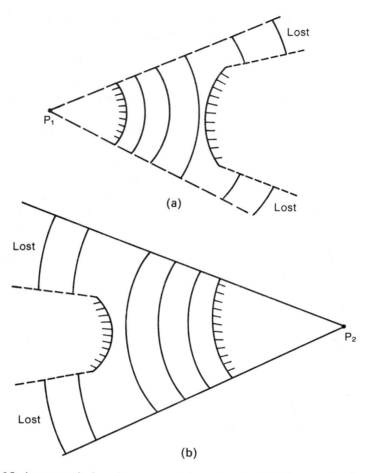

Fig. 3.6.8. A geometrical optics approach to understand the loss mechanism in an unstable resonator: (a) wavefronts for the virtual point source P_1; and (b) wavefronts for the virtual point source P_2.

and

$$\alpha_2 = \frac{\sqrt{1 - 1/g_1 g_2} - 1 + 1/g_2}{2 - 1/g_1 - 1/g_2},$$ (3.6.10)

where

$$g_1 = 1 - \frac{d}{R_1},$$ (3.6.11)

and

$$g_2 = 1 - \frac{d}{R_2}.$$ (3.6.12)

To calculate the equivalent reflection losses, even for perfectly reflecting mirrors, we note that at each reflection part of the beam is completely lost, as shown in Fig. 3.6.8. Denoting the fraction of power reflected by M_1 and M_2 as Γ_1 and Γ_2 we obtain

$$\Gamma_2 = \frac{\text{solid angle of } M_2 \text{ with original } P_1}{\text{solid angle of wave originating at } M_1}$$

$$= \frac{\pi a_2^2/4\pi(\alpha_1 + 1)^2 d^2}{\pi a_1^2/4\pi\alpha_1^2 d^2}, \tag{3.6.13}$$

where a_1 and a_2 are the radii of the mirror apertures. Γ_1 is similarly given by

$$\Gamma_1 = \frac{\pi a_1^2/4\pi(\alpha_2 + 1)^2 d^2}{\pi a_2^2/4\pi\alpha_2^2 d^2}. \tag{3.6.14}$$

Thus the fraction of power which reflects back after a round trip, Γ, is given by

$$\Gamma^2 = \Gamma_1\Gamma_2 = \left[\frac{\alpha_1\alpha_2}{(\alpha_1 + 1)(\alpha_2 + 1)}\right]^2. \tag{3.6.15}$$

Thus the condition for oscillation for the unstable laser will be given by

$$G\Gamma > 1, \tag{3.6.16}$$

where G is the single pass gain and is given by

$$G = \gamma d, \tag{3.6.17}$$

where γ is the lasing gain constant. We note that G has to be quite large for this case if the laser is going to operate with an unstable laser. It is of interest to note that Γ is not dependent on the size of the mirrors. This is because we

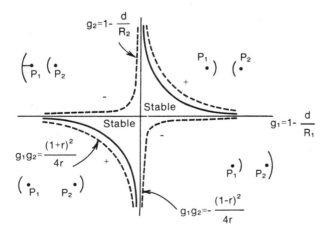

Fig. 3.6.9. Output beam shape for a laser with an unstable laser cavity with two beam slicers and absorbing walls.

have not included the diffraction losses which are, in general, small. To consider this diffraction, we need to solve the integral equation numerically as no analytical solution exists.

Equation (3.6.15) can also be written as

$$\Gamma = \pm \frac{1 - \sqrt{1 - 1/g_1 g_2}}{1 + \sqrt{1 - 1/g_1 g_2}}. \tag{3.6.18}$$

As the resonator is unstable, $g_1 g_2 > 0$ or $g_1 g_2 < 0$. If $g_1 g_2 > 0$, then a positive sign must be chosen as Γ must be less than 1. For $g_1 g_2 < 0$, a negative sign must be chosen.

Thus from (3.6.18) we obtain

$$g_1 g_2 = \frac{(1 + \Gamma)^2}{4\Gamma} \qquad \text{if} \quad g_1 g_2 > 0,$$

$$= -\frac{(1 - \Gamma)^2}{4\Gamma} \qquad \text{if} \quad g_1 g_2 < 0. \tag{3.6.19}$$

For Γ constant, (3.6.19) plots a hyperbola. Thus we can draw equiloss diagrams for a fixed value of Γ, as shown in Fig. 3.6.9. For a cavity with a power loss of 30%, $\Gamma = 0.7$ and $g_1 g_2 = 1.032$ or -0.182. In Fig. 3.6.9, note that the positive branch corresponds to the first and third quadrants, whereas the negative branch corresponds to the second and fourth quadrants.

The advantage of the unstable resonator is that the output beam has a large aperture. It might appear that the output has a hole in the middle and thus difficult to focus; in practice, this hole is of very little consequence. A different arrangement for a laser with an unstable resonator is shown in Fig. 3.6.10. Here it is assumed that the walls of the lasing material container are perfectly

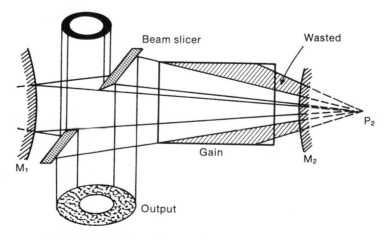

Fig. 3.6.10. Equiloss diagrams for an unstable laser cavity.

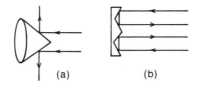

Fig. 3.6.11. (a) Axicon and (b) waxicon.

absorbing. Two mirrors or beam slicers are used for the output beams. It is to be noted that a CO_2 lasing medium has a very high gain and, in general, most of the practical CO_2 lasers do use unstable resonator configuration.

Special mirrors such as waxicons are used to convert the ring mode output of an unstable resonator to a near-Gaussian resonator. Waxicons are W-shaped axicons and are shown in Fig. 3.6.11. Axicons are mirrors and they produce a particular axial distribution of light on reflection.

3.7. Photon, Stimulated, and Spontaneous Emission, and the Einstein Relationship

For a laser to operate, we need a laser cavity and an active medium which acts as a light amplifier. We discussed, in Section 3.3, that light amplification is distributed in nature. Thus we define an amplification constant, γ, which represents the growth of the electric field as the wave travels through the lasing medium given by

$$E(x) = E(0)e^{\gamma x}, \tag{3.7.1}$$

where $E(0)$ is the electric field at $x = 0$. Similarly, we can write the corresponding intensity given by

$$I(x) = I(0)e^{2\gamma x}, \tag{3.7.2}$$

where $I(0)$ is the intensity at $x = 0$. Thus our objective is to derive an expression for this "γ" in terms of physical parameters. However, before we do that, we want to discuss certain aspects of light and its interaction with matter, which is quite different from that which we have considered thus far in this book. It is expected that the reader is familiar with quantum mechanics and its applications to atomic spectra. We will review it very briefly in this section.

Up until now, we have considered electromagnetic waves as waves. However, it turns out that these waves can sometimes be better described by a wave packet or photon. This photon has energy given by

$$E = \hbar\omega = hf, \tag{3.7.3}$$

where h is the Planck constant $= 6.626 \times 10^{-34}$ J s and $\hbar = h/2\pi$. This dual nature of electromagnetic waves, especially for the high-frequency range, is rather complex and somewhat puzzling. However, it is based on experimental data, and it is impossible to explain certain experiments (for example, photo-electric emission) without the photon-like aspects of light waves.

Light is generated when an electron drops from an energy level E_2 to a lower energy level E_1 in matter. The generated photon obeys the conservation of energy given by

$$hf = E_2 - E_1. \tag{3.7.4}$$

Sometimes a photon is absorbed such that the electron transfers from E_1 to E_2. It is customary to denote the energy levels in units of electronvolts (eV). One electronvolt denotes the energy gained by a single electron when it accelerates through a one-volt potential difference. As the electronic charge is 1.6×10^{-19} C and a coulomb-volt is equivalent to a joule, one electronvolt is 1.6×10^{-19} J. Thus when an electron jumps from E_2 ($=2.0$ eV) to E_1 ($=1.9$ eV) it will generate a photon with $\lambda = 1.24$ μm and $f = 2.418 \times 10^{14}$ Hz.

Let us consider a two-level system. The two energy levels are given as E_2 and E_1. Actually, the medium will have many other energy levels. However, for simplicity, let us consider that the only interaction with photons takes place through these two levels. In the equilibrium condition, when this medium is not pumped or energized, the number of electron transitions upward must be equal to the number of transitions downward. Thus no net photons are generated or lost. However, if we somehow increase the number of electrons in the E_2 level beyond the equilibrium value, then there will be an extra supply of electrons in the E_2 level which can make the transition to the lower level. The rate at which this transition takes place will be given by

$$\frac{dN_2}{dt} = -\frac{N_2}{t_{\text{spont}}} = -N_2 A_{21}, \tag{3.7.5}$$

where N_2 is the density of electrons in the upper level. The rate of change of N_2 must be proportional to the number of electrons in level 2 and to a constant A_{21}, which is inverse to the spontaneous emission rate, t_{spont}. Remember that any time a transition takes place, a photon is generated or emitted. Solving (3.7.5), we obtain

$$N_2 = N_{20} + \Delta N_2(0)e^{-t/t_{\text{spont}}},$$

or

$$\Delta N_2 = \Delta N_2(0)e^{-t/t_{\text{spont}}}, \tag{3.7.6}$$

where N_{20} is the equilibrium value of N_2, and $\Delta N_2 = N_2 - N_{20}$ and $\Delta N_2(0)$ is the excess density at $t = 0$.

The light from the spontaneous emission is noise-like and is incoherent. This might seem a little puzzling because the light frequency is given by

$$f_0 = \frac{E_2 - E_1}{h}. \tag{3.7.7}$$

However, this emitted light is really in bursts of decaying exponentials, accord-

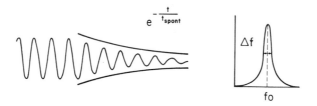

Fig. 3.7.1. A decaying electric field and its Fourier transform.

ing to (3.7.6). Thus, the actual frequency of the light, although centered at $f = f_0$, will have a spread. The spread in frequency is found by taking the Fourier transform of the decaying light, denoted by

$$E = E_0 e^{j2\pi f_0 t} e^{-t/t_{spont}}. \tag{3.7.8}$$

The Fourier transform of (3.7.8) is given by

$$\mathscr{F}\{E\} = E_0 \cdot \frac{1}{1 + (\omega - \omega_0)^2 t_{spont}^2}, \tag{3.7.9}$$

where

$$\omega_0 = 2\pi f_0. \tag{3.7.10}$$

This is plotted in Fig. 3.7.1, and we see that the spontaneous emission will have a frequency spread, Δf, given by

$$\Delta f \approx \frac{1}{t_{spont}} \tag{3.7.11}$$

In contrast to spontaneous emission, there is stimulated emission. The stimulated emission is stimulated by the photons or light wave, and thus they are phase coherent as compared to the spontaneous one. The stimulated rate is also proportional to the number of photons stimulating it. This number of photons is also proportional to the intensity of light; the exact relationship between the two will be derived later. It was Einstein who first pointed out from thermodynamical considerations that a particular relationship exists between these two rates in an enclosed black body which is kept at temperature T in equilibrium.

It is an experimental fact that the blackbody radiation is given by*

$$\rho(f) = \frac{8\pi h f^3}{c^3} \frac{1}{e^{hf/kT} - 1}, \tag{3.7.12}$$

where $\rho(f)$ is the energy density of photons having frequency f, and k is the Boltzmann constant and is equal to 1.381×10^{-23} J/K. Let us consider a medium which has two levels E_2 and E_1 and is placed in this black body as

---

* Note that we represent Boltzmann's constant and propagation constant by the same letter k.

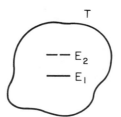

Fig. 3.7.2. Two level systems in a black body.

shown in Fig. 3.7.2. Then, when equilibrium is reached the number of transitions from E_2 to E_1 must be equal to that of E_1 to E_2. The number of photons generated will be given by

$$\text{photons generated per unit time per unit volume} = N_2(W_{21} + A_{21}), \quad (3.7.13)$$

where W_{21} is the stimulated transition rate from level E_2 to level E_1. The term A_{21} is due to the spontaneous emission rate discussed earlier. Similarly, the number of photons lost will be given by

$$\text{photons lost per unit time per unit volume} = N_1 W_{12}, \quad (3.7.14)$$

where W_{12} is the transition rate from E_1 to E_2. The spontaneous emission term, of course, is absent in the photon absorption case.

As mentioned before, W_{21} and W_{12} must be proportional to $\rho(f)$ given by (3.7.12). Thus we can define the so-called B coefficients given by

$$W_{21} = \rho(f)B_{12},$$

and   (3.7.15)

$$W_{21} = \rho(f)B_{21}.$$

Equating (3.7.13) and (3.7.14) and substituting from (3.7.15), we obtain

$$\frac{N_2}{N_1} = \frac{\rho(f)B_{12}}{\rho(f)B_{21} + A_{21}}. \quad (3.7.16)$$

It is also known from thermodynamic considerations that N_2 and N_1 are related by the Boltzmann distribution function, and is given by

$$\frac{N_2}{N_1} = e^{-hf/kT}. \quad (3.7.17)$$

Thus

$$\frac{N_2}{N_1} = \frac{\rho(f)B_{12}}{\rho(f)B_{21} + A_{21}} = e^{-hf/kT},$$

or

$$\frac{B_{12}e^{hf/kT} - B_{21}}{e^{hf/kT} - 1} = \frac{c^3}{8\pi hf^3}A_{21}.$$

The left-hand side of the above equation is dependent on the temperature, T, whereas the right-hand side is independent of temperature. To satisfy this equation at all temperatures, we must have

$$B_{12} = B_{21} = \frac{c^3}{8\pi h f^3} A_{21}. \tag{3.7.18}$$

Thus we obtain for the stimulated emission rate, $W_i = W_{12} = W_{21}$.

$$W_i = \frac{c^3}{8\pi h f^3} A_{21} \rho(f),$$

or

$$(3.7.19)$$

$$W_i = \frac{c^3}{8\pi h f^3} \frac{1}{t_{\text{spont}}} \cdot \rho(f).$$

Then (3.7.18) is the well-known Einstein relationship between the A and B coefficients. To use the Einstein relationship for the case of lasers, we must consider the finite linewidth of the energy levels. One reason for the linewidth was already described in (3.7.11) where spontaneous emission is considered. This type of linewidth is also known as homogeneous linewidth, as it occurs uniformly in all atoms. However, there could be inhomogeneous linewidth broadening, in which case the linewidths are different for different atoms. A special case is for gases which are traveling with thermal velocity. Due to this thermal motion, the emitted frequencies are Doppler broadened and thus are different for different atoms.

Let us denote by $g(f)$ the total linewidth broadening. Then (3.7.18) has to be modified due to this linewidth broadening as follows:

$$W_i = \frac{c^3}{8\pi h f^3} \cdot \frac{1}{t_{\text{spont}}} \rho(f) \cdot g(f). \tag{3.7.20}$$

This is because earlier we considered the position of E_2, and E_1 was given by a delta function. However, for the broadened case we must replace the delta function by

$$\delta(f) \rightarrow \int g(f) \, df. \tag{3.7.21}$$

$g(f)$ for the case of homogeneous broadening is $1/\Delta f$ where Δf is given by (3.7.11). The Doppler inhomogeneous broadening due to thermal motion is given by

$$g(f) = \frac{2(\ln 2)^{1/2}}{\pi^{1/2} \Delta f_d} e^{-[4(\ln 2)(f-f_0)^2/\Delta f_d^2]}, \tag{3.7.22}$$

where $\Delta f_d = 2f_0 \sqrt{2kT/Mc^2} \ln 2$ and M is the atomic mass. The energy density $\rho(f)$ can easily be related to the intensity of the light beam. As shown in Fig. 3.7.3, the intensity of light is given by the energy per unit area per unit time. As the photons move with a velocity, v, in one second the number of photons

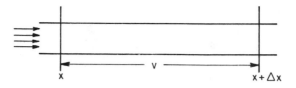

Fig. 3.7.3. Relationship between photon density and intensity.

passing through the area, A, will fill up a cylinder with area A and length v as shown. The volume of this cylinder is

$$V = Av,$$

thus the number of photons in this cylinder is given by

$$N = \frac{Av\rho(f)}{hf}.$$

The corresponding energy per unit area will be the intensity $I(x)$ and is given by

$$I(x) = \frac{Av\rho(f)}{A} = v\rho(f),$$

or

$$\rho(f) = \frac{I(x)}{v}. \tag{3.7.23}$$

Using the above expression for $\rho(f)$ we can rewrite (3.7.20) as*

$$W_i = \frac{v^2}{8\pi hf^3 t_{spont}} I(f)g(f). \tag{3.7.24}$$

3.8. Light Amplifier—Population Inversion

To obtain an expression for γ, we must relate the density of photons to the intensity. This can be done as follows: remember that photons travel with a velocity v, the velocity of light in the particular medium. Let us consider the intensity of the light beam to be $I(x)$ at $x = x$ and $I(x + \Delta x)$ at $x = x + \Delta x$, as shown in Fig. 3.8.1. Let us consider the volume $\Delta x A$, shown in the figure, where A is the cross section. We can write for small Δx

$$I(x + \Delta x) = I(x) + \Delta x \frac{\partial I x}{\partial x}. \tag{3.8.1}$$

---

* Note that in (3.7.24) we have stressed the intensity dependence on the frequency, f, only suppressing its x dependence.

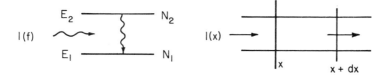

Fig. 3.8.1. Light intensity calculation.

The second term on the right-hand side arises because in the volume $(A\Delta x)$ some photons are generated. However, the number of photons generated per unit time by the stimulated process (remember that stimulated emission is the one which gives rise to coherent laser output, not the spontaneous emission which is like noise to the system) is given by

$$\text{photons generated} = \text{photons emitted} - \text{photons absorbed}$$
$$= A \cdot \Delta x \cdot N_2 W_i - A \cdot \Delta x \cdot N_1 W_i$$
$$= A\Delta x W_i (N_2 - N_1).$$

Thus the power generated in the $A\Delta x$ volume is given by

$$A\Delta x \cdot hf \cdot (N_2 - N_1) W_i,$$

as each photon has an energy equal to hf. Thus,

$$\frac{\partial I(x)}{\partial x} = hf(N_2 - N_1)\frac{v^2 A_{21} g(f)}{8\pi h f^3} I(x), \qquad (3.8.2)$$

or

$$I(x) = I_0 e^{2\gamma x}, \qquad (3.8.3)$$

where

$$\gamma = (N_2 - N_1)\frac{v^2 g(f)}{16\pi f^2 t_{\text{spont}}}. \qquad (3.8.4)$$

Thus we observe that the amplification factor is proportional to $(N_2 - N_1)$. Ordinarily, it is given by (3.7.17).

$$N_2 - N_1 = (e^{-hf/kT} - 1)N_1. \qquad (3.8.5)$$

For light frequencies and at room temperature

$$\frac{hf}{kT} \approx 40.$$

This makes $N_2 - N_1$ negative as N_2 is much smaller than N_1. Thus, in general, we do not have amplification but attenuation of light waves as they pass through the medium. However, if we pump the medium, or energize it such that $N_2 > N_1$, then we can have amplification. This inversion of the electron density from the normal equilibrium condition is absolutely necessary for the lasing medium to lase and is known as a population inversion. In the situation where a population inversion exists, we sometimes define an effective temperature, T_{eff}, given by

$$\frac{N_2}{N_1} = e^{-(hf/kT_{eff})}. \tag{3.8.6}$$

If $N_2 > N_1$ due to pumping,

$$T_{eff} = -\frac{hf}{k}\left[\ln\frac{N_2}{N_1}\right]^{-1}. \tag{3.8.7}$$

Thus, in a sense, the population inversion corresponds to an effective negative temperature if we insist on using (3.7.17).

It is of interest to define the quantity, N_T, called the threshold population inversion density. The quantity N_T denotes the minimum population inversion density needed to start lasing action and then sustain it. Substituting (3.3.12) into (3.8.4), we obtain

$$N_T = (N_2 - N_1)_{Th} = \frac{16\pi f^2 t_{spont}}{v^2 g(f)} \cdot \left(\alpha - \frac{1}{2L}\ln r_1 r_2\right). \tag{3.8.7a}$$

In the expression for N_T, the scattering losses α and the reflection losses of the cavity mirrors r_1 and r_2 appear. It is sometimes convenient to define a quantity called the cavity decay constant, t_c, which is related to the cavity losses. This t_c is the decay constant associated with the decay of an electric field as it bounces back and forth between the mirrors and the lasing medium is not pumped. Thus the electric field $E(t)$ can be represented as

$$E(t) = E_0 e^{-t/t_c}. \tag{3.8.7b}$$

Let us denote by T, the time the electric field takes to make a round trip through the cavity. Thus

$$T = \frac{2L}{v}.$$

If we observe the output at any mirror, then we shall find the decaying pulses of light coming out of the mirror every T seconds. If we start with a value E_0, for the electric field of the starting pulse, then we also know that, because of scattering and reflection losses after a round trip, the electric field $E(T)$ will be given by

$$E(T) = E_0 e^{-2L\alpha} r_1 r_2.$$

Equating the above equation and (3.8.7b) for the value of the electric field at $t = T$ we obtain

$$E_0 e^{-T/T_c} = E_0 e^{-(2L\alpha - \ln r_1 r_2)},$$

or

$$\tau_c = \frac{1}{v}\left[\alpha - \frac{1}{2L}\ln r_1 r_2\right]^{-1}. \tag{3.8.7c}$$

If the losses are small then we can rewrite the above expression as

$$E(T) \approx E_0[1 - 2L(\alpha - \ln r_1 r_2)\dots].$$

Defining the quantity, l_c, the fractional loss per round trip or per pass, we obtain

$$l_c = \frac{E_0 - E(T)}{E_0} = 2L\left(\alpha - \frac{1}{2L}\ln r_1 r_2\right). \tag{3.8.7d}$$

In terms of τ_c and l_c, using (3.8.7a), N_T can be rewritten as

$$N_T = \frac{16\pi f^2 t_{\text{spont}}}{v^3 g(f) t_c},$$

or

$$N_T = \frac{8\pi f^2 t_{\text{spont}}}{v^3 g(f)}\left(\frac{l_c}{L}\right).$$

Of course, t_c and l_c are related by the following equation:

$$t_c = \frac{2L}{v l_c} \quad \text{or} \quad l_c = \frac{2L}{v t_c}.$$

Also, γ_{th} is given by

$$\gamma_{\text{th}} = \frac{l_c}{2L} = \frac{1}{v t_c}.$$

3.9. Different Types of Light Amplifiers and Quantum Efficiency

Until now we have considered only two energy levels in a medium which participate in the stimulated emission. However, the electrons can sometimes be pumped from ground level to a higher level. After some spontaneous emission these electrons may come to the upper lasing level, if it is different from the one to which the electrons were pumped; also, the lower lasing level may or may not be the ground level. In any case, we shall always denote the upper lasing energy level by E_2 and the lower lasing energy level by E_1. Therefore, depending on different circumstances we can have two-level, three-level, or four-level lasers.

A Two-Level Laser

This is the simplest laser where E_1 is the ground level, and the electrons are pumped directly to the E_2 level. For a two-level laser the quantum efficiency, η, is 100%. η is defined as the

$$\eta = \frac{hf}{hf_{\text{pump}}} = \frac{E_2 - E_1}{E_p - E_0} \tag{3.9.1}$$

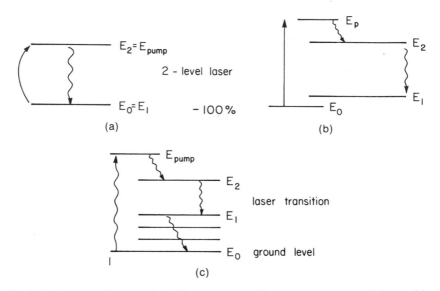

Fig. 3.9.1. Energy diagram for different types of lasers: (a) a two-level laser; (b) a three-level laser; and (c) a four-level laser. E_0 is the ground level and E_1 and E_2 are the lasing levels.

where E_p is the pump level and E_0 is the ground level. Only the $p-n$ junction semiconductor laser can be considered as a two-level laser, and its properties will be discussed in Part IV.

A Three-Level Laser

In a three-level laser, $E_1 \approx E_0$. However, $E_p \neq E_2$. As shown in Fig. 3.9.1(b), f_{pump} is higher than the lasing frequency f. Thus, η is less than 100% and is given by

$$\eta = \frac{E_2 - E_0}{E_p - E_0}. \tag{3.9.2}$$

The ruby laser, the first laser discovered by Maiman, is a three-level laser.

A Four-Level Laser

In a four-level laser, $E_1 \neq E_0$. Thus, as shown in Fig. 3.9.1(c), the quantum efficiency for this laser is even less than that of the three-level laser and is given by

$$\eta = \frac{E_2 - E_1}{E_p - E_0}.$$

We might wonder that if η is so bad for a four-level laser, why we consider it at all. Actually, as will be shown below, in general the four-level laser is relatively easy to pump. Thus nearly all the known lasers are four-level lasers.

Let us consider that N_T is the threshold population inversion density. Then, for a three-level laser, as the ground level is the same as the lower lasing level, we find that

$$N_2 - N_1 = N_T,$$

and

$$N_2 + N_1 = N_0,$$

where N_0 is the total number of electrons participating in the lasing action. Thus, $N_2 = N_0/2 + N_T/2$ at the threshold and, in general, for the laser to begin working

$$N_2 > \frac{N_0}{2} + \frac{N_T}{2} \approx \frac{N_0}{2}. \tag{3.9.3}$$

The last approximation holds because, in general, $N_0 \gg N_T$. However, in a four-level laser, as $E_1 \gg E_0$,

$$N_1 = N_0 e^{-(E_1 - E_0)/kT} \approx 0.$$

This, of course, assumes that $(E_1 - E_0)/kT \gg 1$. For the four-level laser to operate, the population density N_2 must be given by

$$N_2 > N_T. \tag{3.9.4}$$

Thus, we find that

$$\frac{(N_T)_{3\text{-level}}}{(N_T)_{4\text{-level}}} \approx \frac{N_0}{2N_T} = \text{a very large quantity in general.} \tag{3.9.5}$$

Thus it is much easier to pump a four-level laser than the three- or two-level lasers. This is the reason for nearly all the lasers being of the four-level type.

3.10. Rate Dynamics of Four-Level Lasers

Let us consider the general four-level laser shown in Fig. 3.9.1(c). The lasing energy levels are denoted by E_2 and E_1. The other energy levels present participate in pumping and spontaneous transitions. We are interested in the rate of change of the electron density N_2 in level E_2, and that of N_1 for E_1 only.

In the rate equation for N_2 we must consider the increase in the number of electrons by pumping. This pumping rate will be denoted by R_2. R_2 includes all the electrons which arrive at E_2 by spontaneous emission from the upper levels. The decrease in the number of electrons has three components:

(i) The all-important stimulated transition to E_1 which produces the lasing light given by $(N_2 - N_1)W_i$.
(ii) The spontaneous emission to the level E_1 given by N_2/t_{spont}.
(iii) The spontaneous emission to all other lower levels excluding E_1 and given

by

$$\frac{N_2}{t_2}.$$

Thus the rate equation for N_2 can be written as

$$\frac{dN_2}{dt} = +R_2 - (N_2 - N_1)W_i - \frac{N_2}{t_{\text{spont}}} - \frac{N_2}{t_2}. \qquad (3.10.1)$$

We can argue similarly for the E_1 level, to obtain

$$\frac{dN_1}{dt} = R_1 + (N_2 - N_1)W_i + \frac{N_2}{t_{\text{spont}}} - \frac{N_1}{t_1}. \qquad (3.10.2)$$

In the above equation, the unwanted pumping rate, R_1, to the level E_1 is included. The second and third terms of this rate equation are the rates at which electrons are arriving at E_1 from E_2 by stimulated and spontaneous emission, respectively.

The final term denotes the spontaneous emission to all lower levels from E_1 and is characterized by the time constant t_1. In equilibrium, the steady state condition is

$$\frac{dN_2}{dt} = \frac{dN_1}{dt} = 0. \qquad (3.10.3)$$

Or we obtain

$$R_2 - (N_2 - N_1)W_i - \frac{N_2}{t_2'} = 0,$$

$$R_1 + (N_2 - N_1)W_i + \frac{N_2}{t_{\text{spont}}} - \frac{N_1}{t_1} = 0,$$

where

$$\frac{1}{t_2'} = \frac{1}{t_{\text{spont}}} + \frac{1}{t_2}.$$

In general, t_2 is much larger than t_{spont} and thus t_2' can be approximated as

$$t_2' = t_{\text{spont}}.$$

If the medium is pumped, but not lasing, then

$$W_i = 0,$$

or

$$R_2 - \frac{N_2}{t_2'} = 0,$$

and $\qquad\qquad\qquad\qquad\qquad\qquad\qquad\qquad\qquad\qquad\qquad\qquad (3.10.4)$

$$R_1 + \frac{N_2}{t_{\text{spont}}} - \frac{N_1}{t_1} = 0.$$

Solving for N_1 and N_2 from the above, we obtain

$$N_1 = R_1 t_1 + \frac{R_2 t_2' t_1}{t_{spont}}, \tag{3.10.5}$$

and

$$N_2 = R_2 t_2'.$$

Thus, the population inversion density is given by

$$(N_2 - N_1) = R_2 t_2' \left(1 - \frac{t_1}{t_{spont}}\right) - R_1 t_1$$

$$\approx R t_{spont} = (\Delta N)_0, \tag{3.10.6}$$

where R is defined as the effective pumping rate. From (3.10.6) we observe that to have any population inversion

$$t_1 \leq t_{spont}. \tag{3.10.7}$$

This is a very important condition. If this is not satisfied then it does not matter how hard we pump since we can never achieve a population inversion. Physically, it means that the electrons coming to the E_1 level by spontaneous emission must be removed at a rate faster than the arrival rate so that no accumulation takes place. Otherwise, N_1 will increase at a faster rate than N_2, and no lasing action will take place.

Using (3.10.6) and assuming that the condition given by (3.10.7) holds, we can write, in the general case,

$$(N_2 - N_1) = \frac{(\Delta N)_0}{1 + \varphi t_{spont} W_i} = \frac{R t_{spont}}{1 + \varphi t_{spont} W_i}, \tag{3.10.8}$$

where

$$\varphi = \frac{t_2}{t_{spont}} \left[1 + \left(1 - \frac{t_2'}{t_{spont}}\right)\frac{t_1}{t_2'}\right].$$

This equation is obtained by solving for N_2 and N_1 from (3.10.1) and (3.10.2) and taking the difference. In general, $t_2' \approx t_{spont}$ and

$$\varphi \approx 1.$$

The amplification factor, γ, is given by

$$\gamma = (N_2 - N_1) \frac{c^2}{16\pi n^2 f^2 t_{spont}} g(f)$$

$$= \frac{(\Delta N)_0 c^2 g(f)}{16\pi n^2 f^2 t_{spont}} \cdot \frac{1}{1 + W_i t_{spont}}$$

$$= \frac{\gamma_0}{1 + W_i t_{spont}}, \tag{3.10.9}$$

where $\gamma_0 = (\Delta N)_0 c^2 g(f)/16\pi n^2 f^2 t_{spont}$, the gain constant in the absence of feedback (or the mirrors removed such that no lasing action takes place). We know that

$$W_i t_{spont.} = \frac{c^2 g(f)}{8\pi n^2 f^3 h} I(f)$$

$$= \frac{I(f)}{I_s}, \tag{3.10.10}$$

where I_s is the saturation intensity, given by

$$I_s = \left(\frac{8\pi f^2 n^2}{c^2 g(f)}\right) hf = \frac{(\Delta N)_0}{2\gamma_0 t_{spont}}. \tag{3.10.11}$$

Thus, we obtain

$$\gamma = \frac{\gamma_0}{1 + I/I_s}. \tag{3.10.12}$$

We see that the gain constant is initially γ_0 at the point when the feedback is switched on. Then, as the oscillation starts, I increases and γ decreases. Finally, a steady state is reached when

$$\gamma = \gamma_{th}. \tag{3.10.13}$$

This is shown schematically in Fig. 3.10.1. Thus, the intensity of the laser in that case will be given by

$$I = I_s\left(\frac{\gamma_0}{\gamma_{th}} - 1\right). \tag{3.10.14}$$

To obtain laser power we note that in the equilibrium condition, the number of electrons making stimulated transitions to contribute to the lasing action is N_T. Thus, the emitted power is given by

$$P_e = N_T \cdot hf \cdot V \cdot W_i, \tag{3.10.15}$$

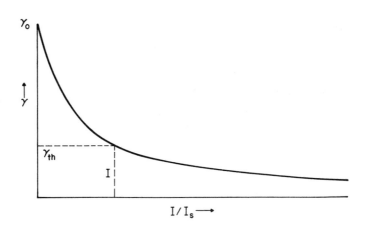

Fig. 3.10.1. Amplifying gain constant as a function of light infinity.

where V is the active volume of the lasing material. A quantity, called critical florescence, P_s, is defined as

$$P_s = N_T hf \cdot V \cdot \frac{1}{t_{spont}}. \tag{3.10.16}$$

P_s really means the amount of spontaneous light which is generated if the lasing material is just at threshold and the laser is not lasing. Thus,

$$\frac{P_e}{P_s} = W_i t_{spont} = \frac{I}{I_s}. \tag{3.10.17}$$

From (3.10.8) we obtain

$$W_i t_{spont} = \frac{R}{N_T} t_{spont} - 1 \tag{3.10.18}$$

or

$$\frac{P_e}{P_s} = \frac{R}{R_0} - 1, \tag{3.10.19}$$

where $R_0 = N_T/t_{spont}$ and physically means the threshold pumping rate. This equation gives us an expression for the lasing power inside the cavity in terms of the pumping magnitude. The harder a laser is pumped, the more power is obtained.

It is of interest to obtain expressions for I, P_e, and W_i in the steady state condition in terms of the decay constant. Including the loss in the cavity, (3.8.2) can be written as

$$\frac{\partial I(x)}{\partial x} = \left(2\gamma - \frac{1}{vt_c} \right) I(x)$$

$$= hf N W_i - \frac{Nhf}{t_c}. \tag{3.10.19a}$$

In the steady state, $N = N_T$ and $\partial I/\partial x = 0$. Thus we have

$$W_i = \frac{1}{t_c}, \tag{3.10.19b}$$

$$I = \frac{N_T hfv}{t_c}, \tag{3.10.19c}$$

and

$$P_e = \frac{N_T \cdot hf \cdot V}{t_c}. \tag{3.10.19d}$$

3.10.1. Optimum Output Power

Equation (3.10.19) gives us an expression for the total laser power inside the cavity. However, the useful power outside, P_0, is less than P_e since only a fraction of P_e is transmitted through the output mirror. To calculate this output power we note that as far as the oscillation condition is concerned, this output power is a loss to the cavity. Thus, if we increase the transmission coefficient of the output mirror, to increase P_0, we inherently reduce P_e, since P_e is a decreasing function of the transmission coefficient. Thus there exists an optimum output mirror transmission coefficient for obtaining maximum output power. To derive this optimum condition, we note that

$$\gamma = \gamma_0 \bigg/ \left(1 + \frac{P_e}{P_s}\right). \tag{3.10.20}$$

Also under the steady oscillation condition

$$\gamma_0 \bigg/ \left(1 + \frac{P_e}{P_s}\right) = \gamma_{th} = \frac{l_c}{2L}, \tag{3.10.21}$$

where l_c, the fractional loss per pass, is defined by (3.8.7d). We thus obtain

$$P_e = P_s \left(\frac{2\gamma_0}{l_c} - 1\right). \tag{3.10.22}$$

In l_c, we have two contributing terms, one due to the useful power output and the other due to the inherent losses. If we define T_0 as the useful mirror transmission and l_i as the inherent losses which are unavoidable, then

$$l_c = T_0 + l_i. \tag{3.10.23}$$

Also the output power, P_0, in this case will be given by

$$P_0 = P_e \cdot \frac{T_0}{T_0 + l_i}. \tag{3.10.24}$$

Using (3.10.22), we obtain

$$P_0 = P_s \frac{T_0}{T_0 + l_i} \left(\frac{2\gamma_0 L}{T_0 + l_i} - 1\right). \tag{3.10.25}$$

Thus, we see, as $T_0 \to 0$, $P_0 \to 0$. However, as T_0 tends to infinity, P_0 also decreases. To obtain the optimum value of P_0 we write

$$\frac{\partial P_0}{\partial T_0} = 0. \tag{3.10.26}$$

Solving (3.10.26) we obtain the optimum output mirror transmission, T_{opt}, given by

$$T_{opt} = P_s + \sqrt{2\gamma_0 L}. \tag{3.10.27}$$

The optimum power output is given by

$$P_{opt} = P_s(\sqrt{2\gamma_0 L} - \sqrt{l_i})^2. \tag{3.10.28}$$

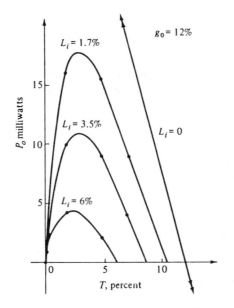

Fig. 3.10.2. Power output, P_0, versus mirror transmission T_0 for various values of l_i for an He–Ne laser. (From P. Laures, *Phys. Lett.*, **10**, 61, 1964.)

It is to be noted that P_s, from (3.10.16), can be rewritten as

$$P_s = \frac{N_T hf V}{t_{spont}} = \frac{8\pi hf^3}{v^2 g(f)} \cdot \left(\frac{V}{L}\right) \cdot l_c$$

$$= \frac{8\pi hf^3}{v^2 g(f)} l_c A, \tag{3.10.29}$$

where A is the cross-sectional area of the lasing mode assumed constant, and related to the spot size given by (3.5.17). Equation (3.10.25) is plotted in Fig. 3.10.2 for representative values for an He–Ne laser. As expected, a maximum is obtained for the optimum value of T_0.

3.11. Properties of Laser Light

Laser light is both spatially and temporally coherent, whereas light from other sources is mostly incoherent. The small time bandwidth, Δf, and the spatial bandwidth, Δf_x, achievable with significant energy by the laser light, are nearly impossible to reproduce by thermal sources. In the following, we discuss these special aspects of laser light, as compared to the incoherent light from other sources with respect to directionality, monochromacity, and statistical phase characteristics. The differences are so dramatic that although both laser light and light from other sources are the same electromagnetic waves, it would be almost correct to say that there is a fundamental difference between them.

Intensity for a fixed bandwidth Δf.

A typical He–Ne laser output is 1 mW, whereas a high power pulsed laser output is on the order of 10^{13} W. If the laser output has a wavelength, λ, and a spread, $\Delta\lambda$, then

$$\frac{\Delta f}{f} = -\frac{\Delta\lambda}{\lambda}, \qquad \text{where} \quad f\lambda = c.$$

For $\lambda \approx 0.5$ μm, $\Delta\lambda \sim 10^{-5}$ μm $= 0.1$ Å, we have $\Delta f \sim 1.2 \times 10^9$ Hz. Each photon at $\lambda = 0.5$ μm has energy $\sim 4 \times 10^{19}$ J. Thus, typical numbers for photons emitted per second will be $0.25 \times 10^{+16}$ to 0.25×10^{22} photons/s. The number of photons emitted per unit Hertz will be approximately 2×10^4 to 2×10^{10}.

If we now compare the laser light with blackbody radiation from (1.8.11) we obtain

$$\text{thermal photon/s} - \text{Hz} = \frac{2}{\lambda^2} \frac{1}{e^{hf/kT} - 1} \Delta A,$$

where the emission takes place from an area ΔA, and T is the blackbody temperature. The Boltzmann factor for our case is given approximately by $e^{-30,000/T}$ and is, in general, very small for ordinary temperatures. For $T = 1000$ K and $\Delta A = 1$ cm^2, we obtain

$$\text{thermal photon/s} - \text{Hz} < 1.$$

Comparing this number with that obtained, even for a 1 mW laser, 2×10^5, we see the enormous difference that exists between thermal light and laser light.

Radiance

The angular spread, $\Delta\theta$, of a typical laser beyond the Rayleigh distance is approximately given by

$$\Delta\theta \sim \frac{\lambda}{d},$$

where d is the aperture diameter. Thus

$$\Delta f_x \sim \frac{1}{d}.$$

The far-field solid angle, $\Delta\Omega$, into which the laser radiation is confined, can be approximated as

$$\Delta\Omega \approx (\Delta\theta)^2 = (\lambda\Delta f_x)^2 \approx \frac{\lambda^2}{A}.$$

Thus the radiance of a laser source (1 mW) will be, for $A = 1$ cm^2,

$$R_{\text{laser}} = \frac{10^{-3} \text{ W}}{\Delta\Omega} = 4 \times 10^5 \text{ W/sr.}$$

As the thermal source radiates over a solid angle of 2π steradian, the radiance will be given by

$$R_{\text{thermal}} = \frac{hf}{\lambda^2} \frac{\Delta f}{e^{hf/kT} - 1}.$$

If we assume, for a thermal source, $\Delta\lambda \sim 10^3$ Å and $T = 10^3$ K, even then

$$R_{\text{thermal}} \sim 4 \times 10^{-16} \text{ W/sr.}$$

This explains why even a 1 mW laser looks "brighter" than a thermal source at $T = 10^3$ K by twenty orders of magnitude.

Brightness

The brightness of a source is given by the power output per steradian of solid angle per hertz of bandwidth. For a laser with output power, P, the brightness is given by

$$B_{\text{laser}} = (P) / \left\{ \left(\frac{\lambda^2}{A} \right) \cdot \Delta f \right\} = \frac{PA}{\lambda^2 \Delta f}.$$

Let us assume that we start with a thermal source of temperature T° K, and filter it spatially as well as temporally to obtain an equivalent brightness for the thermal source equal to the laser source. We obtain

$$B_{\text{thermal}} = \frac{hfA}{(e^{hf/kT} - 1) \cdot \lambda^2} = \frac{PA}{\lambda^2 \Delta f}.$$

To have identical brightness we need a temperature T given by

$$e^{hf/kT} - 1 = \frac{(hf)\Delta f}{P},$$

or

$$T = \frac{hf}{k} \cdot \frac{1}{\ln\left(1 + \dfrac{hf\Delta f}{P}\right)} \approx \frac{P}{k\Delta f}.$$

Using $P = 10^{-3}$, $\Delta f \approx 10^9$, which is equivalent to $\Delta\lambda \sim 10^{-2}$ Å, we obtain

$$T \approx 10^{11} \text{ K.}$$

This is also the reason why we should never look at a laser directly even if it has only 1 mW of power, as it will appear to you as a source with $T \sim 10^{11}$ K.

Monochromacity and Coherence

The bandwidth, Δf, of a good stable laser can be less than 1 kHz compared to a thermal source which is of the order of 10^{14} Hz. To appreciate the differences between these two numbers let us calculate the coherence time and coherence length for both these sources. The coherence length of the laser is $\sim 3 \times 10^{+5}$ m. Thus using a laser source we can do an interference experiment with path lengths ~ 300 km, whereas for white light this distance is only 3 μm. This is also the reason the laser light always has speckle pattern and its intensity appears to fluctuate. Any reflection and scattering, even from far away, can interfere and thus cause change in light intensity. For incoherent light, in the case chosen as an example, any scatterer beyond 3 μm will not cause any fluctuations of intensity. The coherence time of laser light is of the order of milliseconds, whereas for thermal light it is $\sim 10^{-14}$ s.

3.12. Q-Switching and Mode Locking

3.12.1. Single-Mode and Multimode Lasers: Lamb Dip

In this section we will consider the effect of homogeneous and inhomogeneous broadening on the laser output. We shall see that for the case of a homogeneously broadened linewidth, only one mode can oscillate whereas for the inhomogeneously broadened case, multimode oscillations will be common.

Let us consider a lasing medium situated inside a cavity. The pumping is increased slowly from less than threshold to above threshold. The situations are depicted in Fig. 3.12.1, where the gain curve versus the frequency of the lasing medium is shown along with the discrete modes of the cavity. For the much-below-threshold case, spontaneous emission is present. However, because of the cavity, the cavity modes will add up in phase due to reflections and will dominate. As the pumping is increased, the amplitude of the cavity modes increases until the threshold is reached. Let us consider the case where the threshold is increased suddenly to a large value such that quite a large number of modes have gain. For each reflection, these modes grow exponentially. For example, consider the mode with the highest round trip gain of 4. After 10 round trips, its intensity grows by a factor of $4^{20} \sim 10^{12}$ provided the gain remains constant. For a different mode with gain constant 2 only, the increase intensity is by a factor $2^{20} \sim 10^6$. Thus we see that, because of this exponential growth, the mode with the largest gain will dominate as time passes. Also, as this mode increases, the gain of the other modes will start decreasing because of the saturation effect. Eventually, in the stable condition, the threshold condition will be maintained for a single mode and all other modes will be in the spontaneous noise level.

The above scenario of starting the oscillation of a laser is true for the homogeneously broadened case; for the inhomogeneously broadened case,

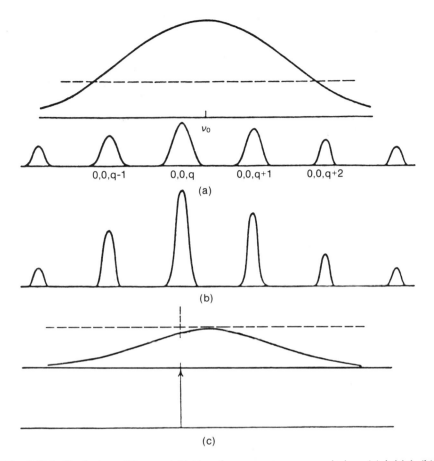

Fig. 3.12.1. Evolution of laser oscillation from spontaneous emission: (a) initial, (b) intermediate, and (c) final.

different modes can oscillate simultaneously—this can be understood by considering Fig. 3.12.2. Below threshold the situation is the same as that discussed previously in connection with Fig. 3.12.1. However, above threshold, the gain of each mode is contributed by a group of atoms whose Doppler frequency matches that of the mode frequency. For the homogeneously broadened case, all atoms match to a single frequency; for the inhomogeneously broadened case, different group of atoms contribute gain to different modes and thus all the modes with gains above the threshold level will oscillate with relative amplitudes determined by their gain constant—this is shown in Fig. 3.12.2. We note that each mode will bring down the gain curve at that frequency to the threshold value, keeping the other parts unaffected. In a sense, each mode will introduce a "hole" in the gain curve. Note that the output power of the mode will be proportional to the corresponding area of the hole.

If the inhomogeneous broadening is due to the Doppler effect of the moving

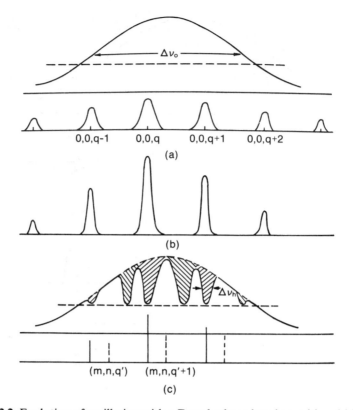

Fig. 3.12.2. Evolution of oscillation with a Doppler broadened transition. (a) The top curve shows the gain versus frequency. The bottom curve shows the amplitude versus frequency of different modes. (b) Larger amplitude of different modes following the gain curve. (c) Gain curve with multimode oscillation. The dashed line corresponds to threshold gain.

atoms, then another interesting phenomenon, the Lamb dip, occurs. Note that for a wave traveling in the $+z$ direction, the Doppler is opposite to that of the wave traveling in the $-z$ direction. For laser oscillation in a particular mode, since the wave must bounce back and forth between the mirrors, it forms a standing wave. The interaction takes place between two sets of atoms with equal and opposite velocities—these introduce holes in the spectra, as shown in Fig. 3.12.3. If only one mode is oscillating, and the cavity mode frequency is shifted or tuned by changing the mirror separation, the output will be given by Fig. 3.12.4 where there is a dip in the middle. Since the cavity mode is near the edge, the power output is low corresponding to point (a); as it approaches the center frequency, corresponding to point (b), the power increases. However, when the mode is at the center, the two holes overlap; and the power is less as the two holes merge, and the overall area decreases.

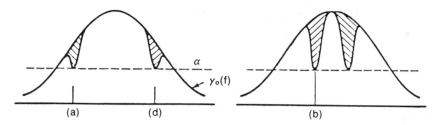

Fig. 3.12.3. Holes in the spectra due to standing waves.

This dip in power is generally referred to as the Lamb dip, since it was predicted by Lamb.

3.12.2. Mode Locking of Multimode Lasers

All of the practical lasers are inhomogeneously broadened and thus, if special care is not taken to raise the threshold level of all but one mode, multimode oscillation takes place. The number of modes oscillating can be quite high. For the case of an He–Ne laser, the linewidth $\Delta\lambda$ may be approximately 0.2 Å corresponding to $(\Delta f)_{\text{Doppler}} \sim 1500$ MHz. Assuming a cavity length of 100 cm, $(\Delta f)_{\text{mode}} = 150$ MHz, and the number of modes excited will be 10. For the case of multimode excitation, the output electric field, $E_{\text{out}}(t)$ will be given by

$$E_{\text{out}}(t) = \sum_{-(N-1)/2}^{(N+1)/2} E_n e^{j2\pi[f_0 t + (\Delta f)_{\text{mode}} n t + \varphi_n]}, \qquad (3.12.1)$$

where E_n is the amplitude for the nth mode, N is the total number of excited modes, and φ_n is an arbitrary phase constant. There is no guarantee that φ_n will be constant. If we can make it a constant by some means, then, the modes are locked. For this case of a mode-locked laser, assuming for simplicity $E_n = E_0 = $ constant, we have

$$E_{\text{out}}(t) = E_0 \sum_{-(N-1)/2}^{(N+1)/2} e^{j2\pi[f_0 t + n(\Delta f)_{\text{mode}} t]}, \qquad (3.12.2)$$

where we have assumed $\varphi_n = $ constant $= 0$.

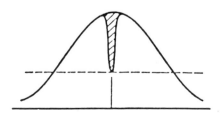

Fig. 3.12.4. Lamb dip.

Equation (3.12.2) is similar to (2.8.16), discussed in connection with diffraction gratings, and can be written as

$$E_{out}(t) = E_0 e^{j2\pi f_0 t} \frac{\sin\{[(\Delta f)_{mode} N t]/2\}}{\sin\{[(\Delta f)_{mode} t]/2\}}.$$ (3.12.3)

The plot of (3.12.3) for N large is very similar to the plot in Fig. 2.8.5; it has a maxima at

$$t = p\left(\frac{1}{\Delta f}\right)_{mode},$$ (3.12.4)

where p is an integer and a pulse width Δt given by

$$\Delta t = \frac{1}{N(\Delta f)_{mode}} = \frac{1}{(\Delta f)_{Doppler}}.$$ (3.12.5)

The maximum amplitude is NE_0. Thus we see that a mode-locked laser produces very short pulses with a repetition rate of $1/(\Delta f)_{mode}$. For an He–Ne laser, the pulse width is of the order of 0.6 ns. For an Nd–glass laser $(\Delta f)_{Doppler} \sim 3 \times 10^{12}$ ($\Delta\lambda \sim 300$ Å) Hz. For this case, $\Delta t \sim 0.3$ ps. For a dye laser $(\Delta f)_{Doppler} \sim 3 \times 10^{13}$ ($\Delta\lambda \sim 1000$ Å) with corresponding $\Delta t \sim 30$ fs. There are other techniques of reducing the pulse width even further to ~ 5 fs. This will be discussed in a later section.

We have not yet discussed how mode locking in a laser is performed. To understand this, consider the standing waves for all the modes. Including the space dependence, (3.12.2) becomes

$$E_{out}(z, t) = E_0 \sum_n \sin\{[2\pi(f_0 t + (\Delta f)_{mode} n t] \sin k_n z,$$ (3.12.6)

where

$$k_n = 2\pi \left\{\frac{f_0 + n(\Delta f)_{mode}}{v}\right\}.$$ (3.12.7)

Using trigonometric identities, we obtain

$$E_{out}(z, t) = E_0 \sum_n \sin\left[(q + n)\frac{2\pi v}{2d} t\right] \sin\left[(q + n)\frac{2\pi v}{2d} z\right],$$

or

$$E_{out}(z, t) = \frac{E_0}{2} \sum_n \cos\left[(q + n)\frac{\pi}{d}\left(t - \frac{z}{v}\right)\right] - \frac{E_0}{2} \sum_n \cos\left[(q + n)\frac{\pi}{d}\left(t + \frac{z}{v}\right)\right],$$ (3.12.8)

where

$$q = f_0 \frac{2d}{v}.$$ (3.12.9)

Finally, we obtain

$$E_{out}(z, t) = \frac{E_0}{2} \cos \frac{q\pi}{d}\left(t - \frac{z}{v}\right) \frac{\sin N \frac{\pi}{2d}\left(t - \frac{z}{v}\right)}{\sin \frac{\pi}{2d}\left(t - \frac{z}{v}\right)}$$

$$- \cos \frac{q\pi}{d}\left(t + \frac{z}{v}\right) \frac{\sin N \frac{\pi}{2d}\left(t + \frac{z}{v}\right)}{\sin \frac{\pi}{2d}\left(t + \frac{z}{v}\right)}. \qquad (3.12.10)$$

The above equation has two pulses, one propagating in the forward direction and one propagating in the backward direction. Thus, if we introduce a shutter inside the cavity with shutter openings at times as shown in Fig. 3.12.5, we will obtain mode locking. This shutter can be a simple acousto-optic or electro-optic modulator with a modulating frequency of $(\Delta f)_{mode}$. This form of mode locking, achieved by forcing the longitudinal modes to maintain a fixed-phase relationship, is called active mode locking. Mode locking can be obtained without any shutters by using a saturable absorber inside the cavity—this form of mode locking is called passive mode locking. Saturable absorbers, mostly dyes in solvents, absorb the light passing through them for low power, but pass light through with no attenuation for high power. Thus a saturable absorber will act as a shutter. Since, in practice, the laser medium itself acts as a saturable absorber for many cases, the laser can mode lock by itself.

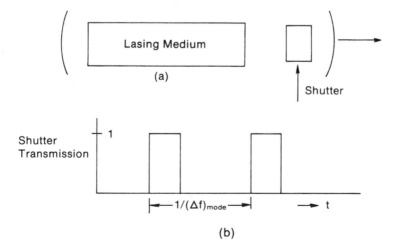

Fig. 3.12.5. Mode locking of a laser using a shutter inside the cavity: (a) configuration; and (b) shutter transmission as a function of time.

3.12.3. Q-Switching

Q-switching is a transient phenomena. This means we can obtain a very large peak pulsed power from a laser even if it can only deliver much lower CW power. Q, or the quality factor of a resonant circuit, is a well-known attribute in radio engineering. It is defined as

$$Q = 2\pi \frac{\text{energy stored at resonance}}{\text{energy lost in a cycle}}. \qquad (3.12.11)$$

If an $L-C$ circuit (Fig. 3.12.6) is used as the resonant system, then we have

$$Q = \frac{\omega_0 L}{R} = \frac{1}{\omega_0 CR}, \qquad (3.12.12)$$

where $L, C,$ and R are the inductance, capacitance, and resistance in the circuit, and the resonance frequency is given by

$$f_0 = \frac{\omega_0}{2\pi} = \frac{1}{2\pi}(\sqrt{LC})^{-1}. \qquad (3.12.13)$$

It can be shown that

$$Q^{-1} = \frac{\text{half width}}{\text{resonance frequency}}. \qquad (3.12.14)$$

For an optical cavity, like a Fabry–Perot interferometer, and using (2.173), we obtain

$$Q = \text{resolving power} = \frac{\pi m \sqrt{F}}{2}. \qquad (3.12.15)$$

Q is also related to the laser linewidth in a laser cavity and is given by (3.3.15c)

$$Q = \mathscr{F}. \qquad (3.12.16)$$

To understand the Q-switching of a laser, we note that a lasing medium can be pumped to a very high population inversion level or high gain when the laser is not oscillating. This can be performed by removing a mirror or

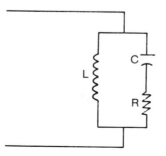

Fig. 3.12.6. $L-C$ equivalent circuit of a cavity.

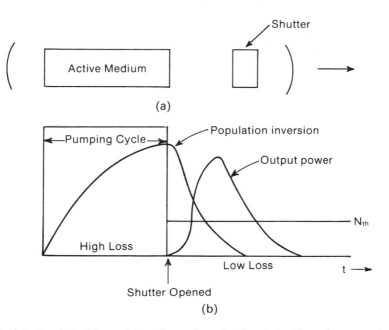

Fig. 3.12.7. Q-switched laser: (a) configuration of a Q-switched laser incorporating a shutter; and (b) population inversion, output power, and shutter transmission as a function of time.

putting some extra loss mechanism in the cavity or, as it is called, by spoiling Q or making it effectively very low. Note that if the laser oscillates in CW or the Q is not spoiled, the high gain or population inversion density comes back to the threshold value in the steady state situation. However, at the beginning, when the Q of the cavity is just restored, the laser output corresponds to the initial large population inversion density. If this continues, the output decreases asymptotically to the steady state value. In Q-switching, this Q-spoiling is done periodically to obtain giant pulses.

A typical Q-switched laser is shown in Fig. 3.12.7 where a shutter is introduced inside the cavity. This shutter might be an acousto-optic or an electro-optic cell or a moving mirror which is mounted on a rotating shaft. The moving mirror acts as one of the cavity mirrors and when they are aligned, Q increases to a very high value. Figure 3.12.7 also shows the pumping of the laser when Q is spoiled, and the optical output when Q is restored.

By including a saturable absorber, discussed in connection with mode locking, we can also obtain Q-switching. We note that initially, the light intensity is low, so the absorption is high and Q is low. However, as the intensity builds up slowly, the absorption saturates, resulting in higher Q and higher pulse power. The difference in operation between Q-switching and mode locking is that the relaxation processes of the saturable absorber are much larger than $1/(\Delta f)_{mode}$ for Q-switching, whereas it must be less for the mode locking. Similar arguments apply for the shutters, also. To obtain mode locking, the shutter operating period must be $1/(\Delta f)_{mode}$. Another way of looking at Q-switching is to consider that large energy is stored by steady pumping to the upper lasing level. This energy is released by the Q-switch in a short time. The situation is very similar to charging a capacitor by a d.c. voltage to a large value to store energy, and then discharging it through a switch to obtain a large spark of current.

To analyze the Q-switching mechanism, we note that since the light pulse duration is very small, we can neglect pumping and changes in population during the pulse. We will also consider that the shutter is instantaneous. The rate of change of the number of photons, φ, in the cavity during pulse is given by

$$\frac{d\varphi}{dt} = \varphi\left(\gamma\frac{vL}{l} - \frac{1}{t_c}\right) \tag{3.12.17}$$

where γ is the exponential growth constant, v is the light velocity in the lasing medium, and t_c is the cavity decay constant. Equation (3.12.17) can by derived by noting that

$$\frac{dI}{dt} = \frac{dI}{dz}\cdot\frac{dz}{dt} = \gamma vI, \tag{3.12.18}$$

where $I = $ intensity $= I_0 e^{\gamma z} = v\varphi$. The factor L/l (L is the length of the active medium and l is the cavity length) comes in because the amplification takes place only within the lasing medium.

Equation (3.12.17) can be written as

$$\frac{d\varphi}{d\tau} = \varphi\left[\frac{\gamma}{\gamma_t} - 1\right] = \varphi\left[\frac{u}{n_t} - 1\right], \tag{3.12.19}$$

where $\tau = t/t_c$,
    $\gamma_t = l/vt_c$ is the threshold gain constant,
    n is the population inversion,
and $n_t = N_t V$ is the total population inversion at threshold.

For every photon generated, an electron makes a transition from the upper level to the lower level. Thus, the change in n is -2 for every photon generated. As the rate at which population difference is changing must be exactly equal to the rate at which photons are increasing, we have

$$\frac{dn}{d\tau} = -2\varphi\frac{n}{n_t}. \tag{3.12.20}$$

Equations (3.12.19) and (3.12.20) give the evolution of a giant pulse in Q-switching. Dividing (3.12.19) by (3.12.20), we obtain

$$\frac{d\varphi}{dn} = \frac{n_t}{2n} - 1. \tag{3.12.21}$$

Solving the above equation we obtain

$$\varphi - \varphi_i = \frac{1}{2}\left[n_t \ln \frac{n}{n_i} - (n - n_i) \right],$$

(3.12.22)

where φ_i and n_i are the initial values. As φ_i is negligible, we obtain

$$\varphi \approx \frac{1}{2}\left[n_t \ln \frac{n}{n_i} - (n - n_i) \right].$$

(3.12.23)

We also note that for t large, $\varphi \to 0$. Thus, denoting by n_f the final value of the population inversion, we obtain

$$\frac{n_f}{n_i} = \exp\left\{ \frac{n_f - n_i}{n_i} \right\}.$$

(3.12.24)

The transcendental equation can be solved numerically to obtain n_i versus n_f. However, a meaningful quantity is the energy utilization factor or the fraction of energy stored in the population inversion that gets converted to a laser pulse. This is given by $(n_i - n_f)/n_i$. A plot of n_i/n_t is shown as a function of $(n_i - n_f)/n_i$ in Fig. 3.12.8. We note that the energy utilization factor approaches 100% as n_i/n_t increases to a large value.

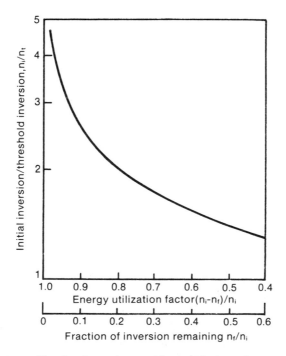

Fig. 3.12.8. Energy utilization factor $(n_i - n_f)/n_i$ and the inversion remaining after the giant pulse. (From W.C. Wagner and B.A. Lengyel, Evolution of the Giant Pulse in a laser, *J. Appl. Physics*, **34**, 1963.)

The laser power output is given by

$$P = hf\varphi = \frac{hf}{2}\left[n_t \ln\frac{n}{n_i} - (n - n_i)\right]. \tag{3.12.25}$$

To find the maximum power we apply the condition

$$\frac{dP}{dn} = 0, \tag{3.12.26}$$

from this we obtain the result that peak power occurs at $n = n_i$. Thus peak power occurs at the beginning of the pulse as expected.

Peak power is given by

$$P_{peak} = \frac{hf}{2t_c}\left[n_t \ln\frac{n_t}{n_i} - (n_t - n_i)\right]. \tag{3.12.27}$$

As $n_t \ll n_i$ in usual practice we have

$$P_{peak} = \frac{n_i hf}{2t_c}. \tag{3.12.28}$$

From (3.10.19d), we note that in the steady state CW operation, power is given by

$$P = \frac{n_t hf}{t_c}. \tag{3.12.29}$$

Thus, the peak power of the Q-switched laser can be one thousand fold more or higher, since n_i can be made very large compared to n_t if the laser is not lasing under a Q-spoiled situation.

The total energy E_{tot}, contained in the pulse is given by the multiplication of maximum energy obtainable $(n_i hf/2)$ and the energy utilization factor

$$E_{tot} = \frac{n_i - n_f}{n_i} \cdot \frac{n_i hf}{2}. \tag{3.12.30}$$

From (3.12.29) and (3.12.30) we can obtain an estimate of the pulse width, Δt, by using the approximation

$$P_{peak}\Delta t \approx E_{tot},$$

or

$$\Delta t = \frac{E_{tot}}{P_{peak}}. \tag{3.12.31}$$

It is to be noted that there is a fundamental difference between the pulsed mode of laser operation and the Q-switched mode. Q-switching can be incorporated in the pulse mode also. Figure 3.12.9 shows the flash lamp output, cavity Q, population inversion, and laser output power versus time. Note that for an ordinary pulsed mode, cavity Q will always be high and independent of time. The laser output will also be a longer pulse with less peak power.

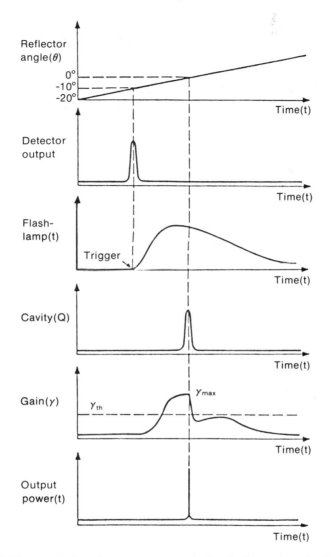

Fig. 3.12.9. Time evolution of laser parameters during the formation of a *Q*-switch the rotating reflector method. (From D.C. O'Shea et al., *Introduction to Lasers and Their Applications*, Addison-Wesley, Reading, MA.)

Gain Switching and Relaxation Oscillation

Many lasers show intensity variation in the form of relaxation oscillation because of the fact that the laser field and the population inversion are interdependent. To explain these oscillations we have to consider the finite time to build up the population inversion and the time needed to build up the oscillation from spontaneous emission. For example, a CO_2 laser can be pumped very fast since the rate is of the order of 10^8 s^{-1} due to collisions with N_2 molecules. However, the spontaneous rate for the lasing frequency is only 0.3 s^{-1}. Thus it takes a much longer time before the lasing output really builds up. As the laser starts building up, the interplay between the laser field and the population generates the relaxation oscillation.

Let us consider the time-dependent four-level laser equation given by

$$\frac{dN}{dt} = R - W_i N - \frac{N}{\tau},$$ (3.12.32)

where τ is the effective decay constant for the upper level excluding the stimulated emission rate. We also know that the stimulated emission rate is related to the number of photons, φ, and is given by

$$W_i = B\varphi,$$ (3.12.33)

where B is constant and is equal to $\gamma V L/l$. The photon generation rate is given by (3.12.19)

$$\frac{d\varphi}{dt} = \varphi BN - \frac{\varphi}{t_c}.$$ (3.12.34)

In the steady state case

$$\frac{dN}{dt} = \frac{d\varphi}{dt} = 0.$$ (3.12.35)

The steady state solution is given by

$$N_0 = \frac{1}{Bt_c}$$ (3.12.36)

and

$$\varphi_0 = \frac{RBt_c - 1/\tau}{B}.$$ (3.12.37)

The threshold pumping rate is obtained by substituting $\varphi_0 = 0$ in (3.12.37). Therefore

$$R_t = \frac{1}{\tau t_c B}.$$ (3.12.38)

Substituting the value for R_t we obtain

$$\varphi_0 = \frac{r - 1}{B\tau}$$ (3.12.39)

where $r = R/R_t$.

Note the difference between the case being considered now and the Q-switching case. In the Q-switched case we neglected the pumping rate. To obtain the relaxation oscillation, we consider small perturbations around the equilibrium value

$$N(t) = N_0 + N_1(t), \tag{3.12.40}$$

$$\varphi(t) = \varphi_0 + \varphi_1(t). \tag{3.12.41}$$

Note that $N_1(t) \ll N_0$ and $\varphi_1(t) \ll \varphi_0$. Substituting (3.12.40) and (3.12.41) into (3.12.32) and (3.12.34) and neglecting higher-order terms, we obtain

$$\frac{dN_1}{dt} = -RBt_c N_1 - \frac{\varphi_1}{t_c} \tag{3.12.42}$$

$$\frac{d\varphi_1}{dt} = RBt_c - \frac{N_1}{\tau}. \tag{3.12.43}$$

Eliminating N_1 from (3.12.42) and using (3.12.43) we obtain

$$\frac{d^2\varphi_1}{dt^2} + RBt_c \frac{d\varphi_1}{dt} + \left(RB - \frac{1}{\tau t_c} \right) \varphi_1 = 0, \tag{3.12.44}$$

or

$$\frac{d^2\varphi_1}{dt^2} + \frac{r}{\tau} \frac{d\varphi_1}{dt} + \frac{1}{\tau t_c}(r - 1) = 0,$$

where $r = RBt_c\tau$. The solution of (3.12.44) is given by

$$\varphi(t) \propto e^{-\alpha t} \cos \omega_m t, \tag{3.12.45}$$

where

$$\alpha = \frac{r}{2\tau}, \tag{3.12.46}$$

and

$$\omega_m = \sqrt{\frac{1}{t_c\tau}(r - 1) - \left(\frac{r}{2\tau} \right)^2}. \tag{3.12.47}$$

As the power output is proportional to $\varphi(t)$, we see that the variation of laser output intensity will be in the form of decaying oscillations or the relaxation oscillation. A typical relaxation oscillation of an Nd laser is shown in Fig. 3.12.10.

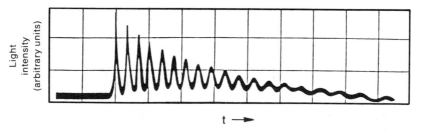

Light intensity (arbitrary units)

t →

Fig. 3.12.10. A typical intensity relaxation oscillation in a laser.

3.13. Lasers

If properly pumped, any matter, gas, liquid, or solid, is capable of lasing. Thus the subject of laser types is a vast one, and it is impossible to cover each and every system. In this section we mention some important and practical ones. These can be subdivided into four different categories: gas, solid, dye, and semiconductor lasers.

3.13.1. The Gas Laser

Before we discuss individual laser systems, it is of interest to point out some common features. For a gas laser to be practical there must be a container of gas, generally made of glass, so that it is transparent to the desired radiation. If a large electric field is applied across this gas, as shown in Fig. 3.13.1, the mobile electrons and ions accelerate and collide with gas molecules. If the voltage is large enough, they produce ionized and excited molecules or atoms. The excited atoms in turn emit light and this whole process is generally known as a gas discharge. Magnetic fields are often used to confine the gases. Radio frequency electric fields are also sometimes used to form the discharge or to aid it in conjunction with a d.c. voltage.

The cavity mirrors can be either inside the gas container or outside. If they are inside (the internal mirror arrangement), then the output light is generally unpolarized. For the outside case, to minimize reflection loss, gas container edges are cut at a Brewster angle. The Brewster angle, θ_B, is given by (see Section 2.12.2)

$$\theta_B = \tan^{-1} \sqrt{\frac{n_2}{n_1}}, \qquad (3.13.1)$$

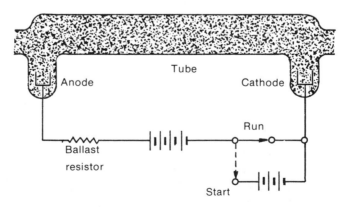

Fig. 3.13.1. Simplified electrical circuit for a gas laser. A larger voltage is needed to start the discharge than to maintain it, so a high-voltage pulse is applied to the gas when the laser is turned on. The ballast resistor serves to limit the current once the discharge is initiated.

where n_2 and n_1 are the refractive indices of the glass and the gas mixture, respectively. It is known that light incident at the Brewster angle, polarized parallel to the plane of incidence, has a transmission coefficient of 1. The passage of light through a glass container at the Brewster angle does not involve any transmission losses and thus the parallel polarization component has a higher effective gain. Thus for the outside mirror arrangement, radiation is plane polarized in general.

The He–Ne Laser

The He–Ne laser is the most popular laser, and was the first laser to be demonstrated operating in CW mode. The energy-level diagram of helium and neon atoms is shown in Fig. 3.13.2. Note that actual laser emission takes place through the neon energy levels. Helium gas is present to provide more efficient excitation. In general, a 10 : 1 mixture of helium and neon is used. As shown in Fig. 3.13.2, the transition from the 5s energy level to the 3p energy level forms the well-known red light ($\lambda = 0.633$ μm) emission. Two other

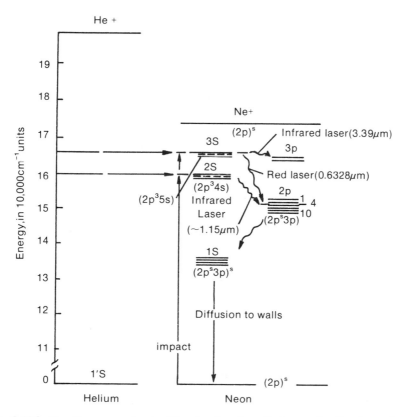

Fig. 3.13.2. He–Ne energy levels. The dominant excitation paths for the red and infrared laser transitions are shown.

transitions (3.39 μm from 5s to 4p and 1.15 μm from 4s to 3p) also produce efficient lasing action if infrared mirrors are used. In general, an He–Ne laser is excited using electric discharge as shown in Fig. 3.13.1. The electrons produced in the discharge collide with helium atoms in the ground state, and excite it to metastable states like the 2^1s and 2^3s energy levels. We note that the 2^1s level of helium has very nearly the same energy as the 5s level of neon. The same applies to the 2^3s level of helium and the 5s level of neon. Thus, in the collision of a 2^1s level excited helium atom and a ground level neon atom, the helium atom loses energy going to the ground state, whereas the neon atom is excited to the 5s level. This resonant collision and subsequent excitation of the neon atom is far more efficient than the direct excitation of neon atoms.

Both the 0.633 μm and 3.39 μm transitions start from the same 5s level. The 3.39 μm transition has a higher gain and thus the laser tends to lase at this frequency unless precautions are taken. This might be to ensure that the cavity mirrors have a very small reflection coefficient at 3.9 μm. Sometimes, small magnets are placed along the length of the laser cavity to create an inhomogeneous magnetic field, which in turn broadens the 3.39 μm line more than the 0.633 μm line. Higher linewidth reduces the amplification factor.

In general, the output power of the He–Ne lasers is in the 0.5–5 mW range. The larger the output, the larger the cavity length and thus the size of the laser. Maximum power is less than 100 mW. Both the outside and inside mirror arrangements are used to manufacture these lasers.

The Ion Gas Lasers (Ar, Kr)

Helium, neon, argon, xenon, and krypton are noble gases and they have electronic states capable of laser transitions. However, except for neon, noble gases are difficult to pump and thus are not of practical interest. However, if these noble gases are first ionized by electron collisions, then they are easy to pump. Actually, they form the highest power visible lasers producing tens of watts. A typical ion laser tube is shown in Fig. 3.13.3. The cathode is coated with a material which emits a large quantity of electrons which in turn produce a very large discharge current (~ 1000 A/cm^2). The current is confined by the magnetic field to a small-diameter bore. This also helps to reduce the collision between the ions/electrons and the glass container. Because of high current density, intense heat is produced. Thus the material for the small-diameter bore is generally either graphite or beryllium oxide. The laser is also water-cooled to dissipate some of the heat.

Typically, a trigger pulse is needed to initiate the discharge. Because the electrons are more mobile than ions, the ion concentration near the cathode builds up which in turn tries to shut off the laser. To remedy this, the bore material has staggered off-axis holes so that the ions can use this as a return path to diffuse toward the anode.

The energy-level diagram of Ar$^+$ is shown in Fig. 3.13.4. An Ar$^+$ ion laser

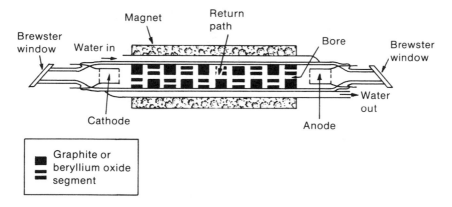

Fig. 3.13.3. Construction of an ion laser tube. The water jacket and magnet windings surround the tube. The return path permits diffusion of ions back to the anode to equalize the pressure caused by a pileup of neutralized ions at the cathode. (From D.C. O'Shea et al., *Introduction to Lasers and Their Applications*, Addison-Wesley, Reading, MA.)

can oscillate simultaneously in many frequencies with typical output powers, as shown in Table 3.13.1. If a single frequency is needed, then either a prism or a diffraction grating is used to select a particular line. The Kr^+ ions can also lase in many frequencies with typical power outputs as shown in Table 3.13.1. Sometimes an Ar^+ and Kr^+ mixture is used as a lasing medium to obtain nearly "white light".

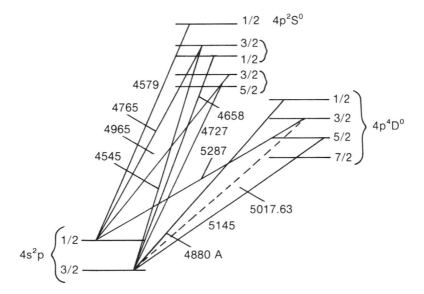

Fig. 3.13.4. Energy level of Ar^+ laser transitions. (From W.B. Bridges, *Appl. Phys. Lett.*, **4**, 1964.)

Table 3.13.1. Table for wavelengths and power output for typical argon and krypton ion lasers.

Laser	Wavelength in Å	Power output
Ar$^+$	5145	3.2 W
	5017	0.3 W
	4965	1.1 W
	4880	3.0 W
	4765	1.1 W
	4727	50 mW
	4657	30 mW
	4579	600 mW
	4545	10 mW
Kr$^+$	6746	
	6471	
	5682	
	5308	
	5208	
	4825	
	4762	
	2619	

The Metal Vapor Lasers

In a metal vapor laser, the solid metal is first vaporized and then brought into the discharge tube to be ionized. The two most successful lasers of this kind are the He–Cd and He–Se laser. In an He–Cd laser, metastable helium atoms are used for the resonant transfer of energy to cadmium ions by collision, in a manner similar to that discussed in connection with the He–Ne laser—this is shown in Fig. 3.13.5. An He–Cd laser oscillates in blue (0.442 μm) and ultraviolet (0.325 μm). The energy-level diagram for an He–Se laser is shown in Fig. 3.13.6. For this case helium ions, by collision with selenium atoms, transfer energy to excite them to the upper lasing levels. An He–Se laser is capable of lasing in 0.46 μm–0.65 μm.

The typical power range of these lasers is 5–25 mW. As mentioned before, in these lasers the metal is heated to vapor and is eventually transported towards the anode. Thus, when the metal is all used up, the laser ceases to operate. To avoid this, either a return path is provided to recover the metal, or a cathode, an anode, and a metal source are added to both ends of the tube. In the latter case, the use of the cathode at one end and the anode at the other end is alternated.

The CO_2 Laser

The CO_2 laser is a molecular laser where the vibrational levels of one carbon atom and two oxygen atoms, bonded by chemical means, are used. Except for

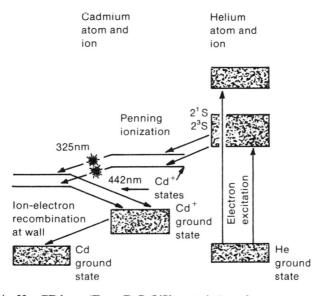

Fig. 3.13.5. An He–CD laser. (From D.C. O'Shea et al., *Introduction to Lasers and Their Applications*, Addison-Wesley, Reading, MA.)

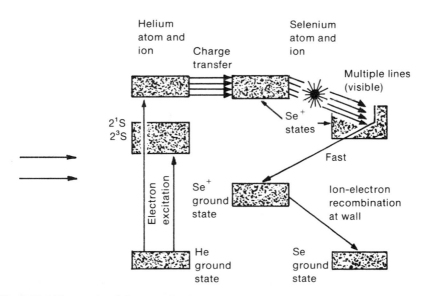

Fig. 3.13.6. Energy-level diagram and important transitions for the He–Se laser. (From D.C. O'Shea et al., *Introduction to Lasers and Their Applications*, Addison-Wesley, Reading, MA.)

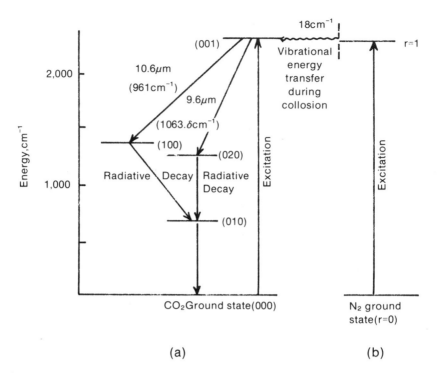

Fig. 3.13.7. (a) Some of the low-lying vibrational levels of the carbon dioxide (CO_2) molecule, including the upper and lower levels for the 10.6-μm and 9.6-μm laser transitions. (b) Ground state ($v = 0$) and first excited state ($r = 1$) of the nitrogen molecule, which plays an important role in the selective excitation of the (001) CO_2 level. (From A. Yariv, *Introduction to Optical Electronics*, Holt, Rinehart and Winston, New York, 1976.)

junction lasers, it has the highest overall efficiency. It can also produce millions of watts CW and even higher amounts in pulsed mode. The energy-level diagram for CO_2 is shown in Fig. 3.13.7. Each vibrational mode contains many rotational levels with small energy separation. Thus they are denoted as a band. The CO_2 laser can oscillate both at 10.6 μm and at 9.6 μm; however, the strongest gain is at 10.6 μm. As for the case of an He–Ne laser, the addition of nitrogen and helium improves the output significantly.

The overall efficiency of CO_2 lasers can be as high as 30%. The output power is proportional to the length of the discharge tube—it can produce approximately 100 W/m of the discharge tube length. However, it takes approximately 12 kV/m for discharge with CO_2 at atmospheric pressure. For a lower pressure, the voltage needed is less, but it produces less power. To solve this problem of a high-voltage supply, two approaches have been very successful. These are:

(1) the Transverse Excitation Atmospheric (TEA) laser, and
(2) the gas dynamic laser.

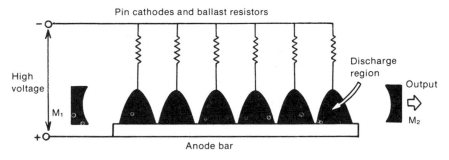

Fig. 3.13.8. TEA laser. The discharge occurs perpendicular to the laser cavity.

For a TEA laser, atmospheric pressure is maintained in the discharge tube; but the gas discharge is not maintained by applying an electric field in the longitudinal direction but rather in the transverse direction, as shown in Fig. 3.13.8. Since the discharge takes place at a critical electric field, less voltage is necessary for transverse excitation. The transverse directions are of the order of 1 cm requiring only 0.12 kV. In general, the discharge is to be maintained uniformly over the whole discharge length, and special care is taken in the design of the anode and cathode. The TEA lasers are the most important of the commercially available CO_2 lasers.

The gas dynamic laser does not use an electrical discharge. For this case, the gas mixture is first heated, then compressed, and finally sent through a nozzle into a region of reduced pressure. The thermodynamic energy stored due to heat and compression is the source of pumping.

The principles of shock tube and rocket technology are applicable for the design of these lasers. They produce enormous amounts of laser output. However, they are bulky and are associated with a tremendous amount of audible roar associated with the high-pressure gas exhaust. Depending on the type of discharge and flow, CO_2 lasers can be of four basic types. These are:

(1) axial discharge with slow axial flow;
(2) axial discharge with fast axial flow;
(3) transverse electron-beam preionization with fast transverse gas flow; and
(4) transverse discharge with transverse fast flow.

A typical axial discharge with a slow axial flow CO_2 laser is shown in Fig. 3.13.9. It generally delivers 50–70 W/m and depends on the efficiency of heat transfer from the gas to the cooling liquid that surrounds the laser tube in a separate jacket or tube—typical tube lengths are 2–3 m. By combining separate discharge tubes with electrical input and gas flow in parallel, power output up to ~ 1 kW can be achieved. The lasers can be pulsed to obtain very high peak power. The tube diameters are generally small, resulting in low-order modes only.

For the case of the axial discharge with a fast axial flow CO_2 laser, the gas mixture is blown through the laser tubes at high speed; also, the mixture

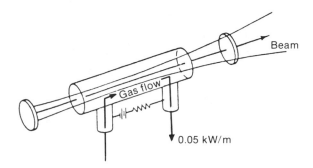

Fig. 3.13.9. Axial discharge–axial flow CO_2 laser schematic.

is recycled through a heat exchanger. The power output is of the order of 600 W/m. The tube diameter may be quite large; thus, unstable and higher-order mode outputs are common—typical outputs are 0.5–10 kW CW.

The transverse electron beam preionization with a transverse gas flow type CO_2 laser uses a lower voltage to sustain the discharge. This is possible by using an electron beam to ionize the gas. The electron beam is generally produced in a high vacuum by thermal emission from a large planar filament cathode; this beam is then accelerated using a high voltage. The high-energy electrons strike a thin metal foil separating the high vacuum of the electron gun and the high-pressure laser cavity. Secondary electrons ejected from the foil produce the actual ionization of the gas. This ionized gas can be maintained in discharge using a lower voltage which is also optimum for population inversion. The power output of this laser is of the order of 10 kW/m and a typical power output is 50 kW CW.

A typical transverse discharge with transverse fast flow CO_2 laser is shown

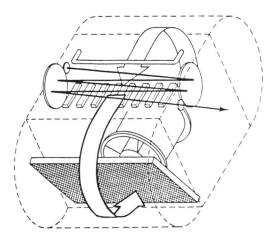

Fig. 3.13.10. Transverse discharge–transverse fast flow CO_2 laser schematic.

in Fig. 3.13.10. The gas flow rate is of the order of 60 m/s and the gas is recirculated. No electron beam preionization is used. The discharge takes place between a hollow water-cooled cathode and a water-cooled segment of the anode individually connected to a ballast resistor. Because of the transverse excitation, relatively low voltage sustains the discharge at quite high current. Typical power output is 600 W/m. However, in general, the beam is folded back and forth through the discharge region five to seven times to achieve a power output of 2.5 kW for a 1.2 m mirror separation.

The Nitrogen Laser

The nitrogen laser is also a molecular laser and is an important source of pulsed ultraviolet radiation at $\lambda = 0.3371$ μm. The lasing action is produced by the transition of electron from state to state, as shown in Fig. 3.13.11. Unfortunately, the lifetime of the upper level is of the order of 5 ns, whereas for the lower level it is in the microsecond range. Thus the laser can only lase for 5 ns or less provided it is pumped in less than a few nanoseconds. This gas discharge of the nitrogen laser is thus generally obtained by discharging a large capacitor consisting of two parallel plates containing nitrogen. The gain of the nitrogen laser is so high that it can oscillate without any feedback in the superradiant fashion; that is, neither of the mirrors in the cavity are needed. One mirror is generally used and the output side contains no mirror at all.

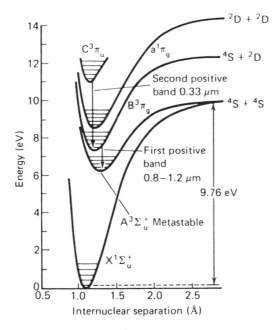

Fig. 3.13.11. Energy-level diagram for N_2 showing the common laser transitions. (From J.T. Verdeyen, *Laser Electronics*, Prentice-Hall, Engelwood Cliffs, NJ, 1981.)

Peak output power can be in the megawatt range. However, pulse widths are in the nanosecond region with a repetition rate of a few hundred hertz.

The Excimer Laser

The excimer laser is also an important source of ultraviolet radiation, and does not have the disadvantage of short duration due to fast lifetime in the upper level. The excimer or excited state dimer is a bounded combination of two atoms in an excited state. Denoting the excited states by an asterisk, the common ones used are Ar_2^*, Kr_2^*, Xe_2^* (noble gases), ArO*, KrO*, XeO* (rare gas oxides), and ArF*, KrF*, XeF* (rare gas halides). Note that in general the noble gases do not form a molecule; however, in the excited state, they can. These excimers can act as either a pump like helium in an He–Ne laser or can be the lasing level itself.

Excimer lasing action can be obtained by an electrical discharge or an electron beam pumping. The electrical discharge is very similar to the case for a nitrogen laser. For the case of electron beam pumping, electrons are accelerated to energies of the order of 1 MeV. These high-energy electrons, in a pulse mode, impinge on a high-pressure gas reaction chamber to excite the excimers.

The Hydrogen Fluoride (HF) Chemical Laser

The HF laser's ingredients are molecular hydrogen and fluoride gas. In general, these molecular species do not react at low temperatures without some external excitation such as ultraviolet radiation, high-energy electron injection, or electrical discharge. The chemical reaction produces hydrogen fluoride in the vibrational excited state. The reaction is highly exothermic and a large quantity of chemical energy is released. This excess energy is the equivalent pumping energy for this chemical laser. It is to be noted that this chemical energy is enormous compared to other forms of pumping energy. This is apparent from the fact that the chemical energy stored in one gallon of gasoline is enough to move a car at high speed to a distance of 100 km. Some of the actual reactions that take place between atomic and molecular hydrogen and fluorine are:

$$F + H_2 \rightarrow HF\ (v \leq 3) + H, \qquad \Delta H = -31.7\ \text{kcal/mol},$$

$$H + F_2 \rightarrow HF\ (v \leq 9) + F, \qquad \Delta H = -97.9\ \text{kcal/mol},$$

v denotes the vibrational levels of the HF molecule. It is to be noted that the end product of the reaction also contains atomic hydrogen and fluoride. Thus, once the reaction starts it continues until all the molecular H_2 and F_2 are consumed, as in the burning of fuel. Chemical lasers have produced the highest levels of total power in a pulsed condition—hundreds of kilo-joules in a sub-microsecond pulse duration.

There are other chemical lasers which have been found useful. These are

HCl, DF, and carbon monoxide lasers. The DF and HCl lasers are very similar to the HF laser, except that hydrogen is replaced by its isotope deuterium in DF, and fluorine is replaced by chlorine in the HCl laser. The CO laser emits in the range between 5–6 μm. A mixture of cyanogen (C_2N_2), helium, and air is passed through an electric discharge. The discharge produces the following chemical reaction:

$$C_2N_2 + O_2 \rightarrow 2CO + N_2 + 127 \text{ kcal.}$$

The vibrationally excited CO molecules participate in the lasing action. Helium simply aids in improving the efficiency. A typical DF laser is shown in Fig. 3.13.12 where F_2 is heated with a carrier gas (helium) and then passed through expansion nozzles, very similar to that in the gas dynamic laser. The chemical reaction takes place after this expansion and in the optical cavity where D_2 is injected.

Chemical lasers have many attractive features, some of which have already been mentioned. These lasers produce the highest output power per unit volume and per unit weight. In general, chemical reactions excite vibrational levels and thus the output wavelength is always in the infrared (1 μm to 12 μm). If one-shot large power is needed as, for example, in a star wars scenario, chemical lasers can produce large amounts of destructive energy without any electrical power.

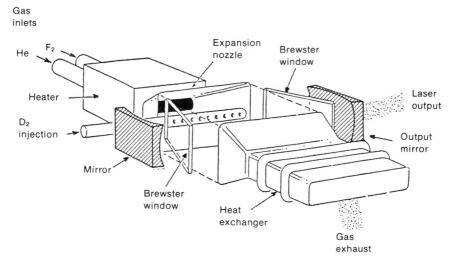

Fig. 3.13.12. Schematic of a chemical laser. One of the chemical reactants (in this case, F_2) is heated with a carrier gas (He) and allowed to expand just before mixing with the second reactant (D_2). The reaction takes place in the region between the two Brewster windows. (The enclosure around this area has been omitted for the sake of clarity.) The output beam is in a direction transverse to the gas flow, as in the gas dynamic laser. (From D.C. O'Shea et al., *Introduction to Lasers and Their Applications*, Addison-Wesley, Reading, MA.)

3.13.2. Solid State Lasers

The first laser demonstrated by Maiman was a ruby laser. The other important solid state lasers are a neodymium YAG laser and a neodymium–glass laser. Because of the importance of semiconductor junction lasers, we shall discuss them separately.

The Ruby Laser

The ruby consists of a crystal of aluminum oxide with chromium (Cr^{3+}) ions as impurities. Actually, the distinctive pink color of ruby comes from the impurity atoms. The energy-level diagram of the ruby laser is shown in Fig. 3.13.13; the blue and green absorption bands of chromium ions are shown in the figure. Because of the large crystal field present in the host aluminum oxide lattice, the degenerate energy levels become split. The lasing action occurs by the transition from this split level ($2E$) to the ground state. The lasing wavelengths are at 0.69430 μm and 0.6927 μm, although the first one dominates. The laser is generally pumped optically using a flash lamp—such as a xenon helical flash lamp—a typical setup is shown in Fig. 3.13.14. Because of the large amount of heat generated by the flash lamp, liquid coolants are often used. The flash lamp is excited by discharging a large capacitor charged with high voltage, since the flash lamp needs a large amount of current. Generally, the operation is pulsed although CW operation is possible.

The radiation from the flash lamp excites electrons to $4F_1$ and $4F_2$ (the green and blue absorption bands) electron states from which, by spontaneous emission, the lasing $2E$ levels are populated. Because of this, it is important that the flash lamp spectra match these absorption bands for better efficiency. Note that the ruby laser is a three-level laser.

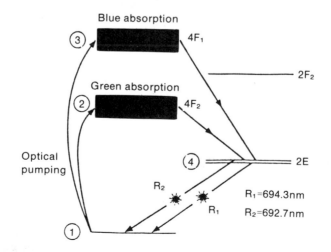

Fig. 3.13.13. Energy-level diagram of a ruby laser. (From D.C. O'Shea et al., *Introduction to Lasers and Their Applications*, Addison-Wesley, Reading, MA.)

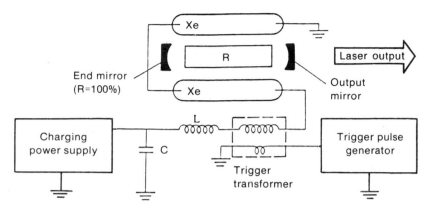

Fig. 3.13.14. Schematic of a simple flashlamp-pumped laser. The trigger pulse generator and transformer provide a high-voltage pulse sufficient to cause the xenon gas (Xe) in the lamps to discharge. The ionized gas provides a low-resistance discharge path to the storage capacitor (C). The inductor (L) shapes the current pulse, maintaining the discharge. The discharge of the lamps optically pumps the laser rod (R).

The laser rod is, in general, placed inside the cavity with two external mirrors. For this case, in general, the laser rod is positioned at the Brewster angle to reduce reflection at the ends. This, however, makes the output light polarized, as explained in Section 3.13.1. It is possible to use a coating at the end of the laser rod so that no external mirrors are needed. However, this is, in general, not used, as the mirror coatings are subject to damage due to high laser power.

Mirrors are also used to focus the radiation from the flash lamp concentrated at the laser rod. A very practical arrangement is to use an elliptical mirror with the cylindrical flash lamp at one focus and the laser rod at the other focus of the ellipse—this is shown in Fig. 3.13.15.

The Neodymium–YAG Laser

The YAG (yttrium–aluminum–garnet, $Y_3Al_5O_{12}$) is a crystal in which Nd^{3+} ions can be used as impurities; these Nd^{3+} ions are responsible for the lasing action. The energy-level diagram is shown in Fig. 3.13.16. The laser emission occurs at 1.0461 μm when electrons make a transition from the upper level $4F_{3/2}$ to the lower level $4I_{11/2}$. Many other laser transitions are possible which are not shown in Fig. 3.13.16—these range in wavelength from 0.94 μm to 1.4 μm. Note that since the lower level is not the ground state, and is ~ 250 meV from the ground state, in general the lower level is nearly empty; thus the Nd–YAG laser is a four-level laser. The operation of the laser is somewhat similar to that discussed in connection with the ruby laser. There are absorption bands between 1.5 eV and 3 eV. The flash lamp excites electrons to these absorption bands from which electrons populate the upper lasing level by spontaneous emission. Because of the four-level nature, the Nd–YAG laser is

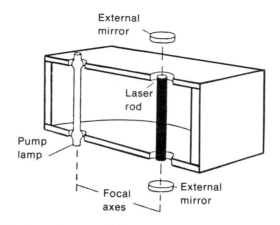

Fig. 3.13.15. Typical continuous solid state laser arrangement employing an elliptic cylinder housing for concentrating lamp light onto a laser. (From A. Yariv, *Introduction to Optical Electronics*, Holt, Rinehart and Winston, New York, 1976.)

very easy to operate in the CW mode. In many commercial operations, the infrared emission of an Nd–YAG laser is frequency-doubled to the visible region using a nonlinear interaction in a crystal.

The Nd–Glass Laser

The Nd^{3+} ions can also be placed in glass as a host material rather than in the YAG crystal. A typical glass is rubidium potassium barium silicate; the energy-level diagram of Nd^{3+} in this glass is shown in Fig. 3.13.17. As with

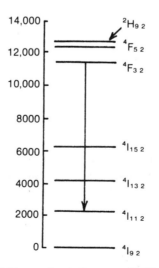

Fig. 3.13.16. Energy level diagram of Nd^{3-} in YAG.

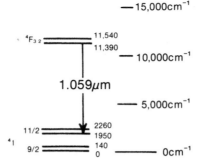

Fig. 3.13.17. Energy-level diagram of Nd_3^+ in rubidium barium silicate glass.

the Nd–YAG laser many lasing wavelengths are possible. The glass being an amorphous material rather than crystalline with fixed periodicity, the energy-level splittings of each individual atom are not identical. This gives rise to a large fluorescent linewidth for the glass. Thus, the glass laser will have a larger threshold than that of the YAG laser. Many times YAG is used as the master oscillator (laser) and Nd–glass lasers (without the feedback mirrors) are used as light amplifiers.

The Nd–glass lasers have produced one of the highest peak pulsed powers produced by any laser. This is possible because of the advantages Nd–glass has over Nd–YAG. First of all, large volumes of glass can easily be fabricated as there are no restrictions due to its crystalline nature and the glass laser can easily be segmented with coolants in between. In general, glass disks with Brewster angle ends are generally used as laser rods for the Nd–glass disk laser amplifier.

3.13.3. Dye Lasers

Dye lasers are liquid lasers where the active material is dye in a host medium of a liquid solvent, such as ethylene glycol. The situation is very similar to solid state lasers when Cr^{3+} or Nd^{3+} is used in a solid host. The advantage of a liquid host is that the concentration of the active ions can easily be changed; the gas lasers have the same advantage. However, the concentration of active ions, and thus the gain, can be much higher for liquid than for gas because of larger concentration.

The dye laser has a unique property which the other lasers do not have, it can be tuned over a broad range. For the case of gas or solid state lasers, the linewidths are very small. Actually, we usually look for narrower linewidths as this makes the threshold power for pumping much smaller; however, the output of the laser can be tuned over the linewidth only. For the case of dye lasers the lines are really bands and they extend not a few angstroms but rather a thousand angstroms. Thus, dye lasers with an external tuning element can be tuned over a very broad range. However, we pay a price for this, the dye

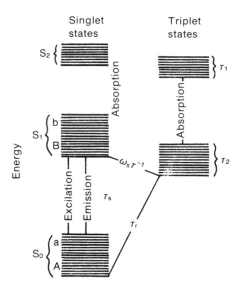

Fig. 3.13.18. Schematic representation of the energy levels of an organic dye molecule. The heavy horizontal lines represent vibrational states and the lighter lines represent the rotational fine structure. Excitation and laser emission are represented by the transitions $A - b$ and $B - a$, respectively. (From A. Yariv, *Introduction to Optical Electronics*, Holt, Rinehart and Winston, New York, 1976.)

laser has a very high threshold power and in general another laser, such as an argon ion laser or a nitrogen laser, is needed to pump it.

The energy band diagram of a typical dye laser is shown in Fig. 3.13.18. The organic dye molecule is known to have two excited states: singlet states denoted by S_0, S_1, and S_2 and triplet states denoted by T_1 and T_2. In the singlet state, the total spin of the excited molecule is zero whereas for the triplet state, it is one. Because of the selection rules singlet–triplet transitions are forbidden. The lasing action occurs, in general, by the transition from lower-lying S_1 levels to different S_0 levels. For different dyes, the output wavelength range and power is shown in Fig. 3.13.19, where an argon ion laser is used as a pump. The most important dye is rhodamine 6G which can be used between 0.57 μm and 0.65 μm with large CW power output. Table 3.13.2 shows the list of organic dyes, their chemical structure, solvents, and range of lasing wavelength.

The dye laser can be pumped by a flash lamp, an argon laser, or a nitrogen laser; for the cases of a flash lamp or nitrogen laser, it is typically pulsed. Both the ultraviolet lines or the visible lines of argon can be used as an effective dye laser pump. If the visible wavelengths are used, then the dye laser wavelength range is 0.56 μm and higher. For shorter wavelengths, ultraviolet pumping is needed. For nitrogen laser pumping using the 0.377 μm line to improve efficiency, a two-step pumping is often used. Two dye molecules are used: one for the lasing action, the other for the efficient absorption of the 0.377 μm

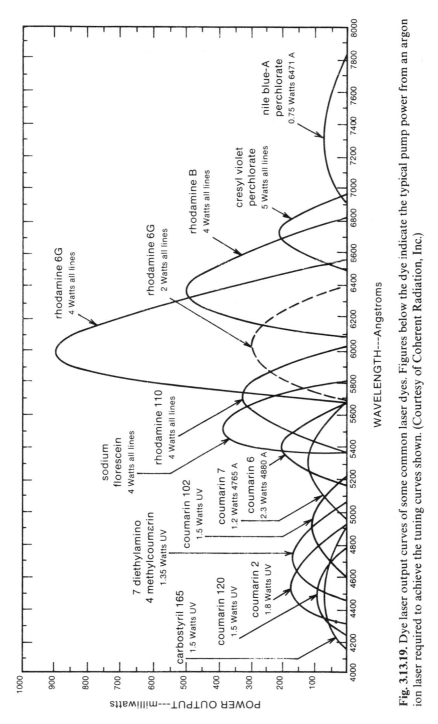

Fig. 3.13.19. Dye laser output curves of some common laser dyes. Figures below the dye indicate the typical pump power from an argon ion laser required to achieve the tuning curves shown. (Courtesy of Coherent Radiation, Inc.)

Table 3.13.2. Molecular structure, laser wavelength, and solvents for some laser dyes. (From B.B. Snaveley, Flash lamp pumped dye lasers, *Proc. IEEE*, **57**, 1969.)

Dye	Structure	Solvent	Wavelength
Acridine red		EtOH	Red 600–630 nm
Puronin B		MeOH H_2O	Yellow
Rhodamine 6G		EtOH MeOH H_2O DMSO Polymethyl-methacrylate	Yellow 570–610 nm
Rhodamine B		EtOH MeOH Polymethyl-methacrylate	Red 605–635 nm
Na-fluorescein		EtOH H_2O	Green 530–560 nm
2, 7-Dichloro-fluorescein		EtOH	Green 530–560 nm
7-Hydroxy-coumarin		H_2O (pH ~ 9)	Blue 450–470 nm
4-Methylem-belliferone		H_2O (pH ~ 9)	Blue 450–470 nm
Esculin		H_2O (pH ~ 9)	Blue 450–470 nm

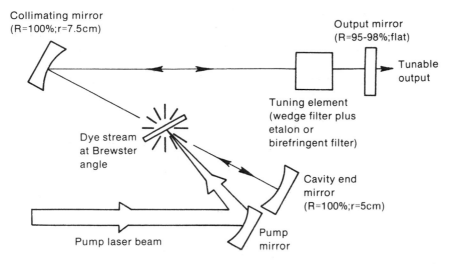

Fig. 3.13.20. Schematic diagram of a laminar-flow dye laser. The dye-laser cavity is formed by the reflector and the output coupler. The other reflector serves to fold the cavity so that the dye-laser output is parallel to the input pump beam. Dye stream flow is perpendicular to the page.

line. This dye fluoresces in the longer wavelength, which produces efficient pumping of the lasing dye.

The gain of the dye lasing medium is very high. Thus, a small volume dye solution is needed to sustain the lasing action in an external cavity, as shown in Fig. 3.13.20. The small volume of dye, however, cannot be sealed in a glass for CW operation, because of the intense heat generation and consequent expansion and inhomogeneity of the lasing medium. In general, the dye solution is pumped through a nozzle which forms a steady stream and a sheet of dye solution at the Brewster angle. The pumped laser is focused to this dye volume using a pump mirror. Between the two cavity mirrors is a wavelength tuning element such as a prism, diffraction grating, or birefringent quartz filter. By adjusting the angles of the prism or the diffraction grating and the optical axis of the quartz filter using a micrometer, we can choose the desired wavelength. The birefringent quartz filter uses one, two, or three quartz plates. The respective linewidths are 300, 100, and 30 Å, respectively. The quartz plates are at Brewster angles in the cavity and their thicknesses are in the ratio of $1:2:4$. The tuning is performed by rotating the plates together.

3.13.4. Semiconductor Lasers

3.13.4.1. The Junction Laser

The most important laser for communication and electronics is the junction or diode laser. This is also the smallest laser we can build—the active area having dimensions of the order of microns. Because of its small size, as in the use of semiconductor diodes for its operation, the junction laser can be integrated with an electronic circuit. The laser output can easily be coupled to the fiber-optic cable, the lasing frequency can easily be modulated in both amplitude and phase electronically, the output power can be adjusted anywhere from microwatts to hundreds of watts, both CW and pulsed, and the wavelength range can be selected from infrared to visible. These are some of the important properties for which the junction laser will probably become the most used laser in the next decade. It will be used extensively in fiber-optics communication and in integrated optical circuits or photonics. In the next section we discuss the light emitting diode (LED) which acts as the light amplifier for the laser diode.

The Light Emitting Diode (LED)

To generate light, electrons must make a transition from an upper-energy level to a lower-energy level. For the case of semiconductors, the upper-energy level is the conduction band and the lower-energy level is the valence band, as shown in Fig. 3.13.21. The bandgap, E_g, determines approximately the wavelength of radiation

$$\lambda \text{ (in } \mu\text{m)} = \frac{E_g \text{ (in eV)}}{1.2394}. \tag{3.13.2}$$

Thus every semiconductor can produce light of a wavelength given by (3.13.2). Table 3.13.3 lists the bandgaps of different semiconductors. Note that we can make a complex semiconductor by combining GaAs and GaP, for example, to make $GaAs_{1-x}P_x$. These ternary compounds have a bandgap which varies smoothly with x. This is very important because this control over the bandgap gives us the opportunity to obtain the laser wavelength desired for some special purpose. For example, most fiber-optic cable has the lowest attenuation at 1.3 μm, and $Ga_{0.27}In_{0.73}As_{0.4}P_{0.6}$ (lattice matched to InP substrate) is used to make laser light for this fiber-optic communication system. We might ask why the most popular semiconductors like silicon and germanium are not listed in Table 3.13.3. The main reason is that silicon and germanium are indirect semiconductors and thus for momentum balance of the electron transition, not only a photon but a phonon is also involved. Thus,

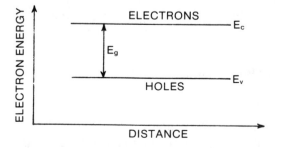

Fig. 3.13.21. Energy-level diagram of a semiconductor.

Table 3.13.3. Bandgap of semiconductors.

	Bandgap in eV at room temperature	Type of gap
Silicon	1.12	Indirect
GaAs	1.43	Direct
GaSb	0.72	Direct
InP	1.35	Direct
InAs	0.36	Direct
CdTe	1.50	Direct
PbSe	0.26	Direct
PbTe	0.32	Direct
$GaAs_xP_{1-x}$	1.43–1.99	Direct
	2–2.26	Indirect
AlGaAs	1.43–1.95	Indirect
	1.95–2.16	Direct
$Ga_xIn_{1-x}As$	0.36–1.43	Indirect
$InAs_xP_{1-x}$	0.36–1.35	
GaSb	0.72	Direct
$Pb_{1-x}Sn_xDe$	—	
InGaP	1.35–2.26	
HgCdTe	0.15–1.50	Direct
GaP	2.26	Indirect
AlAs	2.16	Indirect
InSb	0.17	Direct
AlSb	1.65	Indirect
AlP	2.45	Indirect
$GaAs_{1-x}Sb_x$	0.72–1.42	
$Ga_xIn_{1-x}As_yP_{1-y}$	0.94–1.38	
GaInAsSb	0.73–0.3	
PbEuSeTe	0.19–0.46	
InAsPSb	0.35–0.62	

the transition probability is very low and it is nearly impossible to get any light emission from indirect semiconductors. The semiconductors listed in Table 3.13.3 are all direct semiconductors and they can all generate light efficiently. Figure 3.13.22 shows the E versus K diagram (E is the energy of the electrons and K is its wave vector) for a direct and an indirect semiconductor.

There is some difference between the generation of photons by electron transition in gaseous atoms like He–Ne or argon and that in a semiconductor. In a semiconductor, in general, the valence band is nearly full and the absence of electrons in the valence band is conveniently represented by holes. In a semiconductor, an electron–hole pair recombine to generate a photon of bandgap energy. In an intrinsic semiconductor, or the semiconductor which does not have any impurities, the number of electrons, n, in the conduction band is equal to the number of holes, p, in the valence band and they are given by

$$n = p = n_i = 2\left(\frac{kT}{\hbar^2}\right)^{3/2}(m_e m_h)^{3/4}e^{-Eg/2kT}, \tag{3.13.3}$$

where m_e is the electron effective mass and m_h is the hole effective mass.

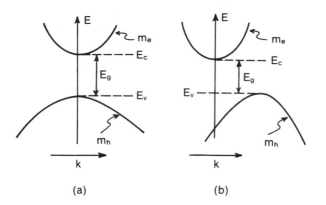

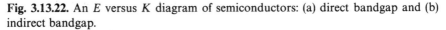

(a)   (b)

Fig. 3.13.22. An E versus K diagram of semiconductors: (a) direct bandgap and (b) indirect bandgap.

Impurities can be added to a semiconductor to change the ratio of electrons and holes. Thus donor impurities, which can release extra electrons, can be added to the semiconductor to make it an n-type. This means that if N_D is the donor impurity density and the position of the donor level in the bandgap, E_d, such that

$$E_c - E_d \ll kT,$$

then

$$n_n \approx N_D \quad \text{and} \quad p_n \approx \frac{n_i^2}{N_D}, \tag{3.13.4}$$

where n_n and p_n represent the number of electrons and holes in the n-type semiconductor, respectively. Similarly, acceptor impurities can be added to the semiconductor to make it a p-type. For a p-type semiconductor, with N_A acceptor density, we have

$$n_p \approx \frac{n_i^2}{N_A} \quad \text{and} \quad P_p \approx N_A. \tag{3.13.5}$$

When a p-type semiconductor and an n-type semiconductor are brought in contact with each other to form a junction diode, an electric field develops in the junction region. In the equilibrium condition, the electron and hole components of current must be individually zero. As there are more electrons in the n region, initially they move to the p region, forming a depletion region of donor atoms charged positively. This continues until the electric field, due to these charged immobile impurities, reduces the flow of electrons from n to p to match those from p to n. Similar arguments hold for the holes, and a depletion layer of acceptor impurities charged negatively forms in the p side. This is shown in Fig. 3.13.23. The depletion widths and the electric field E_0 in the junction region are given approximately by

$$x_{p0} = \left\{ \frac{2\varepsilon V_0}{q} \left[\frac{N_d}{N_a(N_a + N_d)} \right] \right\}^{1/2}, \tag{3.13.6}$$

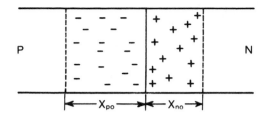

Fig. 3.13.23. A p–n junction showing depletion region.

$$x_{n0} = \left\{ \frac{2\varepsilon V_0}{q} \left[\frac{N_a}{N_d(N_a + N_d)} \right] \right\}^{1/2} \tag{3.13.7}$$

and

$$V_0 = \frac{kT}{q} \ln \frac{N_a N_d}{n_i 2},$$

$$E_0 = -\frac{q}{\varepsilon} N_d x_{p0} \tag{3.13.8}$$

$$= -\frac{q}{\varepsilon} N_a x_{n0}.$$

The above elementary discussion of a p–n junction can be better represented by introducing the concept of the Fermi level in the energy band diagram. Above the Fermi level at absolute zero temperature all the energy levels are empty whereas below it they are full. At room temperature, for typical semiconductors of interest, most of the levels above the Fermi level will be empty. Thus, for an intrinsic semiconductor, the Fermi level is near the middle of the gap and the number of electrons and holes is very small. For a heavily doped or degenerate n-type semiconductor, the Fermi level is in the conduction band itself. Thus, there will be a large number of electrons and very few holes. For p-type semiconductors, the opposite is the case and is shown in Fig. 3.13.24. When a p–n junction is formed under equilibrium conditions, the Fermi levels line up as shown in Fig. 3.13.25, and give rise to an electric field at the junction and the formation of a depletion layer as discussed earlier.

If a voltage, V, is applied to the p–n junction, the equilibrium is disturbed.

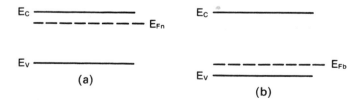

Fig. 3.13.24. Energy-level diagram of doped semiconductors: (a) an n-type semiconductor and (b) a p-type semiconductor. E_{F_n} and E_{F_p} denote the Fermi level position.

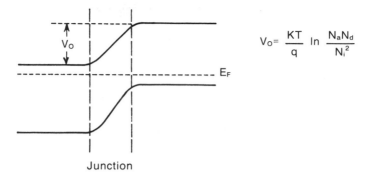

$$V_0 = \frac{KT}{q} \ln \frac{N_a N_d}{N_i^2}$$

Fig. 3.13.25. Energy level diagram of a p–n junction with a bias voltage V_0.

For a forward bias, i.e., a p side connected to the positive and an n side to the negative, a large number of electrons and holes are injected into the depletion region. The number of electrons injected into the p region is given by

$$\Delta n(0) = n_n e^{eV/kT}. \tag{3.13.9}$$

These extra carriers in the depletion region under bias are in a nonequilibrium condition. For this case we can define two quasi-Fermi levels, E_{FC} and E_{FV} for electrons and holes, respectively—these are shown for the forward bias case in Fig. 3.13.26. The quasi-Fermi level, E_{FC}, is related to the electron concentration, n, by the following relationship:

$$n = n_i e^{(E_{FC} - E_i)/kT},$$

and

$$p = n_i e^{(E_i - E_{FV})/kT},$$

where E_i is the intrinsic Fermi level.

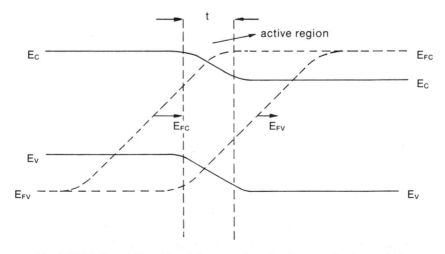

Fig. 3.13.26. Quasi Fermi levels in a p–n junction laser under forward bias.

The injected carriers diffuse and recombine in the depletion region giving rise to diode current. The distribution of electrons in the p region can be shown to be given by

$$\Delta n(x) = n_p + n(0)e^{-x/L_n}, \tag{3.13.10}$$

where L_n is the diffusion length $= \sqrt{D_n \tau_n}$, D_n is the electron diffusion constant, and τ_n is the lifetime of electrons. The lifetime, τ_n, represents the mean time for an electron–hole recombination and thus generation of a photon. Thus, the spontaneous transition probability per unit time, A_{21}, for the junction diode, is given by

$$A_{21} = \frac{1}{\tau_n}. \tag{3.13.11}$$

The excess electrons and holes recombine and diffuse in the depletion region which is also referred to as the active region. For a p–n junction diode for electronic purposes we want τ_n to be as large as possible, so that excess carriers are not lost by recombination. However, for a light emitting diode, the requirements are just the opposite.

The I–V characteristics of the diode are given by

$$I = qA\left(\frac{D_p}{L_p}p_n + \frac{D_n}{L_n}n_p\right)(e^{-qV/kT} - 1), \tag{3.13.12}$$

where A is the cross-sectional area. The emitted spontaneous light has a spectrum containing energies in the range $E_g < hf < E_{FC} - E_{FV}$.

To obtain stimulated emission and eventual laser action we must have a population inversion. In the depletion region of a junction diode under forward bias, population inversion exists. This is because a nonequilibrium condition exists in the narrow region near the junction where the injection takes place. This narrow region is called the active region, and is shown in Fig. 3.13.26. We see that in the active region the band of frequencies defined by

$$E_g < hf < E_{FC} - E_{FV}$$

satisfy the population inversion condition. The width of the active region, t, is approximately equal to $L_p + L_n$.

The Junction Laser

A typical GaAs junction laser is shown in Fig. 3.13.27. The active region and the fundamental mode shape are also shown. The width of the mode, d, is determined by the dielectric waveguide formed by the slightly different index of refraction of the p and n semiconductors. In general, $d \gg t$. However, as we shall discuss shortly, d can be reduced drastically using a heterostructure.

The expression for the gain constant for the semiconductor laser can be written as

$$\gamma(f) = \frac{C^2(N_2 - N_1)/(dl\omega)}{8\pi n^2 f^2 \tau_n} g(f). \tag{3.13.13}$$

Here N_2 and N_1 are the total number of electrons and holes, respectively, and

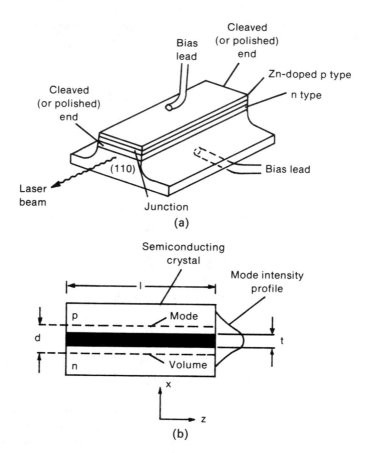

Fig. 3.13.27. (a) Typical p–n junction laser made of GaAs. Two parallel (110) faces are cleaved and serve as reflectors. (b) Schematic diagram showing the active layer and the transverse (x) intensity distribution of the fundamental laser mode.

l and ω are the length and width of the active layer, respectively. If $d < t$, then we should replace d by t in the above expression.

To calculate $(N_2 - N_1)$ as a function of diode current is difficult. However, a simplification can be made if we assume $N_1 \approx 0$ at low enough temperatures. For this case, equating the total number of electrons, injected in the depletion region in time Δt, to the number of spontaneous emissions, we obtain

$$\frac{N_2}{\tau_n} = \frac{I\eta_i}{q}, \tag{3.13.14}$$

where η_i is the internal quantum efficiency. Thus, for the case $N_1 \approx 0$, we obtain

$$\gamma(f) = \frac{C^2 g(f)\eta_i}{8\pi n^2 f^2 qd}\left(\frac{I}{A}\right), \tag{3.13.15}$$

where $A = l\omega$.

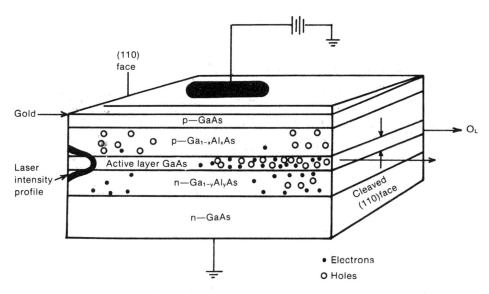

Fig. 3.13.28. A heterojunction laser.

Using (3.13.15) and (3.13.16), we easily obtain the threshold current, I_{th}, for the lasing action to be

$$I_{th} = \frac{8n^2 f^2 q d(f)}{C^2 \eta_i} A \cdot \left(\alpha - \frac{1}{l} \ln r_1 r_2 \right), \tag{3.13.16}$$

where we have used the expression $g(f) = (\Delta f)^{-1}$.

We note that I_{th} is proportional to d. Thus, reducing the mode confinement distance contributes directly to the lowering of the threshold current and to an increase in the power output. To achieve this reduction, the heterostructure junction laser, shown in Fig. 3.13.28, is used. The active layer is a thin GaAs layer which is surrounded on one side by $p - Ga_{1-x}Al_xAs$ and on the other side by n-$Ga_{1-x}Al_xAs$. The difference in the refractive index between a p-GaAs and a p-$Ga_{1-x}Al_xAs$ is much more than that between a p-GaAs and an n-GaAs. Thus the mode confinement is severe and $d \approx t$. Also, the active layer thickness is smaller because of the larger difference in the potential barrier across the junction, since the bandgap energy of GaAlAs is different from GaAs. A typical situation is represented in Fig. 3.13.29 where the lowering of d and t are illustrated.

The power emitted by the stimulated emission, P_e, if the junction diode is biased beyond the threshold condition, is given by

$$P_e = \frac{(I - I_t)\eta_i hf}{q}. \tag{3.13.17}$$

The output power can be written as

$$P_o = \frac{(I - I_t)\eta_i hf}{q} \frac{(1/l) \ln(1/r_1 r_2)}{\alpha + (1/l) \ln(1/r_1 r_2)}. \tag{3.13.18}$$

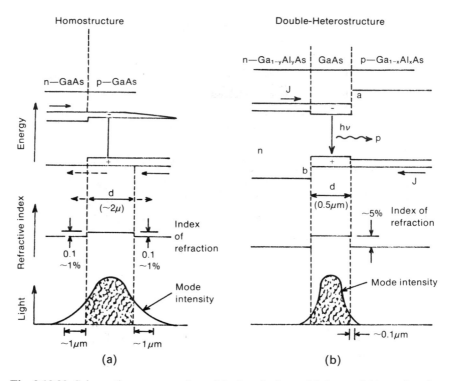

Fig. 3.13.29. Schematic representation of the band edges with forward bias, refractive index changes, and optical field distribution in (a) a homostructure and (b) a double heterostructure diode. (From H. Kressel and H. Nelson, *RCA Review*, **30**, 1969.)

The power efficiency of a junction laser is given by

$$\eta = \frac{P_o}{V} = \eta_i \frac{(I - I_t)}{I} \frac{hf}{qV} \frac{n(1/r_1 r_2)}{\alpha l + \ln(1/r_1 r_2)}. \tag{3.13.19}$$

As the applied voltage is approximately equal to (hf/q), and for $I \gg I_t$,

$$\eta \sim \eta_i. \tag{3.13.20}$$

The internal quantum efficiency, η_i, is very high (0.7–1 in GaAs). Thus the junction laser is the most efficient laser.

As discussed so far, the $p–n$ junction starts emitting light when population inversion is achieved using a high-carrier injection in the forward bias region. This radiation is spontaneous emission without feedback mirrors and these junctions are called LED (light emitting diodes). To make LED lasing, we need to increase the population inversion to the threshold value and add a set of mirrors. External mirrors are not needed, as the reflectivity of the diode–air interface is very high because of the large difference in the refractive index. In practice, the diodes are cleaved along crystalline planes; this guarantees the parallelism of the reflective surfaces without any further polishing of the

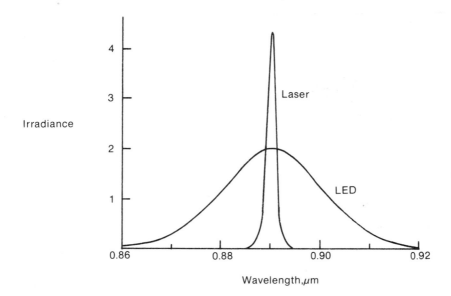

Fig. 3.13.30. Irradiance versus wavelength for an LED and a semiconductor laser.

optical surfaces. Figure 3.13.30 shows the emission spectrum of a semiconductor laser compared to that of an LED.

A typical junction laser cavity has a length of 300 μm with an active region of 3μm wide. Because of the confinement of the light beam to such a small region, the output light has a large beam divergence. Using the diffraction formula, we easily obtain the half-angle of divergence, θ, given by

$$\theta \approx \sin^{-1}\frac{\lambda}{a},\tag{3.13.21}$$

where a is the width. Thus for $\lambda = 0.9$ μm and $a = 3$ μm, $\theta \approx 17°$. Compared to other lasers this is quite large.

As discussed before, there are two types of junction lasers: homojunction and heterojunction. Heterojunction lasers are also of two types: single heterostructure and double heterostructure. In a single heterostructure laser, p-Al$_x$Ga$_{1-x}$A$_s$ is used on n-GaAs substrate. The double heterostructure consists of a p-GaAs sandwiched between a p- and an n-Al$_x$Ga$_{1-x}$As. This triple structure is generally on an n-GaAs substrate with a p-GaAs layer on top for contacts.

Double heterostructure lasers are also referred to as having a large optical cavity (LOC) configuration. The optical cavity is much wider in the double heterostructure, tens of micrometers compared to perhaps a few micrometers. This reduces the danger of damage of the crystal from the radiation. This feature also greatly reduces the diffraction of the beam as it leaves the end of the crystal from an angle of about 2°.

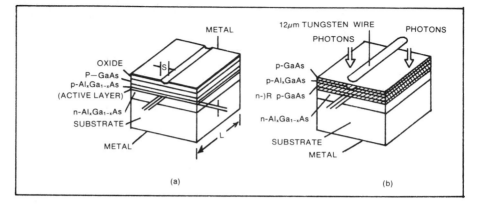

Fig. 3.13.31. Stripe geometry double heterostructure junction lasers: (a) oxide isolation and (b) proton bombarded isolation.

To improve semiconductor laser performance, structures more complex than double heterostructure are often used. One example is the stripe geometry laser shown in Fig. 3.13.31. Heterojunction with stripe geometry reduces the current density and risk of damage due to large radiation fields in the chip. The purpose of the stripe contact is to improve conduction of heat from the active region. $Al_sGa_{1-x}As$ compounds have poorer thermal conductivity than GaAs. Limiting the active region to a narrow stripe allows lateral heat conduction in GaAs to drastically reduce the temperature rise in the active region.

In Fig. 3.13.31(a), the oxide layer isolates all but the narrow stripe contact, restricting the lasing area under the contacts only. In Fig. 3.13.31(b), the stripe geometry laser is fabricated by proton bombardment which produces high resistivity regions. The lasing area is restricted to the unbombarded region. The stripe widths are typically 5–30 μm. The advantages of the stripe geometry are many, these include:

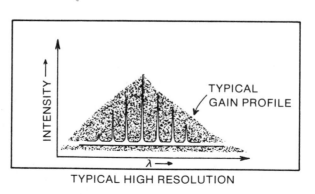

TYPICAL HIGH RESOLUTION

Fig. 3.13.32. Typical high-resolution spectrum with different longitudinal modes of a laser output.

(a) reduction of the cross-sectional area and hence the operating current. Note that lower operating current needs lower heat dissipation and thus room temperature CW operation becomes easier;
(b) elimination of the occurrence of more than one filament (localized high optical intensity area);
(c) improved reliability by removing most of the junction parameter from the surface; and
(d) improved response time.

Many times junction laser oscillates in multimode as shown in Fig. 3.13.32. To calculate the mode separation, $\Delta\lambda$, we note that GaAs is highly dispersive. Thus, to calculate $\Delta\lambda$ between the mth longitudinal mode and its neighbor, we start with (3.3.14), rewritten below,

$$m = \frac{2Ln}{\lambda}, \tag{3.13.22}$$

where L is the length of the cavity and n is the refractive index.
   Differentiating (3.13.22) with respect to λ we obtain

$$\frac{dm}{d\lambda} = -\frac{2Ln}{\lambda^2} + \frac{2L}{\lambda}\frac{dn}{d\lambda}. \tag{3.13.23}$$

Thus we obtain $\Delta\lambda$ as

$$\Delta\lambda = \frac{\lambda^2}{2nL[1 - (\lambda/n)(dn/d\lambda)]}. \tag{3.13.24}$$

It is assumed that m is quite large.
   To obtain a stable single-mode laser, we must make sure that no other modes are excited. This can be done in the following ways:

(1) coupled-cavity;
(2) frequency selective-feedback;
(3) injection locked; and
(4) geometry controlled.

The principle behind coupled-cavity lasers is if the laser light has to travel through additional cavities, the only wavelengths that are positively reinforced (i.e., integral multiples of half wavelengths equal to cavity length), both in the laser's cavity and in the added cavities, are sustained. All other wavelengths are suppressed. The coupling of cavities can be achieved in many configurations, four of which are shown in Fig. 3.13.33. Of these, the cleaved-coupled-cavity (C^3) is of special importance and will be discussed in detail shortly.
   In the *external mirror* approach, the mirror may be flat and parallel to one end facet, but often a slightly concave mirror is used to focus the energy back into the laser's cavity. The air space between the mirror and the laser, whose length is fine tuned by temperature control of the position of the mirror with a resistance heater, is the additional cavity. In the *grooved-coupled cavity* and

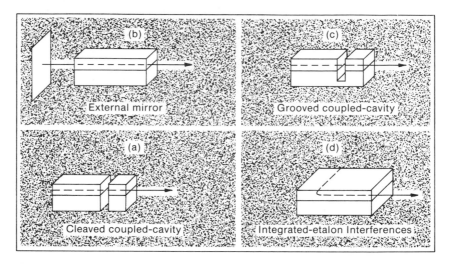

Fig. 3.13.33. Coupled cavity single-frequency lasers: (a) cleaved coupled cavity; (b) external mirror; (c) grown coupled cavity; and (d) integrated etalon interference.

integrated etalon interference, laser light resonates in two active cavities (both laser segments above the threshold). In the latter, the curved segment acts as an etalon between the two straight segments. Unlike temperature and current control, to determine the wavelength in coupled-cavity lasers in a frequency selective feedback approach, the wavelength selection is done by a grating— three possible configurations are shown in Fig. 3.13.34. By proper tilting of the *external grating* we can select the wavelength. If the period of grating is equal to an integral multiple of half of the wavelength desired in the *distributed Bragg reflector*, the Bragg condition is satisfied and only that wavelength resonates in the cavity. In *distributed* feedback, the grating is fabricated directly under or above the laser diode's cavity. The wavelength of the light that resonates is the one reinforced by the period of grating. A low-power He–Ne laser operating at 1.52 μm, by injecting a continuous wave emission of a single wavelength into the laser's cavity, "locks" by stimulated emission only one mode in the laser cavity, as shown in Fig. 3.13.35(a). The *injection locked*

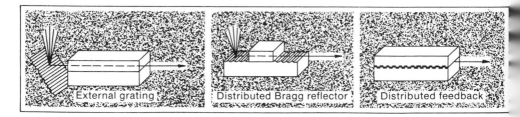

Fig. 3.13.34. Frequency selective feedback lasers: (a) external grating; (b) distributed Bragg reflector; and (c) distributed feedback.

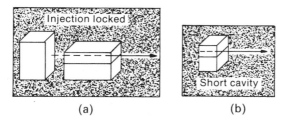

Fig. 3.13.35. Single-frequency operation of the laser: (a) injection locked and (b) short cavity.

lasers are bulky, but have fine stability and spectral purity, even under high modulation frequencies. In *short cavity* lasers (Fig. 3.13.35(b)] and their hybrid forms, the smaller cavity (about 50 μm, about one-sixth the length of other laser diodes) enables the reduction of the number of modes supported by the cavity, and the spacing between adjacent modes is also increased. Thus, this effect, when superimposed with the gain profile of the laser, invariably results in single-frequency operation.

3.13.4.2. The Cleaved-Coupled-Cavity Laser

Figure 3.13.33(a) shows a schematic diagram of a C^3 laser. It consists of two standard Fabry–Perot cavities of 1.3 μm wavelength and GaInAsP laser diodes of 136 μm and 121 μm length, respectively, which were self-aligned and very closely coupled to form a two-cavity resonator. It should be noted that here all the workings and characteristics described for the 1.3 μm laser are equally applicable to the 1.5–1.6 μm laser. The active stripes are separated by < 5 μm; the reflecting facets are formed by cleaving along perfectly parallel crystallographic planes. Complete electrical isolation (> 50 kΩ) between the two individual F–P diodes results.

The basic working principle is illustrated schematically in Fig. 3.13.36. The propagating mode in each active stripe can have a different effective refractive index, N_{eff}, even if they have the same shape, size, and material composition. This is because N_{eff} is a function of the carrier density in the active stripe. This can be varied by varying the injection current below threshold when the junction voltage is not saturated. Thus the mode spacing for active stripes 1 and 2 will be different and given by (3.13.24) as

$$\Delta\lambda_1 \approx \frac{\lambda_0^2}{2N_{eff1}L_1},$$

$$\Delta\lambda_2 \simeq \frac{\lambda_0^2}{2N_{eff2}L_2}.$$

(3.13.25)

Since the two cavities are coupled, those modes from each cavity that coincide spectrally will be the enforced modes of the coupled-cavity resonator. The spectral spacing Λ of these enforced modes will be significantly larger than either of the original mode spacings, depending on $N_{eff1}L_1$ and $N_{eff2}L_2$ as

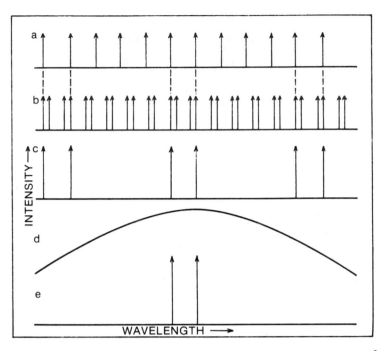

Fig. 3.13.36. Basic working principle of direct frequency modulation in a C^3 laser: (a) modes of cavity (laser); (b) modes of modulator (for two different currents); (c) resultant modes of a C^3 laser; (d) gain profile of the laser medium; and (e) resultant laser spectrum.

given by

$$\Lambda = \frac{\Delta\lambda_1 \Delta\lambda_2}{|\Delta\lambda_1 - \Delta\lambda_2|} = \frac{\lambda_0^2}{2|N_{eff1}L_1 - N_{eff2}L_2|}, \qquad (3.13.26)$$

if we assume $\Delta\lambda_1 \simeq \Delta\lambda_2$.

Thus when the enforced modes are superimposed on the gain profile, the adjacent enforced modes are suppressed with an enforced mode near the gain maximum only being present. Now, if laser 1 is biased with an injection current I_1 above the lasing threshold, it acts as a laser. Laser 2 is biased with some current I_2 below the threshold, thus acting as an etalon. Under these conditions, the situation is described by solid lines in Fig. 3.13.36. If I_2 is increased to I_2', keeping I_1 the same, a change in the carrier density in the active region 2 will cause a decrease from N_{eff2} to N_{eff2}'. This results in a shift of the modes of laser 2 towards shorter wavelengths, as shown by the dashed lines in Fig. 3.13.36. As a result of such changes, the modes from laser 1 and etalon 2 that originally coincided become misaligned, and the adjacent mode on the shorter wavelength sides comes into play. Figure 3.13.37 shows the various spectra obtained with different current levels applied to the modulator diode. This new mechanism, which is called cavity-mode enhanced frequency modula-

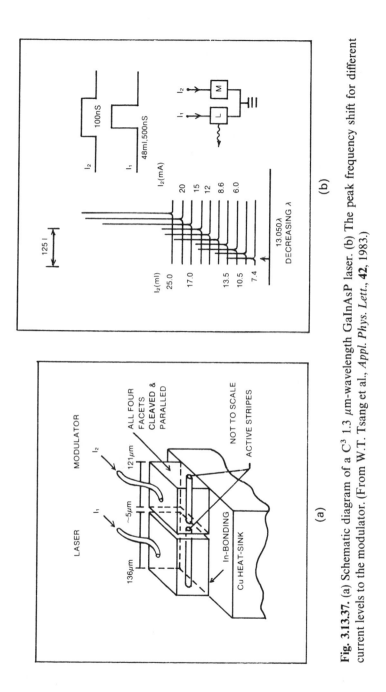

Fig. 3.13.37. (a) Schematic diagram of a C^3 1.3 μm-wavelength GaInAsP laser. (b) The peak frequency shift for different current levels to the modulator. (From W.T. Tsang et al., *Appl. Phys. Lett.*, **42**, 1983.)

tion (CEM–FM), results in a very large frequency-tuning rate (expressed in MHz/mA) and a very wide frequency-tuning range, at least half of the spectral width of the gain profile, i.e., ≥ 150 Å. The range can be further increased by temperature control.

3.13.5. Free-Electron Lasers and Cyclotron Resonance Masers

The lasers we have discussed so far use a material in which electrons make transitions from a higher-energy level to a lower-energy level to produce stimulated emission. Electrons can also radiate when they are accelerated in free space. The interaction of a proper electromagnetic field and a beam of moving electrons will accelerate the electrons in such a way that they will radiate coherently. In both free-electron lasers and cyclotron resonance masers, a magnetic field is used to accelerate the electrons. Free-electron lasers generally work at wavelength regions ranging from millimeter to ultraviolent. Cyclotron resonance masers are efficient in the region of centimeter to millimeter wavelengths.

3.13.5.1. Free-Electron Laser

The free-electron laser (FEL) uses a totally new concept for generating coherent radiation, and offers a variety of advantages over the conventional lasers discussed so far in this chapter. In place of solid, liquid, or gas as the gain medium, FELs use a high-energy electron beam in a magnetic field. A FEL is shown schematically in Fig. 3.13.38. It consists of an accelerator to produce the electron beam, a wiggler magnet to force the electrons to oscillate and radiate, and an optical system to form the laser beam. The wiggler magnet consists of a series of alternating magnetic poles which form a magnetic field directed up and down along the length of the wiggler. As the electrons pass through this magnetic field, they are deflected alternately left and right. Because of this transverse motion, the electrons emit radiation at the wiggler frequency. Due to the relativistic effects, the radiation is strongly forward

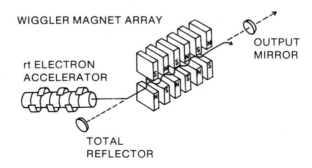

WIGGLER MAGNET ARRAY

rt ELECTRON ACCELERATOR

OUTPUT MIRROR

TOTAL REFLECTOR

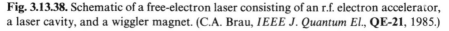

Fig. 3.13.38. Schematic of a free-electron laser consisting of an r.f. electron accelerator, a laser cavity, and a wiggler magnet. (C.A. Brau, *IEEE J. Quantum El.*, **QE-21**, 1985.)

directed and appears at a frequency which is Doppler shifted to a much shorter wavelength.

It can be shown that the wavelength of light is given approximately by

$$\lambda = \frac{\lambda_w}{2\gamma^2}\left[1 + \left(\frac{qB\lambda_w}{2\pi mc}\right)^2\right], \tag{3.13.27}$$

where λ_w is the wiggler wavelength, γ is the ratio of the electron beam energy to the electron rest energy (0.511 MeV), B is the rms wiggler magnetic field strength, and m is the rest mass of the electron.

The force on an electron in the presence of a magnetic field is given by the $\mathbf{V} \times \mathbf{B}$ term where $\mathbf{V}$ is the velocity of the electron. This interaction causes a

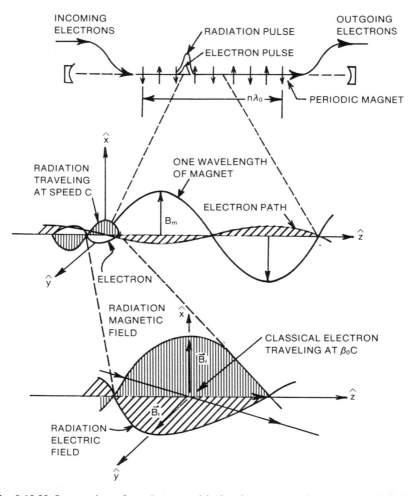

Fig. 3.13.39. Interaction of an electron with the electromagnetic wave: (a) a full view; (b) an expanded view showing one period of the magnet; and (c) a close-up view of the radiation field and electron pulse. (From W.B. Colson, *Physics of Quantum Electronics*, vol. 5 (ed. S.F. Jacobs), Addison-Wesley, Reading, MA, 1978.)

trapping wave* which causes electrons to bunch in the axial direction. The trapping wave bunches the electrons by decelerating some and accelerating others. The $\mathbf{V} \times \mathbf{B}$ force involves the electron wiggle velocity, which is typically much less than the axial velocty and the strength of the radiation magnetic field.

When the coherent optical field from a laser (even the FEL itself) is superimposed on the electrons, the magnetic field of the optical beam interacts with the electrons. At resonance, when the laser wavelength satisfies (3.13.27), the interaction becomes strong and the electrons are accelerated or decelerated slightly by the optical field, depending on whether the electrons are oscillating in phase or out of phase with the local magnetic field. As a result, the faster electrons catch up with the slower ones and form bunches spaced at the optical wavelength. The electrons then radiate coherently with respect to each other and with respect to the optical field. The electron emission then adds coherently to the optical beam and amplifies it as in a conventional laser. Figure 3.13.39 shows schematically the interaction of the electron, optical, and wiggler magnetic field.

3.13.5.2. Cyclotron–Resonance Masers

A beam of electrons traveling with velocity v injected in a magnetic field B (as shown in Fig. 3.13.40) will gyrate with a frequency, ω_c, given by

$$\omega_c = \frac{qB_0}{m} = \frac{qB_0}{m_0\gamma} = \frac{\omega_{c0}}{\gamma}, \tag{3.13.28}$$

---

* Also called the pondermotive wave.

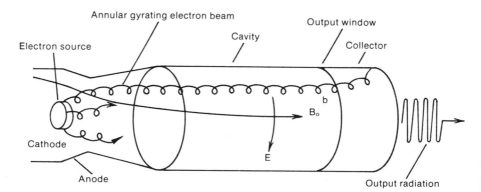

Fig. 3.13.40. A cyclotron–resonance–maser oscillator, in schematic view. The electron source is a magnetron injection gun. The cathode emits an annular beam that gyrates about an applied magnetic field B_0 as it propagates through a cavity. The cavity operates in a transverse-electric mode near its cutoff frequency. The spent electron beam is collected, and radiation is emitted through an output window. (From P. Sprangle and T. Coffey, *Physics Today*, March 1984.)

where ω_c is known as the cyclotron frequency and γ is related to the transverse velocity only.

The heart of the cyclotron–resonance maser is a beam of nearly mono-energetic electrons injected in a magnetic field such that it gyrates at the cyclotron–resonance frequency. An electromagnetic field with frequency very near the cyclotron–resonance frequency is also present in the structure, as shown in Fig. 3.13.40. The electron source is generally a cathode whose design is very similar to that of a magnetron injection gun. Magnetron injection guns can produce several amperes, and electron energies as high as 100 keV. The cathode emits an annular beam of electrons which propagates through the electromagnetic cavity. The cavity operates in the transverse electric (TE) mode near cutoff. The electrons gyrate and radiate giving rise to the output.

Similar to the free-electron laser, electron bunching takes place when the radiation frequency slightly exceeds the gyrating frequency. Note that high-frequency operation requires a large ratio of transverse to longitudinal velocity. This ratio is typically 1 : 3 and efficiency can be as high as 60%.

For a wavelength of 3 mm (94 GHz), ω_{c0} demands that the magnetic field strength be 34 kiloGauss. To obtain such a high magnetic field, we generally need a superconducting magnetic field. Typical cyclotron maser outputs are

$$\text{peak power} \sim 1 \text{ MW,}$$

$$\text{pulse duration} \sim 1\text{–}5 \text{ ms,}$$

$$\text{bandwidth} \sim 5\%.$$

References

[1] A.E. Siegman, *An Introduction to Lasers and Masers*, McGraw-Hill, 1971.
[2] A. Yariv, *Optical Electronics*, 2nd ed., Holt, Rinehart, and Winston, 1976.
[3] D.C. O'Shea, W.R. Callen, and W.T. Rhodes, *Introduction to Lasers and Their Applications*, Addison-Wesley, 1977.
[4] J.T. Verdeyen, *Laser Electronics*, Prentice-Hall, 1981.
[5] S. Martellucci and A.N. Chester, eds., *Free Electron Lasers*, Plenum Press, 1983.
[6] J.T. Luxon and D.E. Parker, *Industrial Lasers and Their Applications*.

PART IV

Applications

4.1. Introduction

There are too many applications of optics and lasers in engineering, and we have already mentioned some of them in the course of this book. In this part, we will consider some of these applications in detail. Section 4.2 considers only conventional optical engineering, i.e., the camera, the microscope, the telescope, etc. Fiber-optics and integrated optics are elaborated on in Section 4.3. This is followed by optical signal processing and the different industrial and medical applications of lasers. The final section (4.6) includes three topics, i.e., optical interconnection, optical computing, and Star War.

4.2. Optical Instruments

In this section we will discuss some commonly used optical instruments; these are lens magnifers, telescopes, binoculars, compound microscopes, spectrometers and cameras. For each instrument, the operation is first explained using geometrical optics; this is then followed by the effect of diffraction on the operating limits of the instruments.

4.2.1. The Lens Magnifier

To understand a lens magnifier, we should consider the optics associated with a normal human eye. As shown in Fig. 4.2.1, the focusing system consists of a cornea, c, an adjustable iris, I, a lens, L, and the retina, R. The light incident on the eye is refracted by the cornea which separates a liquid, M_1, with a refractive index of 1.34 from air. The lens consists of a material with varying refractive index, the values being 1.42 at the center and 1.37 at the edges. The radii of curvature of the lens can be adjusted by tension which results in a change in the focal length of the lens. The lens focuses the images on the retina. The main body of the eye, between the retina and the lens contains a jelly-like "vitreous humor", M_2, which has a refractive index of 1.34. The iris, I, adjusts

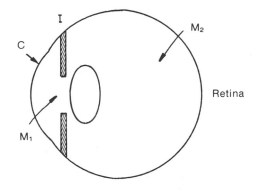

Fig. 4.2.1. Focusing system of the eye. Most of the refraction occurs at the first surface, where the cornea, C, separates a liquid N_1 ($n = 1.34$) from the air. The lens, L, which has a refractive index varying from 1.42 in the center to 1.37 at the edge, is focused by tension at the edges. The main volume contains a jelly-like "vitreous humour" ($n = 1.34$). The image is formed on the retina R. The iris, I, adjusts the aperture according to the available illumination.

the aperature according to the intensity of illumination. For a normal eye, proper focusing cannot be obtained for distances shorter than a specified value, D, and the value of D varies from individual to individual. For design purposes, the norminal value is taken to be $D = 25$ cm. The angular resolution of the eye is dependent on the photosensitive element separation in the retina. It turns out that this resolution also matches the diffraction limited angular resolution due to the iris. As the angular resolution is fixed, the linear resolution of the eye is highest for objects situated at a distance D from the eye, as this is the nearest object distance for a properly focused image on the retina. If an object is brought any nearer, the eye cannot focus it. However, using a simple lens with focal length f, situated as shown in Fig. 4.2.2, we can bring the effective object for the eye at a distance, D. Thus we have the following

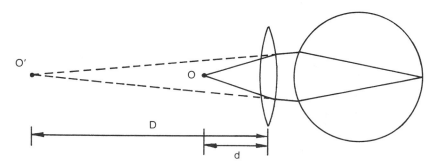

Fig. 4.2.2. Simple lens used as a magnifier. An object at O close to the eye can be focused by the eye as though it were at a more distant point O'.

relationship:

$$\frac{1}{d} - \frac{1}{D} = \frac{1}{f}, \tag{4.2.1}$$

where d is the distance of the actual object from the lens which is very near the eye. The magnification, M, of this simple lens magnifier is given by

$$M = \frac{D}{d} = \frac{D}{f} + 1. \tag{4.2.2}$$

To obtain high M, we need very short focal length lenses. Using a single lens, it is very difficult to obtain it, if the system is to be free of aberration. A compound microscope is the solution for this problem.

4.2.2. The Telescope

We have already discussed telescopic systems in Section 1.4.2. The object of the telescope is to have angular magnification, p_α, as large as possible, this is given by

$$p_\alpha = -\frac{f_1}{f_2}. \tag{4.2.3}$$

A more meaningful expression for the angular magnification can be obtained by considering Fig. 4.2.3. The light from an object on the optical axis will have wavefront perpendicular to the optical axis as shown by w_0. An object at an angle θ_1 with respect to the optical axis will have a wavefront denoted by w_1. Both the wavefronts w_0 and w_1 are shown as they enter the telescope and as they exit. Due to the angular magnification, the exit angle is θ_2. If the aperture of the front element of the telescope is given by y_1, then the path difference between w_1 and w_0, l, is given by

$$l = y_1 \theta_1. \tag{4.2.4}$$

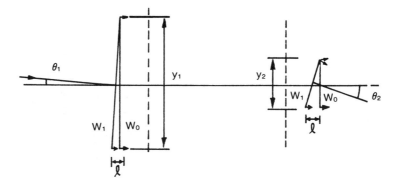

Fig. 4.2.3. Wavefronts through the telescope for the understanding of angular magnification.

Similarly, at the exit window of the telescope we obtain

$$l = y_2 \theta_2. \tag{4.2.5}$$

Note that the path differences must be the same as the light passes through the telescope.* Thus, we must have

$$p_\alpha = \frac{\theta_2}{\theta_1} = \frac{y_1}{y_2}. \tag{4.2.6}$$

From (4.2.6), we observe that we need to have a large angular magnification, $y_1 \gg y_2$. This can be achieved in various configurations using mirrors and lenses, such as:

    the astronomical telescope,
    the Galilean telescope,
    the Newtonian telescope,
    the Cassegrain telescope,
    the Gregorian telescope,
    the Herschel telescope.

An astronomical telescope consists of two biconvex lenses with positive focal lengths, as shown in Fig. 4.2.4(b). The first lens, f_1, is generally called the objective and the second lens, f_2, is called the eyepiece. The separation between the lenses is given by

$$d = |f_1| + |f_2| \tag{4.2.7}$$

and the image is inverted.

In the Galilean telescope the objective is a biconvex lens with positive focal length, whereas the eyepiece is a biconcave lens of negative focal length. For this case

$$d = |f_1| - |f_2| \tag{4.2.8}$$

and the magnification is positive.

The Newtonian telescope uses a concave mirror as the objective. The reflected light from the mirror is reflected again by a flat mirror before it passes through the objective lens. The situation is very similar to the astronomical telescope. Note that a small portion of incident light is lost due to the placement of the flat mirror.

The Cassegrain telescope also uses a concave mirror as the objective. However, its eyepiece is also a mirror, a convex mirror. The objective mirror has a hole at the center through which the output light passes. The situation here is very similar to the Galilean telescope.

The Gregorian telescope is similar to the Cassegrain telescope in structure except a concave mirror is used as the eyepiece. Thus, for this case, the image is inverted.

---

* For a discussion on path difference, see Sec. 2.10.2.

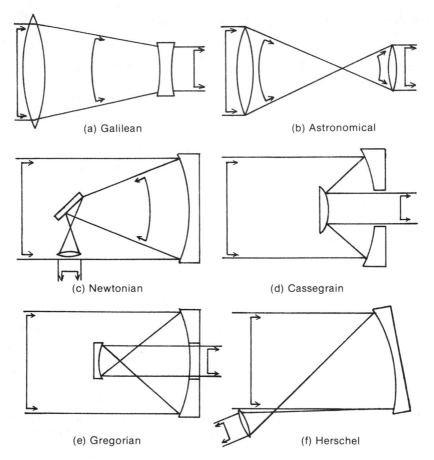

Fig. 4.2.4. The reduction in width of a wavefront in various types of telescope. The telescopes are all adjusted for direct viewing of the emergent beam; the emergent wave could instead be made convergent, so that a photographic plate could be placed at the focus.

The Herschel telescope uses a concave mirror and a lens; however, no incident light is lost in this case. This is possible by tilting the concave mirror such that it focuses light outside the input aperture, as shown in Fig. 4.2.4(c). Otherwise, this telescope is very similar to the Newtonian one and has the advantage that no flat mirror is needed.

To compare the relative advantages of the different telescopes, it is important to note that mirrors have no chromatic aberration and they reflect nearly all the light. However, the mirror has a disadvantage in terms of distortion because of temperature and gravity. The mounting of mirrors needs to be superior, compared to that of lenses, as the first-order effect on the optical path length is zero for the lens. For the case of the mirror, if it bends 1 μm, the path difference is 2 μm.

If a graticule is to be placed as a reference at the image position, the Galilean and Cassegrain telescopes are unsuitable as they do not have a real image. This is important for survey and position measurement equipment.

4.2.3. Binoculars

Binoculars are really two telescopes, one for each eye. Typically, an astronomical telescope is used with the objective lens system corrected for chromatic aberration. The eyepiece also generally contains two lenses, the first one being used as a field lens to increase the field of view by reducing or eliminating vignetting.

To understand the design of binoculars, consider Fig. 4.2.5 where we desire a magnification of 20 (M). To obtain a field of view of 2.5°, the output angle must be 50°. To accommodate this large angle, the diameter of the eye lens and the field lens must be about 15 mm diameter. This also determines the focal length of the objective, as we have

$$\frac{0.75\,\text{cm} \times 180}{2.5° \times \pi} = f_{\text{obj}} \approx 17\,\text{cm}. \tag{4.2.9}$$

The focal length of the eyepiece must be 0.85 cm. Note that as the size of the eye-pupil is about 4 mm, the objective need only be 80 mm (D) in diameter. The specification of this telescope will be noted as 20 × 40 or $M \times D$.

Note that the separation between the eyepiece and the objective is ~ 18 cm. This being quite large for practical purpose, two prisms (as shown in Fig. 4.2.6)

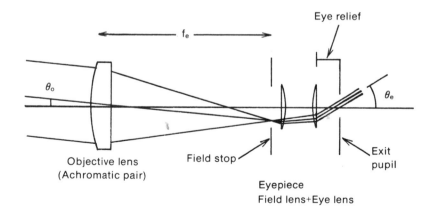

Fig. 4.2.5. Astronomical telescope, as used in the binocular telescope:

$$\text{magnification } M = \frac{\theta_e}{\theta_o} = \frac{f_o}{f_e},$$

$$\text{field of view } \theta_o = \frac{\text{field lens diameter}}{\text{objective focal length}}.$$

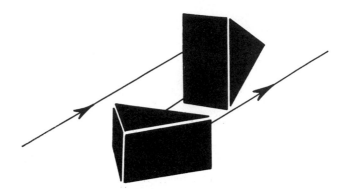

Fig. 4.2.6. A pair of erecting prisms, as used in the binocular telescope.

are used to shorten the size of the binoculars. This also erects the image from the inverted one expected from the astronomical telescope.

For small magnifications, a Galilean telescopic arrangement can be used to obtain a direct image without inversion; this is customary for opera glasses. Note that large magnification is not possible because of the length of the telescope without prism folding.

4.2.4. Compound Microscopy

A compound microscope is capable of achieving very large magnification by using a telescoping system in conjunction with a magnifying glass. Actually, the magnifying glass and the objective of the telescope are combined together followed by an eyepiece—this is shown schematically in Fig. 4.2.7. The objective lens system is the most important part of the microscope and this must have as small an aberration as possible for large angles. The magnification is given by

$$M = \frac{g}{f_1} \cdot \frac{D}{f_2}, \tag{4.2.10}$$

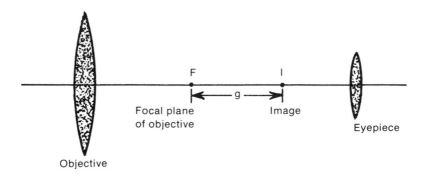

Fig. 4.2.7. Geometry for calculating the magnification of a compound microscope.

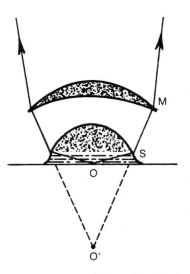

Fig. 4.2.8. Microscope objectives. Oil immersion objective in which the object O and virtual objective O' are on the aplanatic surfaces of a sphere S. The wavefront curvature is again reduced by a series of meniscus lenses M.

where f_1 is the focal length of the objective, f_2 is the focal length of the eyepiece, g is the distance of the intermediate image from the focal plane of the objective lens, and D is the nearest distance of distinct vision.

The length, g, is generally known as the optical tube length of the microscope. For the highest magnification, many times the oil immersion is used. This is shown in Fig. 4.2.8 where the oil has a refractive index very near the value of the first lens which has a very short focal length. The object, O, is placed just beyond the focal plane in oil and it forms the virtual image, O', further away. The rays from the virtual image are collected by the meniscus lens, M, whose first surface is on the sphere with O' as the center.

In a reflective objective compound microscope, the equivalent of the Cassegrain telescope is used—this is shown in Fig. 4.2.9.

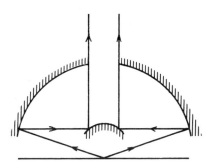

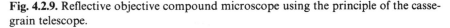

Fig. 4.2.9. Reflective objective compound microscope using the principle of the cassegrain telescope.

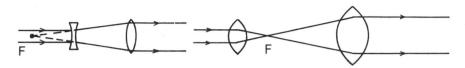

Fig. 4.2.10. Transmissive beam expanders.

4.2.5. Beam Expanders

In many laser applications, the small laser beam needs to be expanded to a larger diameter. Thus, the requirement is just the opposite of a telescope, i.e., the incident light must be on the eyepiece and the output is through the objective—this is shown in Fig. 4.2.10 for the astronomical and the Galilean telescopes. The advantage of the astronomical case is that by using a pinhole at the focus, the laser beam can also be spatially filtered, if needed. The disadvantage is that for high-power lasers, the focused spot may give rise to air breakdown and the transmission of large power through the lenses may not be desirable due to absorption losses.

Reflective beam expanders are shown in Fig. 4.2.11. Note that the reflective expanders may be built more compactly. In all cases, the output beam diameter is related to the input diameter by the following equation:

$$W_2 = \frac{f_2}{f_1} W_1. \tag{4.2.11}$$

4.2.6. Photographic Lens Systems

The design of lens systems for photographic purposes dates back to the 1850s, when portrait and landscape lenses were developed for early versions of the camera. It is interesting to note that the prototype of the modern camera, a device known as the "camera obscura", did not utilize a lens system at all. This device consisted of a black box with a small hole in one wall; light passing through this hole formed an inverted image on the opposite wall.* The camera

---

* The first permanent photograph was made by Joseph Nicéphore Niépce in 1826 using such a camera; his subject was a rooftop scene in Châlon sur Saône, France.

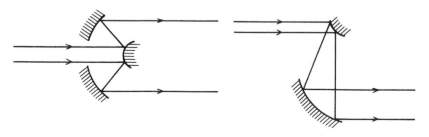

Fig. 4.2.11. Reflective beam expanders.

obscura formed a surprisingly well-defined image over a wide angular field of view, due to its high depth of focus. In addition, the absence of a lens made the pinhole camera's image practically distortion-free, and there was no focusing system to limit the clarity of the final image. Instead, the quality of the image was limited by diffraction effects; for maximum sharpness, the pinhole diameter should be proportional to the distance from the image plane. However, the camera obscura required extremely long exposure times, even with the most sensitive films; as the art and science of photography became more advanced, the need for distortion-free camera lenses (to extend the versatility and speed of photography) became apparent.

The general requirement of a photographic lens is its ability to form a uniformly sharp, distortion-free, real image of the object over the entire field of view. Furthermore, we require that the image be flat and uniformly bright over the whole image area. A good lens system should also permit relatively short exposure times, on the order of a fraction of a second in medium sunlight (or slightly longer under low illumination), which implies a large relative aperture or f number. Under such requirements, a single positive lens element makes a very poor camera lens, since it is generally not possible to correct for the inherent abbreviations of a single lens. Instead, it is necessary to combine several different lens elements to form a lens system in which aberrations may be suitably corrected.

In view of the many possible lens combinations, and the rather strict requirements of a photographic lens, it is not surprising that lens design remains an active field of study. Many contemporary camera objectives are based on a few well-known successful forms, although slight variations are numerous. Because of the many exceptions and unusual designs, it has historically been difficult to classify camera lenses. In 1946, Kingslake developed a classification system based on the number of components in a lens; his system was implemented at the Eastman Kodak Company, and has since become an industry standard (Table 4.2.1). More recently, researchers such as Hoogland have proposed a revised classification based on the degree of complexity of a lens system. Any comprehensive attempt at classification is bound to encounter its share of exceptions and borderline cases, however.

This section will present a review of several photographic lens systems which form the basis of most modern camera lenses. Designs such as the wide-angle, telephoto, and zoom lenses will be discussed, as well as the Cooke triplet, double Gauss, Tessar, and Petzval portrait lenses. Finally, a summary of recent developments in computer-aided lens design will be presented.

4.2.6.1. The Wide-Angle Lens

For a normal camera lens, the diagonal of the field of view is roughly equal to the focal length of the lens. Thus, the angular field of view, ψ, is defined as the angle subtended at the lens by the diagonal of the film area (Fig. 4.2.12). The angular field of view can be thought of as relating to the fraction of the object scene included in the photo. Since the film diagonal is approximately

Table 4.2.1. Lens classification system. (Developed by R. Kingslake for the Eastman Kodak Company, circa 1946.)

Several common types of lenses are defined and classified according to the number of components a given lens contains.

1. Singlet	A single lens element.
2. Doublet	Two lens elements.
3. Triplet	Three lens elements.
4. Quadruplet	Four lens elements.
5. Petzval	Two thin, positive components, widely separated, designed to give high aperture over a narrow field.
6. Telephoto	A positive front member widely separated from a negative rear member, such that the distance from the front vertex to the focal plane is much less than the focal length.
7. Reversed telephoto	A telephoto lens with the negative member in front (wide-angle view).
8. Zoon lens	Lens with a continuously variable focal length, in which the image is held constantly in focus by mechanical means.
9. Special types	Viewfinders, mirror systems, etc.

(From R. Kingslake, A classification of photographic lens types, *J. Opt. Soc. Amer.*, **36**, 1946.)

equal to the focal length of the lens, the angular field of view is commonly between 40° and 60° for a standard lens of focal length 50–58 mm.

As the object moves closer to the camera (object distance decreases), the image distance must increase for a fixed focal length lens system. This, in turn, decreases the field of view. In the opposite situation, the image distance decreases and a larger angle of view is required. If the film size (i.e., the diagonal of the image space) is kept constant, then the angular field of view may be

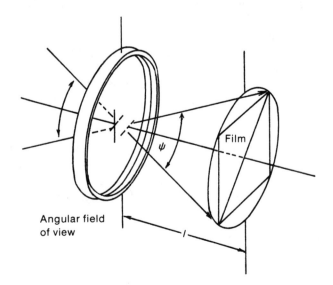

Fig. 4.2.12. Angular field of view for a simple lens: ψ = view angle, f = focal length of the lens. (From E. Hecht and A. Zajac, *Optics*, Addison-Wesley, Reading, MA, 1974.)

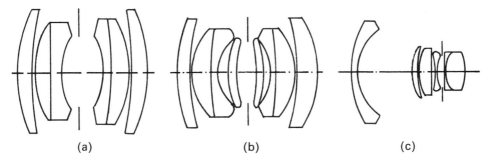

Fig. 4.2.13. Wide angle lenses: (a) variation of double Gauss, the 35-mm Summaron f/2.8; (b) variation of double Gauss, the 35-mm Summicron f/2.0; and (c) inverted telephoto design, the 35-mm Skoparex f/3.4. (From H.H. Brandt, *The Photographic Lens*, The Focal Press, New York, 1960.)

increased by reducing the effective focal length of the lens. Thus, wide-angle lenses may be designed with focal lengths ranging from 40 mm down to as low as 6 mm; their corresponding field of view is increased to between 70° and 80°. Special purpose wide-angle systems have even been designed whose field of view goes beyond 180°, although some distortion is unavoidable in these systems.

Typically, a wide-angle lens may be derived from several other common lens types, such as the Tessar or Gauss designs (Fig. 4.2.13). Such designs help to overcome the more prevalent problems of wide-angle lenses, such as distortion at high fields of view and loss of illumination and sharpness towards the edges of the image. For ultra-wide-angle lenses ($\psi > 90°$), vignetting is also a persistent problem. However, suitable designs based on common lens types, as discussed previously, have made it possible virtually to eliminate aberrations, and result in high-quality, professional camera lenses.

Another common practice for achieving wide-angle effects is the use of an inverted telephoto design, as shown in Fig. 4.2.13. These designs are characterized by a strongly negative lens group followed by a smaller positive lens group. The design has the advantage of a long-back focal length; however, for $\psi > 100°$, barrel distortion is an inevitable result. Such distortion provides a more uniformly illuminated image, by concentrating light in the usually weaker edges of the image. A major design consideration is pupil aberration, since the location of the entrance pupil differs from the paraxial location at large field angles. This difficulty has been overcome, with some degree of success, by using computer-aided lens design techniques.

Wide-angle lenses are capable of taking in a greater object field from a given viewpoint than standard camera lenses. Thus, they are indispensable whenever an object must be fully recorded in a single photograph and it is impossible to use a viewpoint sufficiently far away. These lenses have found applications in architectural photography (photographs in narrow streets or of room interiors) as well as aerial photography and photogrammetry.

4.2.6.2. The Telephoto Lens

It is well known that if a single lens is used to observe an object which is very far away (i.e., the object distance, u, tends to infinity), then the image distance, v, will be approximately equal to the focal length of the lens, f. The lateral magnification is then given by

$$m_x = -\frac{v}{u} \simeq -\frac{f}{u}.$$

So, if a larger image is desired without changing the distance between the object and the camera, one solution is to use a lens of long focal length. This tends to be impractical, however, since a long focal length implies that the lens must be positioned far away from the film.

An alternative way to achieve the same effect is through the use of a telephoto lens. Basically, the telephoto lens consists of a converging front component placed some distance in front of a diverging rear component (Fig. 4.2.14). As can be seen from the figure, for this configuration the principle planes H (and H') are located distances D (and D') from the first (and second) lens elements (here depicted as single lenses). The image distance, v, is now measured from H'; so, as H' moves far away,

$$v = |D'| + (a = \text{distance to film}).$$

This is in contrast to the single lens case, where a is the focal length of the single lens. Using this new expression for v gives a lateral magnification of

$$m_x = -\frac{v}{u} = -\frac{|D'| + a}{u}.$$

So, if $|D'|$ is greater than the separation between the front and rear elements, the lenses may be mounted close to the film while still achieving the desired image magnification. This makes for a much more convenient and compact camera system. Typically, a is of the order of 50 mm, while $|D'|$ is approximately equal to 500 mm; a telephoto lens commonly has a focal length greater than 80 mm.

Modern telephoto lenses employ two separated sets of compound lenses, each independently corrected for aberrations. Early telephoto designs allowed the user to vary the separation between the converging and diverging elements, thereby adjusting the power of the lens. Although such attachments had been abandoned in favor of the fixed-focus telephoto lens, recent advances in lens production technology have renewed interest in the so-called "telephoto zoom lens" design (see Section 4.2.6.3).

It is important to distinguish between the true telephoto lens design and the more conventional long-focus lens, sometimes called the tele-Gauss design. While both are designed for the photography of distant objects, a long-focus lens is simply a variation on the design of a common Gauss lens which produces larger images. Here, the front-half of a Gauss symmetrical derivative is placed in front of a simple converging lens; an aperture stop or variable

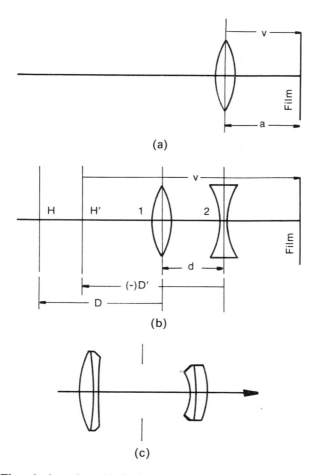

Fig. 4.2.14. The telephoto lens: (a) single camera lens; (b) basic telephoto design; and (c) the 180-mm Tele Xenar f/5.5. (From M.V. Klein and T.E. Furtak, *Optics*, 2nd ed., Wiley, New York, 1986.)

shutter is often placed between the two (Fig. 4.2.15). It is also interesting to note that a reversed telephoto lens may be used to achieve wide-angle effects (see Section 4.2.6.1). In connection with this application, note that since the focal length of a telephoto lens is at least twice as long as the diagonal of the image field, the field of view for a telephoto lens drops off rapidly with increasing focal length, often decreasing to only a few degrees at focal lengths greater than 1000 mm.

A quantity known as the "telephoto effect" is defined as the ratio of the focal length of the combined system to the approximate focal length of an ordinary lens used at the same camera extension. Modern telephoto lenses have telephoto effects of two to three, which is sufficient for most applications. An early design problem for telephoto systems was high pincushion distortion,

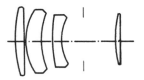

Fig. 4.2.15. The long focal length or tele-Gauss lens; the 135-mm Tele Travenar f/3.5.

since both the converging and diverging components tended to produce this type of distortion. Although such problems were considered to be unavoidable for some time, in 1926 Lee succeeded in producing a distortion-free telephoto lens by separating the components of the diverging lens and utilizing the astigmatism in both surfaces to correct the pincushion distortion (Fig. 4.2.16). Contemporary telephoto lenses are distortion-free over a wide range, and of very high quality.

4.2.6.3. The Zoom Lens

If a lens system consists of two elements (or groups of elements) it is possible to change the focal length of the system by varying the separation between the elements. If two elements of focal lengths f_1 and f_2 are separated by a distance, d, then the total focal length, f, of the combination is given by

$$\frac{1}{f} = \frac{1}{f_1} + \frac{1}{f_2} - \frac{d}{f_1 f_2}.$$

Thus the total focal length becomes shorter as the distance between the elements is reduced. The displacement of lens groups along the optic axis is referred to as "zooming", and lens systems of this type are called zoom lenses. Such lenses are useful if it is desired to image an object plane of variable size into a constant size image.

A zoom lens design consists of three basic parts, the focusing, zoom, and relay elements (Fig. 4.2.17). The focusing part is a single group of lenses which performs focusing of the incident light and presents a fixed virtual object position to the zoom part. The zooming part or "zoom kernel" consists of two lens groups which may be mechanically displaced along the optic axis. The kernel changes the lateral magnification when zooming; behind the zoom kernel, the image size remains constant for transfer to the film. It is the relay part, usually composed of two lens groups, which transfers the image to the

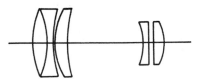

Fig. 4.2.16. Distortion-free telephoto lens design by Lee (f/5.0).

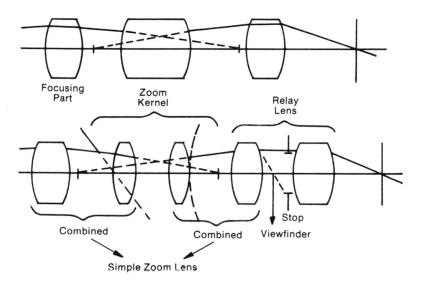

Fig. 4.2.17. The zoom lens: (a) standard construction of a zoom lens; and (b) practical zoom lens, the 36–82-mm Voigtlander Zoomar f/2.8. (From H.H. Brandt, *The Photographic Lens*, The Focal Press, New York, 1960.)

photographic film. Thus, as shown in Fig. 4.2.17, there are a total of five lens groups within a zoom lens design. The variable focal length property makes possible the continuous adjustment of image size, within certain limits, while maintaining a given distance between the object and the image.

In practice, the focusing part and zoom kernel, together with the first group of the relay part, constitute an afocal zoom attachment to the rear group of the relay part. Zoom lenses are classified based on the form of this afocal attachment, which now consists of only four lens groups. Two of these groups must have positive powers, the other two negative powers. Thus, there are four basic types of zoom lenses, which differ from one another in the arrangement of their positive and negative powers (Table 4.2.2).

Table 4.2.2. Basic types of zoom lenses.

	Five-group zoom lenses
"+ − − +" type	Vignetting behavior is the best of all the five-group zoom types; low sensitivity to misalignment.
"+ − + −" type	Poor vignetting behavior; difficult to correct aberrations over a wide zoom range.
"− + + −" type	Poor vignetting behavior; low sensitivity to misalignment; not easily adapted to wide-angle behavior.
"− + − +" type	Good vignetting behavior; difficult to correct aberrations.
	Four-group zoom lens
"+ − +" type	Some vignetting, may be readily corrected; poor aberration correction.
"− + −" type	Poor vignetting behavior; aberrations may be readily corrected.

It is also possible to design simpler zoom lenses, with a more limited zoom range, using a total of four lens groups rather than five. This design may be characterized by an afocal zoom attachment consisting of only three lens groups. This is done by combining either the focusing part or front group of the relay part with the zoom kernel, effectively reducing the size of a normal five-group zoom lens by one group (Fig. 4.2.17). In this new design, the afocal attachment may once again be classified by the arrangement of positive and negative powers. Only two of the possible combinations have come to be of practical significance; these are summarized in Table 4.2.2.

Any two of the three lens groups must be movable, while the third remains fixed; this leads to three subclassifications for each possible arrangement of positive and negative powers. These subclassifications are, however, fairly similar in their basic principles, and will not be treated in detail here.

The design of a zoom lens is much more difficult than the design of fixed focus lenses. In particular, the behavior of a zoom lens is sensitive to the mechanical mechanism used to translate lens groups, including the possible tilting or misalignment of individual lens elements. Most of the conventional aberrations may be well corrected in the zoom kernel; modern lens designs have even proposed the use of aspheric lenses to improve overall performance. Some degree of vignetting is generally acceptable, provided that it remains constant throughout the zoom range.

It is also desirable to design a close focusing capability into a zoom lens. For general photography applications, a zoom lens must be able to maintain an image in focus for a range of object distances, at any zoom position. Typically, the lens is focused by moving the front or focusing part, as discussed previously. However, aberration correction in the focusing part is complicated by two conflicting requirements. First, it is desirable to have large entrance pupil diameter at the longest focal length position of the lens. Second, a larger field of view is required at the shortest focal length position. As a compromise between these two objectives, the minimum focal distance for such lenses has been on the order of 2 m.

This difficulty was first addressed in 1972, with the introduction of a telephoto–zoom lens (also known as a "telezoom lens") with macro focusing capability. The similarities in general design between a telephoto and zoom lens (Figs. 4.2.14 and 4.2.17) suggest that it may be advantageous to combine features of both into a single lens. In addition, this new lens has the capability of moving the zooming group along a mechanical track which is separate from the means used to achieve the zooming effect. Now, the zooming group can be used to achieve close focusing at distances of less than 2 m; the focusing group is used only to focus at distances greater than this. The new close focusing or "macro" mode greatly expanded the capabilities of the telezoom lens. Of course, since the zooming group was used for close focusing, no zooming effect was possible in the macro mode. Still, lateral magnifications on the order of $0.5 \times$ could be achieved with this design.

Although zoom lens designs may become quite complex (Fig. 4.2.18), lens

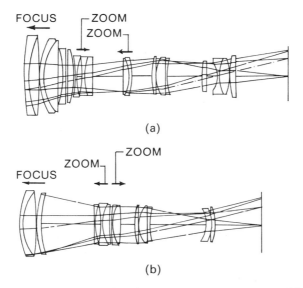

(a)

(b)

Fig. 4.2.18. The telezoom lens: (a) the Vivitar 90–180-mm telezoom with macro mode for continuous close focusing, f/4.5; and (b) the Vivitar 100–200-mm telezoom with macro mode for continuous close focusing, f/4.0. (From E. Betensky et al., Continuous close focusing telephoto zoom lenses, *Proc. OSA/SPIE*, **237**, 1980.)

systems have been produced which are well corrected for aberrations, and mounting mechanisms have been developed to minimize the effects of mechanical displacement. In addition to their use as camera lenses, zoom lens systems have also found applications in motion picture and television camera systems.

4.2.6.4. Basic Lens Configurations

A. The Double Gauss

One of the most common photographic lens configurations in use today is based on a telescope objective designed by K.F. Gauss, the famous German mathematician and astronomer. The so-called "double Gauss" lens was first implemented by P. Rudolph at the Carl Zeiss Co. in 1896. These designs owe their enduring popularity at least in part to their anastigmatic properties, and the fact that they can be well corrected for most other types of aberrations as well.

The basic double Gauss design consists of two outer positive elements and two inner negative elements; the space between the two elements in each half of the lens takes the form of a diverging meniscus (Fig. 4.2.19). Both halves of the lens are usually spaced symmetrically about a central stop or diaphragm. The two inner elements are often replaced by cemented glasses to render the lens achromatic. With slight modifications, all aberrations (with the possible

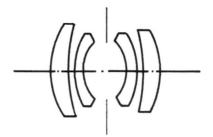

Fig. 4.2.19. The double Gauss lens.

exception of coma) may be well corrected. Variations on this design are numerous, and it has rapidly become one of the mainstays of modern optical design.

B. The Cooke Triplet and Tessar

Another important basic lens form is the Taylor–Cooke triplet, originally designed in 1894 by Mr. H. Dennis Taylor for the optical firm of T. Cooke and Sons, York, England. This lens is asymmetrical, in contrast to the double Gauss, and consists of three lens elements. The front and rear elements are positive, while a negative element is placed between them; the three components are separated by air spaces (Fig. 4.2.20). A lens stop is located near the central element, between it and the rear positive element. This comparatively simple construction is not only easy to manufacture, but may be well corrected for most aberrations (including coma).

The Cooke triplet constitutes an inexpensive anastigmatic lens design. Its performance has been improved in recent years by splitting up the various single lens elements and replacing them with cemented components. Thus, numerous derivatives of the Cooke triplet have appeared, and have found application both as photographic and projection lenses. One important variation is the Tessar lens, developed by the Carl Zeiss Co., in which the back positive component is replaced by a cemented doublet (Fig. 4.2.21).

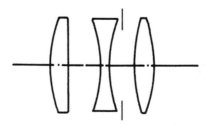

Fig. 4.2.20. The Taylor–Cooke triplet.

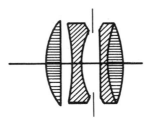

Fig. 4.2.21. The tessar.

C. The Petzval Portrait Lens

One of the oldest photographic lenses still in use today is the Petzval portrait lens, first designed in 1840 by Professor J. Petzval of Vienna for the firm of Voigtländer and Son.* This unsymmetric design consists of four elements, combined into two independently achromatic pairs, separated by a large air space (Fig. 4.2.22). A lens stop is positioned between the two lens pairs. The front component is composed of two cemented lens elements, while the rear component contains two uncemented elements.

Since the lens is fairly long, the astigmatic surfaces (i.e. the image field) could be flattened over an area large enough for portrait photography. Because it was relatively fast for its time (on the order of $f/4$) it soon became a standard photographic lens. The design posessed high vignetting, however, which limited the field of view to about 20° in practice. Eventually, it was replaced in cameras by higher speed anastigmats, although it is still used almost exclusively as a motion picture and slide projection lens.

4.2.6.5. Computer-Aided Lens Design

The lens design process has historically been a complex, time-consuming task, involving a large amount of numerical calculations in order to arrive at an

---

* The Petzval lens was also the first mathematically calculated lens design; prior to the 1840s, lens systems were designed on a purely empirical basis.

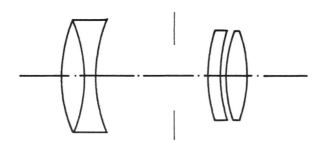

Fig. 4.2.22. The Petzeval portrait lens.

optimal design. Over the past twenty years, developments in computer programming and technology have made computer-assisted lens design both a productive and economical alternative. This section presents a brief review of some recent developments in the field, and their impact on modern optical design.

First, consider the lens design process itself. The practical purpose of optical design is the development of a procedure which, if properly implemented, will lead to a working, economical solution to some design problem. Thus, the design process consists of identifying a need, determining the feasible technical solutions (subject to such constraints as cost limits, available materials, and time limitations), and then using some optimization process to select the "best" of the allowed solutions. Clearly, the design process is as much an art as it is a science; the digital computer can serve as a useful design tool, however, subject to certain inherent limitations.

Originally, large computers operating in high-level languages such as FORTRAN were favored for design purposes because of their high speed. Such systems, however, tend not to be very user-friendly. A recent trend is the use of small microcomputers, or desk-top computers, to perform design functions. Although these machines are slower than larger computers, this is more than compensated for by their highly interactive, user-friendly format. Microcomputers have also become much less expensive in recent years. However, perhaps the single most important feature of microcomputer optical design is the introduction of interactive computer graphics.

Since a wide range of desk-top computers are available (IBM, Apple, Hewlett-Packard, etc.) the focus has shifted to computer software for lens design. A functional program should be high speed (about 300 ray traces per second), comprehensive, versatile, user-friendly, and well documented. In this way, the high accuracy, memory, and graphics capabilities of a given system may be fully exploited. Numerous optical design software packages are available, among them Code V, OSLO, Accos V, and others.

4.3. Fiber-Optics and Integrated Optics

4.3.1. Introduction

Until now we have mostly considered light propagation in free space or in a homogeneous medium, where diffraction plays the most important role. However, we can have guided optical waves. Guided wave propagation, in contrast to free space propagation, deals with propagation in an inhomogeneous medium, specially fabricated so that most of the light energy is transported along a prescribed path nearly unattenuated. If the guiding medium is in the shape of a fiber or cable, we call it fiber-optic, whereas if the guiding medium is planar, like the surface of a substrate, we generally refer to it as integrated optics.

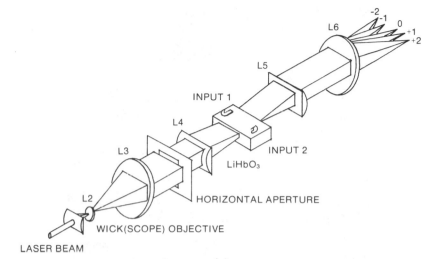

(a)

(b)

Fig. 4.3.1. (a) Optical processor for real-time spectrum analysis; (b) photograph of the optical processor; (c) photograph of translational stage assembly needed for acousto-optic interaction. (From P. Das and D. Schumer, *Ferroelectric*, **10**, 1976.)

(c)

Fig. 4.3.1 (*continued*)

To understand the basic concepts behind integrated optics, let us consider the example shown in Fig. 4.3.1. It consists of a real-time r.f. spectrum analyzer using the principle of Bragg diffraction in an acousto-optic device. Figure 4.3.1(a) shows the implementation using a gas He–Ne laser, a bunch of lenses to collimate the beam, an acousto-optic device, a spatial filter to separate different diffraction orders, and a photodetector array, the output of which is connected to an amplifier. A picture of a typical implementation of this (in the author's laboratory in 1972) is shown in Fig. 4.3.1(b). It is obvious that, compared to ordinary electronic equipment, the optical implementation is somewhat cumbersome because of the mechanical structures needed for the lenses, etc. However, an elegant solution of this problem is the use of integrated optics, and this implementation is shown in Fig. 4.3.2. In this figure, the whole

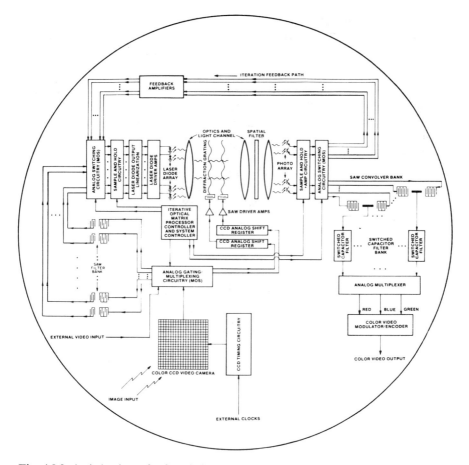

Fig. 4.3.2. Artist's view of a futuristic system on a single GaAs wafer using different signal processing devices.

device is on a GaAs substrate. GaAs is a piezoelectric semiconducting material and has unique properties. As discussed in Section 3.13.4, the laser can be fabricated on GaAs using some of its derivatives, such as GaAlAs. The light is guided onto the substrate on which lenses also can be formed. The acousto-optic device itself can also be fabricated directly onto it, as the photodetector array and the amplifier needed for its output. Thus, we see that this integrated optics implementation makes the optical implementation of the r.f. spectrum analyzer mechanically rugged, easy to mass produce (because of planar configuration), and the size is reduced to very small dimensions. The situation is very similar to the ordinary electronics of the past, where discrete elements like resistors, capacitors, inductors, and transistors were used to build electronic circuits, rather than integrated circuits. Actually, the words "integrated optics" were coined because of this similarity.

Fiber-optics have many applications. However, the communications use is probably the most important one. Fiber-optics communication is like cable television rather than free-space radio or television. We might wonder why fiber-optics has become so important in communications. In the following, we will discuss some of the advantages of using a fiber-optic cable.

(1) Wide bandwidth. The higher the bandwidth of a channel, the more information can be communicated using that channel. A telephone coversation takes approximately 4 kHz of bandwidth. If one million people want to talk through the telephone over a long distance, we need a million channels which can handle 4 kHz each, or one fiber-optic channel which can handle a 4 GHz bandwidth. In general, only a fraction of the carrier center frequency can be used as a useful bandwidth; thus, if the carrier is 1 MHz, we might be able to use a 100 kHz bandwidth. As the center frequency of light is $\sim 10^{15}$ kHz, the achievable bandwidth is enormous, $\sim 10^{12}$ or more. Thus a single, properly designed, fiber-optic cable can, in principle, replace all the telephone, microwave, and satellite channels between New York and Los Angeles.

(2) Smaller size and lighter cables. Most of the electronic cables used for telephones, etc., are made of copper. Compared to these copper cables, fiber-optic cables (made of glass fibers) are much lighter weight-wise and much smaller in diameter. This is of great importance in an aircraft or a submarine, where the changeover from copper cable to fiber-optic cable achieves significant weight reduction, in addition to a significant reduction of space. Also, copper is significantly more expensive than sand (the basic ingredient of fiber-optic cables).

(3) Fiber-optic cables have nearly negligible cross talk when a bundle is formed, and they are highly immune to r.f. interference. Optical fibers do not pick up electromagnetic interference (caused by lightning and other electrical noise generators, such as electric motors, relays, etc.), as they do not act as antennas for these disturbances. Because of this, optical fibers also provide greater security through an almost total immunity to wire tapping. As light is mostly confined to the optical fiber and does not radiate outside the cable, there is no way to eavesdrop without actually tapping directly into the fiber.

(4) Fiber-optic cables can be laid throughout chemical plants, coal mines, etc., where explosive gases exist, without the fear of causing fire. This is because, even if the fiber-optic cable is damaged, no spark is produced. Also, fiber-optic cables, being made of glass in contrast to metal, have a higher tolerance to temperature extremes as well as corrosive gases and liquids. Thus, a longer lifespan for fiber-optics cables is predicted.

(5) Transmission losses are, in general, lower in fiber-optic cables, as compared to coaxial cables. Thus, in a telephone communications system, longer distances between repeaters are possible. Repeaters are needed to boost the signal strength by using amplifiers to take care of the transmission loss. A lesser number of repeaters means greater reliability and ease of maintenance. Overall, fiber-optic systems have the potential to be significantly cheaper than coaxial cable telephone systems.

Because of these advantages, fiber-optic systems are being used increasingly in telephone systems, cable television, computer links, military communication needs, etc. There are other sophisticated uses of fiber-optic cables; for example, they can be used as sensors for temperature, pressure, and rotation; and very sensitive hydrophones and gyroscopes have also been built. These sensors can be used in the most hazardous atmospheres such as nuclear reactors, pit furnances, power stations, etc.

4.3.2. Guided Light

Light can be guided in a fiber-optic cable or on a planar substrate; guiding on a planar substrate is also called a slab waveguide and is used for integrated optics. Because of its simplicity, we will discuss the slab waveguide first.

4.3.2.1. The Slab Waveguide

A typical slab waveguide is shown in Fig. 4.3.3. For the guiding of light, n_1 must have a value higher than both n_2 and n_3. If $n_2 = n_3$, then the guide is called a symmetric guide. To understand guiding in a simple fashion, consider the total internal reflection for the case of a symmetrical guide. A ray is incident on the face of the guide at an angle θ and refracts at an angle θ_1. Thus

$$\sin \theta = n_1 \sin \theta_1.$$

The refracted light is incident on the n_1–n_2 interface at an angle $90 - \theta_1$. Let us denote the critical angle for the total internal reflection by θ_c. Then

$$n_1 \sin \theta_c = n_2. \tag{4.3.1}$$

If $90 - \theta_1$ is larger than θ_c, then the ray will go through total internal reflection. This will continue whenever the ray meets the n_1–n_2 interface and the ray will be trapped or guided. Thus, all the incident rays with cone angles extending from 0 to θ will be confined if

$$n_1 \sin(90 - \theta) = n_2,$$

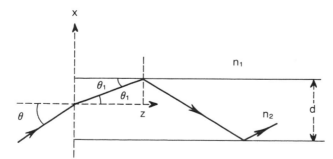

Fig. 4.3.3. Ray diagram showing the guiding of waves, due to total internal reflection.

or

$$\sin \theta = \sqrt{n_1^2 - n_2^2}. \tag{4.3.2}$$

In general, the difference between n_1 and n_2 is small. So we can approximate this angle to be

$$\theta \sim \sqrt{n_1^2 - n_2^2}, \tag{4.3.3}$$

this θ is also known as the numerical aperture.

To properly solve the slab waveguide problem, we should solve the Maxwell equations with the appropriate boundary conditions. The complete derivation of this problem, although straightforward, is beyond the scope of this book. In the following we discuss only the important results.

Light propagates in different discrete modes within the waveguide. These modes can be divided into transverse electric (TE) or transverse magnetic (TM). The TE mode means that there is no electric field component in the direction of propagation (i.e., $E_z = 0$). Similarly, for TM, $H_z = 0$. For a symmetric waveguide, the modes can be further subdivided into even and odd modes. The electric field for the TE even mode is given by

$$E_y = A \cos \alpha_1 x e^{-jK_z z}, \qquad |x| \le d,$$

$$= \frac{\alpha_1}{\alpha_2} A e^{-\alpha_2 (x-d)} e^{-jK_z z}, \qquad |x| \ge d, \tag{4.3.4}$$

where

$$\alpha_1 = \frac{\omega^2}{c^2} n_1^2 - K_z^2, \tag{4.3.5}$$

$$\alpha_2 = K_z^2 - \frac{\omega^2}{c^2} n_2^2, \tag{4.3.6}$$

and K_z is the propagation constant, A is the constant representing the field strength, and d is the width of the guide.

Note that there is a finite electric field in the medium n_2 although there is total internal reflection. However, this electric field decays exponentially as we move away from the interface. α_1, α_2, and K_z are determined from the characteristic equation given by

$$\tan \alpha_1 d = \frac{\alpha_2}{\alpha_1}. \tag{4.3.7}$$

Similar results are also obtained for the odd modes whose electric field is given by

$$E_y = A \sin \alpha_1 x e^{-jK_z z}, \qquad |x| \le d. \tag{4.3.8}$$

It is customary to define a quantity, R, given by

$$R = \frac{\omega^2 d^2}{c^2} (n_1^2 - n_2^2). \tag{4.3.9}$$

For $R < \pi/2$, only two modes can propagate, one even and one odd. For a multimode slab waveguide

$$R \gg 2\pi,$$

or

$$d \gg \frac{\lambda_0}{\sqrt{n_1^2 - n_2^2}} = \frac{\lambda_0}{NA}, \qquad (4.3.10)$$

where NA is the numerical aperture.

Note that the propagation constant, K_z, is a function of ω. Thus, in general, the waves are dispersive. It can be shown that for a length, L, of the multi-mode guide, the difference in propagation delays (between the low- and high-frequency light waves) is given by

$$\Delta\tau_g = \frac{L}{c}(n_1 - n_2), \qquad (4.3.11)$$

where the material dispersion has been neglected. Figure 4.3.4 plots the electric field variation for the first four TE modes as a function of x.

4.3.2.2. Fiber-Optic Cables

Fiber-optic cables are cylindrical waveguides. There are two basic types: the graded index fiber and the stepped index fiber. For the step index fiber, the refractive index has an abrupt discontinuity at the interface between the core

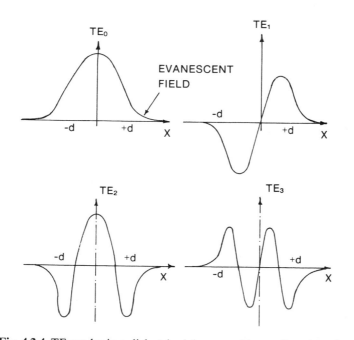

Fig. 4.3.4. TE modes in a dielectric slab waveguide as a function of x.

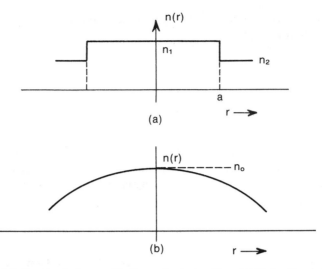

Fig. 4.3.5. (a) Refractive index profile for a step index fiber. (b) Refractive index profile for a gradient index fiber.

and the cladding given by

$$n = n_1, \qquad r \le a,$$
$$= n_2, \qquad r > a,$$
$$n_1 < n_2, \tag{4.3.12}$$

this is shown in Fig. 4.3.5(a). For the graded index fiber whose refractive index profile is shown in Fig. 4.3.5(b), we have

$$n(r) = n_0 \left(1 - \frac{r^2}{2b^2} \right), \tag{4.3.13}$$

where b is a constant. Although other functional dependence is possible, the parabolic variation is the most common one. In the graded index, the refractive index gradually decreases as a function of the radius. It can be shown that the modes in the gradient index fiber, with refractive index variation given by (4.3.13), can be written as

$$E_{mn}(x, y, z) = E_{mn0} e^{-x^2 K_0/2b} H_m \left\{ \left(\frac{k_0}{b} \right)^{1/2} x \right\} H_n \left\{ \left(\frac{k_0}{b} \right)^{1/2} y \right\} e^{-jK_{m1n}z} \tag{4.3.14}$$

where

$$k_{mn} = k_0 \left[1 - \frac{2}{bk} (m + n + 1) \right]^{1/2},$$

$$k_0 = \frac{\omega}{c} n_0, \tag{4.3.15}$$

k_{mn} = propagation constant for the (m, n) mode,

and E_{mn0} is the field strength for the (m, n) mode, and m, n represent the mode numbers and can be any integer. Note the similarity between (4.3.14) and (3.6.2) which represents the laser modes.

Note that for the graded index fiber, we do not have a total internal reflection in a particular interface. What happens is that the rays bend continuously until the x component of the propagation constant is reversed. In a different way, we can consider that the light is periodically focused along the fiber as it propagates. The electric field decays in a Gaussian fashion, as a function of radius, with the effective spot size given by

$$\omega \approx \sqrt{\frac{\lambda}{\pi b}}. \tag{4.3.16}$$

Note that, theoretically, the gradient index fiber must extend to infinity along

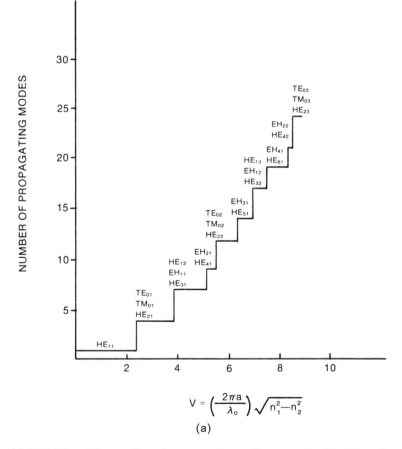

Fig. 4.3.6. (a) Plot of the number of propagating modes versus the fiber V number. (b) Normalized intensity plots for several LP modes for frequencies far away from cutoff and near cutoff. (From T. Okoshi, *Optical Fibers*, Academic Press, New York, 1982.)

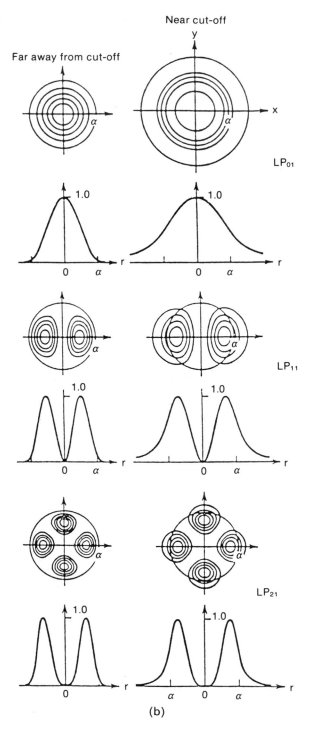

Fig. 4.3.6 (*continued*)

r. However, for practical purposes, the radius r_0 needs to be a few multiples of ω. For a fixed r_0 it can be shown that the maximum mode number, m_{max}, which can be guided is given by

$$2m_{max} + 1 = \frac{r_0^2}{\omega^2}. \tag{4.3.17}$$

The step index fiber is similar to the slab waveguide; however, it can propagate the hybrid modes (EH and HE) over and above the TE and TM modes. Figure 4.3.6 shows different propagating modes as a function of $V = R$, where R is defined by (4.3.9). Thus, for a single-mode fiber, we propagate only the HE_{11} mode; this is true if $V < 2.405$. The r dependence of the electric field for the HE_{11} mode is given by

$$E_y = A\frac{J_0(\alpha_1 r)}{J_0(\alpha_1 a)}, \qquad r < a,$$

$$= A\frac{K_0(\alpha_2 r)}{K_0(\alpha_2 a)}, \qquad r > a, \tag{4.3.18}$$

where α_1 and α_2 are defined in (4.3.5) and (4.3.6) and K_0 represents the modified Bessel function of order 0 and represents the alternating electric field as a function of r. Figure 4.3.6 shows the normalized intensity plots of the three modes including HE_{11}. Note that the linearly polarized (LP) modes are linear combinations of the EH and HE modes for the weakly guiding case, i.e., $n_1 \approx n_2$.

The approximate total number of modes that can exist in a step index fiber is given by

$$N = \frac{4v^2}{\pi^2}. \tag{4.3.19}$$

4.3.3. Integrated Optics

4.3.3.1. Guide, Couplers, and Lenses

Planar waveguides were discussed in the last section; however, for device purposes, channel waveguides are more important. In channel waveguides guiding takes place in two dimensions, as shown in Fig. 4.3.7. All four sides of the channel must be surrounded by the refractive index of a lower value than that of the channel. In Fig. 4.3.7 four possible types are shown, they are:

(a) the embossed or rib waveguide;
(b) the ridge waveguide;
(c) the embedded strip waveguide; and
(d) the stripline guide.

The exact analysis of these channel guides is very complex. However, simplified approximate analysis predicts results very similar to the guides discussed in Section 4.3.2. Figure 4.3.8(a) shows the intensity pictures for the first six modes of a square channel with parameters as shown in Fig. 4.3.8(b).

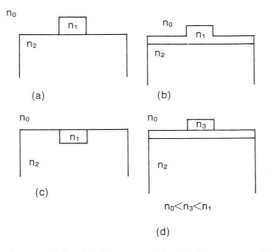

Fig. 4.3.7. Channel waveguides: (a) rib waveguide; (b) ridge waveguide; (c) embossed strip waveguide; and (d) stripline waveguide.

If the source is also on the substrate, like that shown in Fig. 4.3.2, then we can directly couple light to the waveguide. Light from an external light source can be coupled to optical waveguides using the following couplers:

(a) the prism coupler;
(b) the grating coupler;
(c) the tapered filter coupler; and
(d) the butt coupling.

The prism coupling configuration is shown in Fig. 4.3.9, where the prism of higher refractive index is placed on the waveguide with small airgap. The incident light gets total-internally reflected at the interface. However, the evanescent field at the interface couples light to the guide. The coupling efficiency can be as high as 100% if we have a proper beam profile and airgap width.

For a grating coupler as shown in Fig. 4.3.10, the diffracted light couples energy to the waveguide provided the diffracted K vector matches the K vector of any mode of the waveguide. From a practical point of view, grating couplers are convenient and their coupling efficiencies can be designed to be quite high. The output tapered film coupler is shown in Fig. 4.3.11. The length of the tapered part is generally of the order of tens of wavelengths. The film is made of the waveguide material itself. For light propagating in the film, because of the taper at some point, the ray hits the film substrate interface at less than the critical angle and thus refracts out. An input coupler can also be fabricated similarly.

Butt coupling is achieved by directly placing the laser at the edge of the waveguide. Conceptually, this is the simplest one although it requires proper mechanical polishing and gluing for efficient coupling.

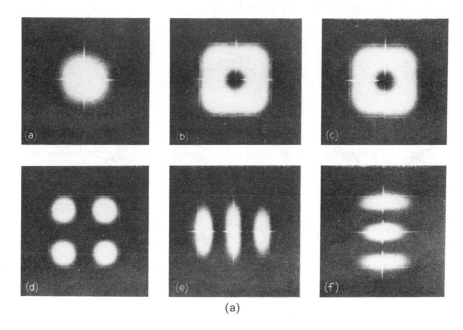

(a)

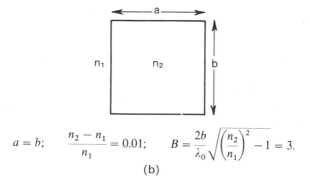

$$a = b; \qquad \frac{n_2 - n_1}{n_1} = 0.01; \qquad B = \frac{2b}{\lambda_0} \sqrt{\left(\frac{n_2}{n_1}\right)^2 - 1} = 3.$$

(b)

Fig. 4.3.8. Intensity pictures for the first six modes for a square channel waveguide: (a) intensity pictures and (b) parameters for the square channel waveguide. (From J.E. Goell, Rectangular Dielectric Waveguides, in *Introduction to Integrated Optics* (ed. M.K. Barnowski), Plenum Press, New York, 1973.)

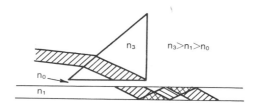

Fig. 4.3.9. Prism coupler.

Fig. 4.3.10. Grating coupler.

The planar optical elements needed for signal processing are lenses, mirrors, and beam-splitters. Waveguide lenses can be fabricated using three techniques, and they are known as:

(i) the Lunenberg lens;
(ii) the geodesic lens; and
(iii) the grating lens.

A Lunenburg lens is a sphere having a refractive index profile given by

$$n(r) = \sqrt{2 - r^2}. \tag{4.3.20}$$

For an incident plane wave, the focus is at the rim of the sphere, as shown in Fig. 4.3.12(a). By modifying the profile, we can focus at any distance away from the rim, as shown in Fig. 4.3.12(b). For integrated optics, the lens is fabricated by sputtering a material of higher refractive index on the waveguide surface. The index profile is obtained by using a properly shaped mask through which one makes vaccum deposition by sputtering.

The geodesic lens is very similar to the Lunenburg lens. However, for this case, the path length variation along the radial direction is obtained by a depression formed on the substrate, before a waveguide is fabricated. For a depression, with dimensions shown in Fig. 4.3.13, the focal length is given by

$$f = \frac{R_0}{2(1 - \cos\theta)}. \tag{4.3.21}$$

Unfortunately, the lens has strong spherical and other aberrations, and to correct them, we form an aspheric surface which corrects them. Using this aspheric surface and diamond turning technique to fabricate them suc-

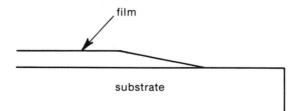

Fig. 4.3.11. Tapered film coupler.

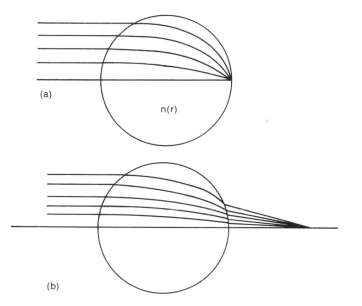

(a)

n(r)

(b)

Fig. 4.3.12. Lunenburg lens. (a) Focus at the rim of the sphere. Radial refractive index profile $n(r) \propto \sqrt{2 - r^2}$. (b) Focus outside the sphere.

cessfully near diffraction limited focusing has been obtained on $LiNbO_3$ waveguides.

Grating lenses are formed using chirp gratings on the waveguide substrate. A particular case is shown in Fig. 4.3.14, where permanent gratings are formed. We can use ultrasonic gratings which are electronically controllable using a surface acoustic wave (SAW). If we need a totally reflecting mirror, we can use a large number of gratings, called the distributed Bragg reflector.

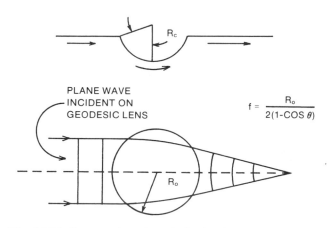

Fig. 4.3.13. Cross section of geodesic thin-film waveguide lens.

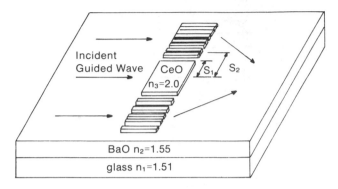

Fig. 4.3.14. Grating lens using chirp grating.

4.3.3.2. Modulators, Deflectors, and Directional Couplers

Modulators and deflectors using electro-optic effects are very similar to those discussed in Section 4.4.2 in connection with regular optics. Of course, with integrated optics, there are practical advantages such as planar structure, ease of fabrication, compact devices with no adjustment, etc. However, the analysis is rather complex because of the two-dimensional variation of the electric field and the light wave. A typical deflector is shown in Fig. 4.3.15, where electrodes are deposited so that the electric field can be applied to form an electro-optic prism. The three-electrode system forms two prisms in series, doubling the deflection. Note that the electric fields in the two prisms are of opposite polarity.

An electro-optic modulator is shown in Fig. 4.3.16, where application of the voltage across the electrodes causes the refractive index to change within the guide and thus modulates the light. The directional coupler is a versatile device and has been used extensively in microwave technology. It is equally important in integrated optics and can be used as the basic building block for modulators, switches, filters, multiplexers, and demultiplexers. Basically, the

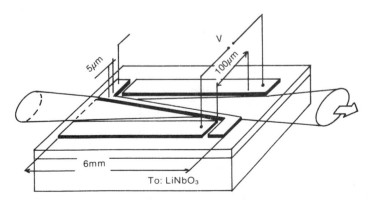

Fig. 4.3.15. Thin film electro-optic prism deflector.

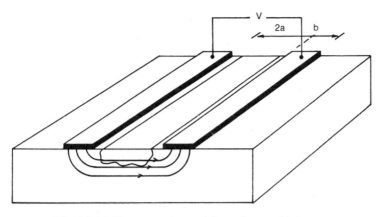

Fig. 4.3.16. Electro-optic modulator for a guided wave.

waveguide in integrated optics is leaky and if another waveguide is near it, the evanescent field between the two couples gives rise to coupled-mode equations which form the basis for understanding these devices.

Figure 4.3.17 shows the schematic of a directional coupler of length L. For the purpose of analysis, consider the section between x and $x + \Delta x$. Light propagates from input ports 1 and 2 to output ports 3 and 4, and their electric fields and intensities are denoted by E_i and $|E_i|^2$ where i varies from $1 \to 4$. The directional coupler is characterized by the scattering coefficients S_{ij} which obey the following equations:

$$E_3 = S_{13}E_1 + S_{23}E_3, \tag{4.3.22}$$

$$E_4 = S_{14}E_1 + S_{24}E_2. \tag{4.3.23}$$

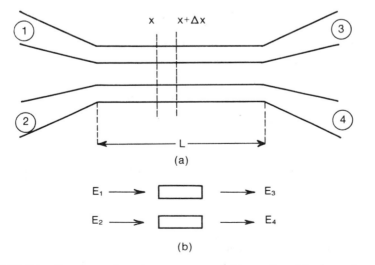

Fig. 4.3.17. Directional coupler using evanescent field coupling: (a) schematic and (b) small section of the coupler.

The conservation of energy demands that

$$S_{13}S_{23}^* + S_{14}S_{24}^* = 0. \tag{4.3.24}$$

We also note that

$$E_3 = E(x + \Delta x) = E_1 + \frac{dE_1}{dx} \cdot \Delta x + \frac{d^2E_1}{dx^2} \frac{(\Delta x)^2}{2} + \cdots \tag{4.3.25}$$

as $E_1(x) = E_1$.

Similarly,

$$E_4 = E_2 + \frac{dE_2}{dx} \cdot \Delta x + \frac{d^2E_2}{dx^2} \cdot \frac{(\Delta x)^2}{2!} + \cdots. \tag{4.3.26}$$

Substituting these equations into (4.3.22) and (4.3.23) we obtain

$$\frac{dE_1}{dx} = \frac{S_{13} - 1}{\Delta x} \cdot E_1 + \frac{S_{23}}{\Delta x} E_2 - \frac{\Delta x}{2!} \frac{d^2E_1}{dx^2} \tag{4.3.27}$$

and

$$\frac{dE_2}{dx} = \frac{S_{24} - 1}{\Delta x} E_2 + \frac{S_{14}}{\Delta x} E_1 - \frac{\Delta x}{2!} \frac{d^2E_2}{dx^2}. \tag{4.3.28}$$

We also know that for $\Delta x \to 0$, S_{13} and S_{24} can be written as

$$S_{13} = e^{jK_1\Delta x} \approx 1 + jK_1\Delta x, \tag{4.3.29}$$

$$S_{24} = e^{jK_2\Delta x} \approx 1 + jK_2\Delta x, \tag{4.3.30}$$

where K_1 and K_2 are propagation constants of the two guides, respectively. From (4.3.24), and noting that $S_{13} \approx S_{24}^* \approx 1$, we obtain

$$S_{14} = -S_{23}^* = j\alpha\Delta x, \tag{4.3.31}$$

where α is defined as the coupling coefficient per unit length and depends on the overlap of the two evanescent fields.

Using the values S_{ij} derived above and using (4.3.27) and (4.3.28), we finally obtain the coupled-mode equations for this distributed system

$$\frac{dE_1}{dx} = jK_1E_1 + j\alpha E_2, \tag{4.3.32}$$

$$\frac{dE_2}{dx} = jK_2E_2 + j\alpha E_1. \tag{4.3.33}$$

To solve these equations, we eliminate E_1 from (4.3.33) by differentiation to obtain

$$\frac{d^2E_2}{dx^2} - j(K_1 + K_2)\frac{dE_2}{dx} + (\alpha^2 - K_1K_2)E_2 = 0. \tag{4.3.34}$$

Assuming α to be real, we obtain

$$E_2 = e^{j[(K_1+K_2)/2]x}[A \sin x\theta + B \cos x\theta], \tag{4.3.35}$$

where

$$\theta = \left\{\alpha^2 + \left(\frac{K_1 - K_2}{2}\right)^2\right\}^{1/2} = \left[\alpha^2 + \left(\frac{\Delta K}{2}\right)^2\right]^{1/2}. \qquad (4.3.36)$$

Similarly, E_1 is given by

$$E_1 = \frac{K_1 - K_2}{2\alpha} E_2 + e^{j[(K_1 + K_2)/2]} \cdot \theta[-A \cos \theta x + B \sin \theta x]. \qquad (4.3.37)$$

If the boundary conditions are given by

$$E_1(0) = 1, \qquad (4.3.38)$$

$$E_2(0) = 0, \qquad (4.3.39)$$

we have

$$B = 0, \qquad (4.3.40)$$

and

$$A = \frac{j\alpha}{\theta}. \qquad (4.3.41)$$

For this case, we obtain

$$E_1 = e^{j[(K_1 + K_2)/2]x}\left[\cos \theta x + j\frac{K_1 - K_2}{2\theta} \sin \theta x\right], \qquad (4.3.42)$$

and

$$E_2 = je^{j[(K_1 + K_2)/2]} \cdot \alpha \cdot \frac{\sin \theta x}{\theta}. \qquad (4.3.43)$$

The coupler efficiency, η, for a length, L, of the coupler is given by

$$\eta(L) = \left|\frac{E_2(L)}{E_1(0)}\right|^2 = \alpha^2 \frac{\sin^2 L\theta}{\theta^2}. \qquad (4.3.44)$$

Thus the maximum power transfer amount is given by α^2/θ^2 and it occurs for $L\theta = (n + \frac{1}{2})\pi$. Note that for $\Delta K = 0$ and $\alpha L = (n + \frac{1}{2})\pi$, full power transfer occurs and $\eta = 1$.

ΔK can be changed by applying the electric field to one of the guides, as shown in Fig. 4.3.18; this arrangement can thus be used as a switch or modulator. An interesting case arises when the coupling is between a TE mode and a TM mode; and this only happens in the presence of the electro-optic effect. However, for this case, ΔK is never zero. We can apply a periodic electric field to simulate the effect, as if ΔK were zero. This is true for the device shown in Fig. 4.3.19 provided

$$K_{TE} - K_{TM} = \frac{2\pi}{\Lambda}, \qquad (4.3.45)$$

where K_{TE} and K_{TM} are the propagation constants of the TE mode and the TM mode, respectively, and Λ is the period of the applied electric field. The device in Fig. 4.3.19 can also be called a mode converter.

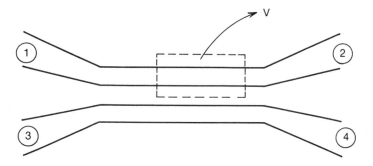

Fig. 4.3.18. Electro-optic switch or modulator using directional coupler.

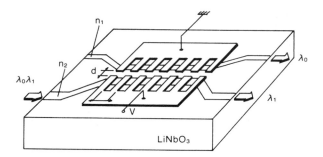

Fig. 4.3.19. Electro-optic mode converter.

The main element of the electro-optic A/D converter is the Mach–Zender interferometric modulator shown in Fig. 4.3.20, implemented in the integrated optics fashion using LiNbO$_3$ as the substrate. The input light is divided equally between two guides and then recombined again forming an interferometer. In one path two metal plates are added, so that the velocity of the guided wave can be changed by applying the electric field due to the electro-optic interaction. The output light intensity is given by

$$I = I_0 \cos^2\left(\frac{\varphi}{2} + \frac{\psi}{2}\right),$$ (4.3.46)

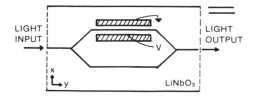

Fig. 4.3.20. Schematic drawing of an integrated optical Mach–Zender interferometer modulator fabricated from single-mode channel waveguides in LiNbO$_3$.

where I_0 is the incident light intensity, φ is the phaseshift caused by the electro-optic effect, and ψ is any phaseshift present due to an imbalance between the two arms of the interferometer. Note that if the light from the two branches is not in phase, then the two recombined lights form a second-order mode and cannot propagate through the output guide. From Section 2.12.7, we have

$$\varphi = 2\pi L \frac{\Delta n}{\lambda} = \pi \frac{V}{V_\pi}, \tag{4.3.47}$$

where L is the length of the modulator metal strip and V is the applied voltage.

Equation (4.3.46) can be rewritten as

$$\left(I - \frac{I_0}{2}\right) = \frac{I_0}{2} \cos\left[2\pi \frac{V}{V_\pi} + 2\psi\right]. \tag{4.3.48}$$

Thus the frequency of the periodic variation of the output, measured with reference to half of the input intensity, is proportional to the applied voltage. This fact is used to form the A/D converter. For details see reference [32].

4.3.4. Fiber-Optic Cables

In this Section we discuss some practical aspects of fiber-optic cables; these are attenuation, dispersion, and pulse propagation.

The main loss mechanisms for optical fibers are:

the intrinsic material absorption loss;
the Rayleigh scattering loss;
the waveguide scattering loss; and
the microbending loss.

As most of the fibers are made of high-silica glass, they have a very high absorption rate, due to direct photon absorption to create an electron-hole pair at an approximate energy bandgap of 8.9 eV, corresponding to $\lambda = 0.14~\mu\text{m}$. On the other hand, in the infrared region, the absorption is due to molecular vibration. A typical loss curve is shown in Fig. 4.3.21, where we note that above 1.5 μm the intrinsic loss is less than 0.5 db/km. In general, the intrinsic loss is very small in the wavelength region of 0.8–1.5 μm. In this region, however, the vibrations of the dopants and other impurities (i.e., OH^-, Fe^{2+}, Cu^{2+}, etc.) give rise to losses which are highly dependent on the particular fiber. In Fig. 4.3.21, OH^- absorption is shown to be predominant at ~ 1.4 eV.

Rayleigh scattering loss arises due to thermal vibration and compositional variations. It can be shown that

$$\alpha_R = \frac{8\pi^3}{3\lambda^4}\left[(n^2 - 1)\beta KT + 2n\left(\frac{\partial n}{\partial c}\right)^2 \overline{\Delta C^2}\delta v\right], \tag{4.3.49}$$

where α_R is the Rayleigh scattering loss coefficient, β is the isothermal compressibility, C is the dopant concentration, and $\overline{\Delta C^2}$ is the mean squared

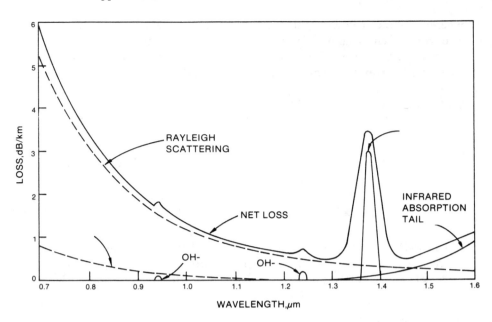

Fig. 4.3.21. Typical loss curve for high-silica glass.

fluctuation of C over volume δv. In general, α_R is dependent on the manufacturing process and a typical one is shown in Fig. 4.3.21. Due to manufacturing variability and imperfections, there is always a loss term due to waveguide formation and microbends in the fibers. However, with proper care it can be significantly reduced.

From the point of view of optical signal processing, the bandwidth limitation and its effect on pulse propagation is more important. The main reason for pulse broadening is due to modal dispersion (only when multimode is used) and basic material dispersion, or chromatic dispersion. Modal dispersion can easily be eliminated using single-mode fibers and, as discussed in Section 4.2, this can be done over a very large $\sim 10^3$ GHz bandwidth if material dispersion is neglected. It is convenient to expand the propagation constant $\beta(\omega, E)$ as follows:

$$
\begin{aligned}
\beta(\omega, E) = {} & \beta_0 + \left(\frac{d\beta}{d\omega}\right)_{\omega_0} (\omega - \omega_0) + \frac{1}{2}\frac{d^2\beta}{d\omega^2}\bigg|_{\omega_0} (\omega - \omega_0)^2 \\
& + \frac{1}{6}\frac{d^3\beta}{d\omega^3}\bigg|_{\omega_0} (\omega - \omega_0)^3 + \frac{\pi}{\lambda} n_2 |E|^2 \\
= {} & \beta_0 + \beta_1(\omega - \omega_0) + \frac{1}{2}\beta_2(\omega - \omega_0)^2 \\
& + \frac{1}{6}\beta_3(\omega - \omega_0)^3 + \frac{\pi}{\lambda} n_2 |E|^2,
\end{aligned}
\qquad (4.3.50)
$$

it is assumed that ω_0 is the center frequency. The constant term, β_0, introduces a simple phaseshift, the second term, β_1, causes the pulse delay, and the third term, β_3, causes pulse distortion. The last term is the nonlinear interaction term due to the Kerr effect discussed in Section 2.12.7, and n_2 is related to the Kerr coefficient.

The group velocity, $v_g(\omega)$, is given by

$$\frac{1}{v_g(\omega)} = \beta_1 + \beta_2(\omega - \omega_0) + \tfrac{1}{2}\beta_3(\omega - \omega_0)^2. \tag{4.3.51}$$

If the length of the fiber is L, then neglecting the contribution of the third term in (4.3.50), the delay time t_d is given by

$$t_d(\omega) = \frac{L}{v_g(\omega)}. \tag{4.3.52}$$

If the spectral width is $\Delta\omega$, then the uncertainty in time delay, t_d is given by

$$\Delta t_d \approx L\beta_2 \Delta\omega. \tag{4.3.53}$$

Defining the chromatic dispersion, D, as

$$D = \frac{1}{L}\frac{dt_d}{d\lambda}, \tag{4.3.54}$$

we have

$$D = \frac{2\pi c}{\lambda^2}\beta_2, \tag{4.3.55}$$

where D has units of ps/nm-km. Equation (4.3.53) can also be written as

$$\Delta t_d = LD\, d\lambda = \frac{2\pi cL}{\lambda^2}\beta_2\, d\lambda. \tag{4.3.56}$$

A typical D versus λ curve for two optical fibers is shown in Fig. 4.3.22. To obtain the minimum pulse width, τ_{in}, at the input, which can be transmitted for a length, L, we note that the output pulse width is given by

$$\tau_{out} = \tau_{in} + \Delta t_g = \tau_{in} + L\beta_2 \frac{1}{\tau_{in}}, \tag{4.3.57}$$

where we approximated $\tau_{in} \sim 1/\Delta\omega$. Differentiating (4.3.57) with respect to τ_{in}, and equating to zero, we obtain the minimum transmitted pulse, τ_{min}, given by

$$\tau_{min} = 2\sqrt{L\beta_2} \propto \lambda\sqrt{L|D|} \tag{4.3.58}$$

for

$$\tau_{in} = \sqrt{L\beta_2}. \tag{4.3.59}$$

In (4.3.58) the absolute value of D is used, as D can be positive or negative.

A typical value of D at $\lambda = 1.55\ \mu$m is 10 ps/nm-km, which predicts

$$\tau_{min} \propto 14\sqrt{L(\text{in km})}\ \text{ps}. \tag{4.3.60}$$

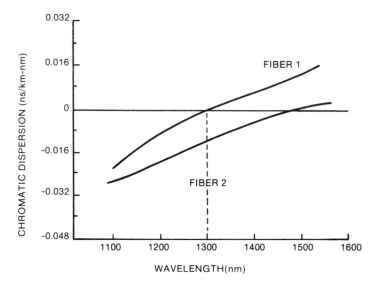

Fig. 4.3.22. Chromatic dispersion curve of optical fiber.

From Fig. 4.3.22 we note that D is zero at 1.3 μm. It can be shown that the pulse propagation for this case gives rise to pulse distortion in the form of ringing. The shortest propagated pulse for this case can be shown to be given by

$$\tau_{min} \approx 3(\beta_3 L)^{1/3}. \tag{4.3.61}$$

Using the value of $D' = dD/d\lambda = 0.08$ ps/nm²-km for a typical fiber, it is estimated that*

$$\tau_{min} \approx 1.2(L \text{ (in km)})^{1/3} \text{ ps}. \tag{4.3.62}$$

Comparing (4.3.60) and (4.3.62) it is obvious that it is desirable to use fiber-optic cable at $\lambda = 1.3$ μm where $D(\lambda) = 0$; however, attenuation is minimum at $\lambda = 1.55$ μm. So we have a somewhat ideal fiber, from the pulse propagation point of view, if the chromatic dispersion curve can be somehow modified to resemble fiber 2 shown in Fig. 4.3.22. There are two ways this can be done. The simplest approach is fiber concatenation or the use of two fibers with $D(\lambda)$ equal, but opposite in sign. This can be done by using different doping. It is known that GeO_2 doping shifts λ_0 to longer wavelengths whereas BeO_2 shifts it to a shorter wavelength. The second approach is to have a doubly clad fiber with a refractive index variation as shown in Fig. 4.3.23(a). The figure also shows the predicted chromatic dispersion for the values given in the figure.

The nonlinear term in (4.3.50) causes what is known as self-phase modula-

---

* Note that D' is related to β_3.

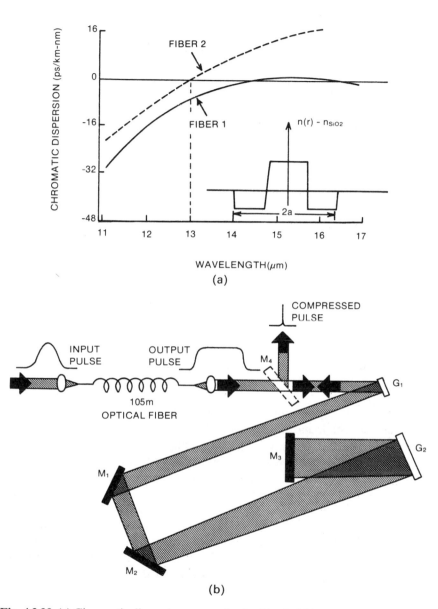

Fig. 4.3.23. (a) Chromatic dispersion curve of a double clad fiber with index variation shown in the inset. (b) Schematic drawing of an optical pulse compressor. The dispersive delay line consists of gratings G_1 and G_2 and mirrors M_1, M_2, and M_3. The compressed pulse is deflected out by mirror M_4. (From A.N. Johnson et al., *Appl. Phys. Lett.*, **44**, 1984.)

tion. The Kerr effect can be written as

$$\Delta n = \tfrac{1}{2} n_2 E_{\text{peak}}^2 = \bar{n}_2 I, \tag{4.3.63}$$

where I is the intensity. A typical value is $n_2 \sim 3.2 \times 10^{-6}$ cm^2/W, and for 1 W of propagating power $\Delta n \sim 10^{-10}$. Although this change in refractive index is very small, it causes frequency chirp in the pulse spectrum.

This phenomenon has been used successfully in forming femto-second pulses by using pulse compression. The light pulse, generally from a mode-locked dye laser in the pico-second range, is passed through a fiber-optic cable to have the frequency spectrum chirped. This chirped pulse is sent through a diffraction grating to obtain pulse compression. A typical setup is shown in Fig. 4.3.23(b).

4.3.5. Applications

The largest application of fiber-optics and integrated optics is in communications systems. A block diagram of a fiber-optic communication system is shown in Fig. 4.3.24. The source consists of either an LED or a semiconductor laser; for a coherent system, however, we must use a laser. The transmitter consists of modulators, different multiplexing switches, and amplifiers. Different modulators have been discussed in other parts of this book. The modulation scheme can be either analog (amplitude or frequency) or digital (pulse code modulation). The digital systems are of more use. The main element of the receiver is the detector, in conjunction with other electronics, to decode the modulated data.

Further details of fiber-optic communications are shown in Fig. 4.3.25. This system uses a broadband single-mode fiber, and double star network architecture for local distribution. There are many other possible distribution, networking, and switching schemes. There is a broadband switch at the local central office and a second switch at a remote electronics terminal; in this terminal, the distribution of the selected channels to the subscriber is made.

Any fiber-optic system must use a light source, and there are two options. In one case, the use of a light source for each local network; the other alternative being the use of a high-power laser in a central station, splitting the power N ways as shown in Fig. 4.3.26. This figure shows a local distribution network architecture, using centrally located shared lasers and external modulators to create a two-way transmission system with no lasers at the

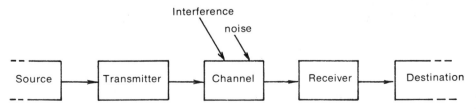

Fig. 4.3.24. Block diagram of a communication system.

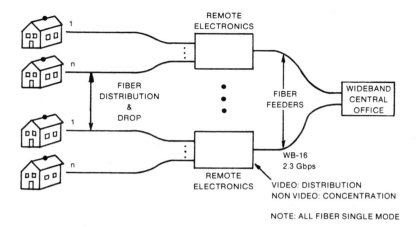

Fig. 4.3.25. Broadband single-mode fiber double-star network architecture for local distribution. (From L.R. Linnell, *IEEE J. Selected Area Comm.*, **SAC-4**, 1986.)

network terminal. The high-power laser sources, s_1 and s_2, are split m ways by a star coupler. Then each channel is modulated externally by the modulators denoted as M_{ij}.

To obtain two-way communication we need two channels, denoted as upstream and downstream channels. Wavelength multiplexing and demultiplexing are used to distinguish the two channels. The optical detectors are denoted by D_{ij}. Typically, it is projected that m can be 100–1000 and can have single-mode fibers with 100 GHz/km bandwidth distance product.

The main advantage of a coherent transmission system is the theoretical high-sensitivity detection, limited by photoelectron statistics or quantum noise. Another advantage is the inherent multifrequency capability, the ability to do optical frequency division multiplexing. In a low-loss single-mode fiber

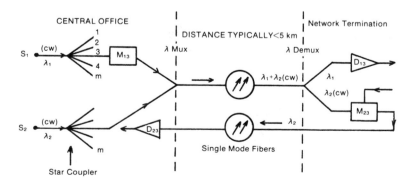

Fig. 4.3.26. Local distribution network architecture using centrally located shared lasers and external modulators to create a two-way transmission system with no lasers at the network termination. (From S.S. Cheng et al., A distributed star network architecture for inter office applications, *J. Lightwave Tech.*, **LT-4**, 1986.)

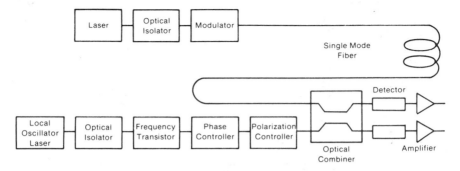

Fig. 4.3.27. Typical coherent optical fiber transmission system illustrating the use of integrable waveguide components.

in the 1.55 μm wavelength band there is $\Delta\lambda \sim 120 \times 10^{-9}$ m available. This corresponds to a bandwidth of 15,000 GHz. A coherent transmission system, however, needs a more elaborate receiver including a local oscillator whose phase, frequency, and polarization must be controlled. Integrated optics is well suited for fabricating this receiver. A typical coherent optical fiber transmission system block diagram, illustrating the integrable waveguide components, is shown in Fig. 4.3.27. A particular receiver using LiNbO$_3$ substrate is shown in Fig. 4.3.28. The receiver performs the functions of polariza-

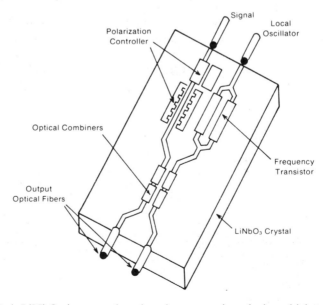

Fig. 4.3.28. A LiNbO$_3$ integrated optic coherent receiver device which performs the functions of polarization correction, optical frequency tracking, and signal and local oscillator combining. (From W.A. Stallard et al., Novel LiNbO$_3$ integrated-optic component for coherent optical heterodyne detection, *Electron. Lett.*, **21**, 1985.)

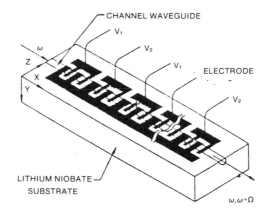

Fig. 4.3.29. An electro-optic frequency shifter. (From H.F. Taylor, Application of guided-wave optics in signal processing and sensing, *Proc. IEEE.*, **75**, 1987.)

tion correction, optical frequency tracking, and signal and local oscillator combining.

Other integrated optics functional subsystems have been built; and here we discuss just one of them. A frequency shifter is a useful system both for coherent transmission and for optical fiber gyroscopes. Both electro-optic or acousto-optic effects can be used for this purpose. Figure 4.3.29 shows an electro-optic frequency shifter. This uses the mode conversion of a TE mode to a TM mode by the periodic metal fingers across which proper voltages are applied. The fingers produce an electro-optic grating whose period matches the TE–TM beat length. Thus, part of the light coupled into the TE mode emerges in the TM mode, shifted by the frequency of the electro-optic modulating voltage. An acousto-optic frequency shifter is shown in Fig. 4.3.30. It uses two surface acoustic wave (SAW) interdigital transducers. The

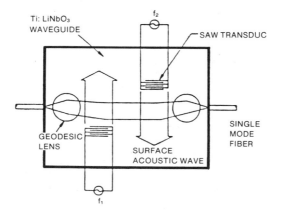

Fig. 4.3.30. An acousto-optic frequency shifter for use with single-mode fibers. (From H.F. Taylor, Application of guided-wave optics in signal processing and sensing, *Proc. IEEE.*, **75**, 1987.)

SAW is an elastic wave which is mostly confined to the surface and is thus ideal for guided-wave acousto-optic interaction. The single-mode input and output optical fibers have relative frequency shifts of difference frequency $(f_1 - f_2)$, where f_1 and f_2 are the frequencies of the input voltages to the SAW transducers.

4.4. Optical Signal Processing

4.4.1. Introduction

Optical signal processing involves devices or systems which improve the performance of a signal. The signal to be processed is, in general, electric in nature; thus, the natural question arises as to why we want to do optical signal processing—because the electrical signal has first to be converted to an optical signal before processing. To answer this question, we note that very efficient high bandwidth and high time–bandwidth product modulators are available through acousto-optic or electro-optic interaction or through the direct modulation of the lasers, especially the junction laser; but more important are the following points:

(i) For digital processing we need a very fast A/D converter for analog signals. At present, A/D converters, beyond the sampling rates of tens of megahertz, are not easily available at a reasonable price. Even for analog signals, high bandwidth requirements can easily be met by acousto-optic or electro-optic devices which are very difficult to accommodate by technologies other than optical. Thus, the instantaneous power spectra of an electrical signal with bandwidths exceeding 1 GHz can, at present, be performed easily and probably only by acousto-optic devices.

(ii) The signal to be processed is not always electrical. With the increasing use of fiber-optic communication cables in the near future, there will probably be more signals on the optical carrier. In this case, we are better off processing the signals directly by optical signal processing, than by first converting the optical signal back to an electrical signal by a detector and then processing it. A typical example might be the data from the ultrasonic detectors all over the submarine which are brought to the optical processor through fiber-optic cable.

(iii) The more obvious case is that the signal itself is in the optical domain, such as an optical image or transparency.

(iv) Inherent parallelism with optical signal processing is probably the biggest asset of optical signal processing. Imagine trying to connect an electrical signal to a million discrete points simultaneously. This can easily be done in optics if a junction laser, placed at the focal point of a lens, is modulated with the electrical signal. Note that this is the fundamental reason for the interest in and importance of optical signal processors. Another example is the correlation or matched filtering of images in parallel.